MIGRATION
OF MACROSCOPIC
INCLUSIONS IN SOLIDS

STUDIES IN SOVIET SCIENCE

1973

Motile Muscle and Cell Models
   N. I. Arronet

Densification of Metal Powders during Sintering
   V. A. Ivensen

The Transuranium Elements
   V. I. Goldanskii and S. M. Polikanov

Pathological Effects of Radio Waves
   M. S. Tolgskaya and Z. V. Gordon

Gas-Chromatographic Analysis of Trace Impurities
   V. G. Berezkin and V. S. Tatarinskii

A Configurational Model of Matter
   G. V. Samsonov, I. F. Pryadko, and L. F. Pryadko

Complex Thermodynamic Systems
   V. V. Sychev

Central Regulation of the Pituitary-Adrenal Complex
   E. V. Naumenko

Crystallization Processes under Hydrothermal Conditions
   A. N. Lobachev

Migration of Macroscopic Inclusions in Solids
   Ya. E. Geguzin and M. A. Krivoglaz

STUDIES IN SOVIET SCIENCE

MIGRATION OF MACROSCOPIC INCLUSIONS IN SOLIDS

Ya. E. Geguzin

A. M. Gorkii Khar'kov State University
Khar'kov, USSR

and

M. A. Krivoglaz

Institute for Metal Physics
Academy of Sciences of the Ukrainian SSR
Kiev, USSR

Translated from Russian by
Albin Tybulewicz
Editor, *Soviet Physics - Semiconductors*

SPRINGER SCIENCE+BUSINESS MEDIA, LLC

Geguzin, Iakov Evseevich.
Migration of macroscopic inclusions in solids.

(Studies in Soviet science)
Translation of Dvizhenie makroskopicheskikh vkliucheniĭ v tverdykh telakh.
Bibliography: p.
1. Solids. 2. Diffusion. I. Krivoglaz, Mikhail Aleksandrovich. II. Title. III. Series.
[DNLM: 1. Diffusion. 2. Physics. QC176.8.D5 G299d 1974]
QC176.8.D5G3813   530.4'1   73-83894
ISBN 978-1-4757-5844-3   ISBN 978-1-4757-5842-9 (eBook)
DOI 10.1007/978-1-4757-5842-9

Yakov Evseevich Geguzin was born in 1918 in the town of Donetsk. A 1941 graduate of the A. M. Gor'kii State University in Khar'kov, he holds the degree of Doctor of Physicomathematical Sciences. Professor Geguzin, after years of industrial and theoretical work in many areas of solid state physics, now holds the Chair of Physics of Crystals at Khar'kov State University. He is a member of Scientific Councils on Solid-State Physics and on Physics of Surface Phenomena of the Academies of Science of the USSR and the Ukrainian SSR.

Mikhail Aleksandrovich Krivoglaz was born in 1929 in Kiev. He was graduated from the Physics Department of Kiev University. Professor Krivoglaz holds the degree of Doctor of Physicomathematical Sciences, is professor at Kiev University, and heads a division at the Institute of Metal Physics of the Academy of Sciences of the Ukrainian SSR.

The original Russian text, published by Metallurgiya Press in Moscow in 1971, has been corrected by the authors for the present edition. The translation is published under an agreement with Mezhdunarodnaya Kniga, the Soviet book export agency.

DVIZHENIE MAKROSKOPICHESKIKH VKLYUCHENII V TVERDYKH TELAKH
Ya. E. Geguzin and M. A. Krivoglaz

ДВИЖЕНИЕ МАКРОСКОПИЧЕСКИХ ВКЛЮЧЕНИЙ В ТВЕРДЫХ ТЕЛАХ
Я. Е. ГЕГУЗИН И М. А. КРИВОГЛАЗ

© 1973 Springer Science+Business Media New York
Originally published by Publishing Corporation, New York in 1973

Preface to the American Edition

We welcome the initiative of Plenum Press to acquaint English-speaking and, particularly, American specialists with our book. The subject of migration of macroscopic inclusions in solids is only about 15 years old. Nevertheless, it has attracted many investigators and is developing very rapidly.

Much new work has been published since the appearance of the Russian edition. We have been able to deal only partially with this work in the American, revised, edition.

We shall be very happy if the book is found useful by the American and other English-speaking scientists.

Ya. E. Geguzin and M. A. Krivoglaz

Preface

Extensive theoretical and experimental investigations have recently been made of macroscopic inclusions in crystal matrices (solid particles of the second phase, liquid inclusions, pores, and gas-filled cavities). The size of such inclusions exceeds greatly the atomic spacings and they can migrate as a whole under the influence of external forces. The mechanism responsible for such migration is the diffusion of the matrix (host) matter between the "front" and "rear" parts of an inclusion. Thus, the migration of inclusions is one of the manifestations of diffusion processes in real crystals.

The diffusional migration of macroscopic inclusions in solids can have a considerable influence on many processes and phenomena which are important in modern technology. For example, the mobility of inclusions may determine the kinetics of the high-temperature creep of dispersion-hardened alloys and composites (i.e., the heat resistance of these materials); the kinetics of the recrystallization of heterogeneous systems; the kinetics of the swelling of fuel elements in nuclear reactors; certain features of the kinetics of the densification of porous diodes during their firing, etc.

The present book is the first attempt ever to review the experimental and theoretical investigations of the migration of inclusions in solids. One of the authors (Geguzin) is responsible for the review of the experimental results presented in Chaps. I and II and for the whole of Chap. III. The other author (Krivoglaz) is a theoretician: he is responsible for the first two chapters, with the exception of the sections dealing with experimental results.

Contents

CHAPTER III

Influence of the Migration of Inclusions on
High-Temperature Processes in Solids

One-Component Solids

1. MIGRATION MECHANISMS

When external forces are applied to a solid or a liquid second-phase particle or to a bubble of gas in a liquid, the particle or the bubble begins to move at a velocity determined by the Stokes formula. The moving inclusion forces the liquid apart and sets up viscous flow.

Phenomenologically similar processes may occur also in crystals containing inclusions* and subjected to external forces which act either on the inclusion as a whole or on some atoms in the inclusion or in the host crystal. However, the microscopic nature of the migration of inclusions in crystals and liquids must differ considerably because in a crystal an inclusion is bound tightly to the host lattice and in the absence of plastic flow this lattice is immobile (we shall always assume, unless otherwise specified, that there is no plastic flow).

In spite of the strong binding to its host crystal, an inclusion can move relative to the lattice because of the directional diffusion of atoms in the field of an external force, which can be quite rapid at high temperatures. Such a directional or ordered diffusion flux results in a gradual loss of atoms from layers in the host lattice which are on one ("front") side of the inclusion and the gradual "etching away" of these layers. Simultaneously, new atomic layers of the host lattice appear on the other ("rear") side

---

*An inclusion is a second-phase crystalline or liquid particle or a microcavity which can be empty or filled with gas.

of the inclusion. Consequently, a microcavity containing the inclusion (or filled with a gas) can move relative to the host lattice.

There are three possible mechanisms of the diffusional migration of inclusions in solids. The first is due to a diffusion flux of vacancies $\vec{I}_v$, which appears in a crystal matrix under the action of external forces and which gives rise to a flux of atoms of the same magnitude but opposite direction: $\vec{I} = -\vec{I}_v$.[†] Far from an inclusion the flux has a constant value $\vec{I}_\infty$, but near an inclusion the lines of flow are distorted and the flux $\vec{I}(\vec{r})$ depends on the radius vector $\vec{r}$. In particular, the lines of flow of vacancies or atoms may run from the boundaries of a grain or from other sources to the inclusion, or they may run from one part of the surface of the inclusion to another. Since the vacancy fluxes reaching the front and rear parts of the inclusion are different, atoms are gradually transferred from the layers in front of the inclusion to the layers behind it. These atoms fill the vacancies behind the inclusion and the inclusion as a whole moves relative to the crystal matrix.

We shall assume that the other two mechanisms of diffusional migration are unimportant and that, in particular, the host atoms do not penetrate the inclusion. Then, the velocity $\vec{v}'(\vec{r}_s)$ of an element of the surface S of an inclusion whose radius vector is $\vec{r}_s$ can be found by equating the volume $\vec{v}'(\vec{r}_s)\vec{n}(\vec{r}_s)\,dtdS$ swept through by the element dS of the surface S of the inclusion during its motion over a distance $\vec{v}'(\vec{r}_s)dt$ in a time $dt[\vec{n}(\vec{r}_s)$ is a unit vector of the outer normal to the surface S] to the volume $\omega\vec{I}(\vec{r}_s)\vec{n}(\vec{r}_s)dSdt$ (ω is the atomic volume in the host matrix) of the atoms which diffuse from a surface layer of area dS during a time dt (Fig. 1).[‡] This equality can be written in the form

$$\vec{v}'\left(\vec{r}_s\right)\vec{n}\left(\vec{r}_s\right) = \omega\vec{I}\left(\vec{r}_s\right)\vec{n}\left(\vec{r}_s\right). \tag{1.1}$$

[†] If the dominant mechanism is the interstitial rather than the vacancy diffusion, the flux of atoms is simply equal to the flux of defects and the signs of both are the same. All the results obtained in the present chapter are derived phenomenologically; they apply also to the interstitial diffusion.

[‡] By definition of the flux $\vec{I}$ the product $\vec{I}\vec{n}$ is equal to the number of atoms passing, per unit time, across a unit area of a plane perpendicular to the vector $\vec{n}$.

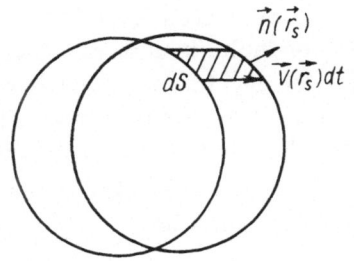

Fig. 1. Change in the volume of an inclusion resulting from the motion of a surface element of area dS over a distance $\vec{v}(\vec{r}_S)dt$; here $\vec{n}(\vec{r}_S)$ is a unit vector along the outer normal, drawn at the point $\vec{r}_S$.

If a crystal containing an inclusion is subjected to external forces, the velocities $\vec{v}'(\vec{r}_S)$ vary with time during an interval t comparable with the vacancy relaxation time

$$\tau \propto R^2/D_v \qquad (1.2)$$

(R is the characteristic dimension of the inclusion and D_v is the vacancy diffusion coefficient). In the case of a micropore, the velocities of different parts of the pore surface will be different during such a period. Therefore, the external forces not only cause translational motion of the pore but also alter its shape and volume. Such forces may also rotate (reorient) the pore relative to the host crystal. The velocity of translation of a pore $\vec{v}'$ may be defined as the average velocity of its different parts:

$$\vec{v}' = \frac{1}{S} \int \vec{v}'\left(\vec{r}_S\right) dS. \qquad (1.3)$$

The rate of change of the volume Ω of a pore is equal to the surface integral of the normal components of the velocities of different parts of the pore:

$$\frac{d\Omega}{dt} = \int \vec{v}'\left(\vec{r}_S\right)\vec{n}\left(\vec{r}_S\right) dS = \omega \int \vec{I}\left(\vec{r}_S\right)\vec{n}\left(\vec{r}_S\right) dS. \qquad (1.4)$$

After a time interval which is long compared with the relaxation time τ the relaxation processes are completed and a quasi-equilibrium distribution of the vacancies is established around an inclusion so that the velocities $\vec{v}'(\vec{r}_S)$ cease to depend on time. Simultaneously, the processes of reorientation and change of shape of the pore are completed and the pore now moves without a change in its shape (however, the volume can still change).

The velocities $\vec{v}'(\vec{r}_S)$ and $\vec{v}'$ defined by Eqs. (1.1) and (1.3) represent the motion of parts of the surface of an inclusion or of its center relative to the lattice of the host crystal [the fluxes $\vec{I}(\vec{r})$

are also defined relative to the system of coordinates linked to the lattice]. In considering the migration of inclusions due to the diffusion of vacancies in the bulk of the host crystal, we must bear in mind that vacancy fluxes also cause the host atoms to move within the lattice. Far from an inclusion, this flux of atoms relative to the lattice is $\vec{I}_\infty$. Consequently, the host matter far from the inclusion moves relative to the lattice at a velocity

$$\vec{v}_a = \frac{\vec{I}_\infty}{N_0} = \omega \vec{I}_\infty \, , \tag{1.5}$$

where $N_0 = 1/\omega$ is the number of atoms per unit volume of the host crystal. In particular, when a single crystal or a grain in a polycrystalline matrix is subjected to a uniform field of external forces, the boundaries of the single crystal or of the grains move at the velocity defined by Eq. (1.5). Consequently, the planes in the host lattice move relative to the crystal or its ends at a velocity $-\vec{v}_a$.

In addition to the velocities $\vec{v}'(\vec{r}_S)$ and $\vec{v}'$ in a system of coordinates linked to the host lattice, we can also consider the velocity $\vec{v}(\vec{r}_S)$ of an element of the surface of an inclusion and its transitional velocity $\vec{v}$ in a system of coordinates linked to the matter of the host crystal (far from the inclusion), or to the ends of the crystal. This system of coordinates is more natural because the position of an inclusion can then be specified easily by measuring the distances from the boundaries of a crystal or a grain. The velocities in these two coordinate systems are related by the following self-evident expressions:

$$\vec{v} = \vec{v}' - \vec{v}_a, \tag{1.6}$$

$$\vec{v}\left(\vec{r}_S\right) = \vec{v}'\left(\vec{r}_S\right) - \vec{v}_a, \quad \vec{v}\left(\vec{r}_S\right)\vec{n}\left(\vec{r}_S\right) = \omega \left[\vec{I}\left(\vec{r}_S\right) - \vec{I}_\infty\right]\vec{n}\left(\vec{r}_S\right). \tag{1.7}$$

In the diffusional migration of inclusions, the velocity of an inclusion should depend strongly on the state of the inclusion−host interface which determines the boundary conditions applicable to the vacancy diffusion fluxes in the host crystal. Let us assume that there are no vacancy sources or sinks at the boundary of a crystalline inclusion or in the bulk of the host crystal (in this case, the inclusion has a coherent boundary with an absolutely perfect host crystal). Let us also assume that the diffusion coefficient of the matter within the inclusion is considerably smaller than the diffusion coefficient of the host crystal. Then, the vacancy fluxes flow around the inclusion and, in the absence of other migration mechanisms, the inclusion is bound rigidly to the lattice and its velocity $\vec{v}'$ relative to the lattice is zero. Since the flux of vacancies flowing around the inclusion transports the host atoms and displaces the boundaries of the host crystal (or grain), the inclusion travels relative to the matter of the host crystal far from the inclusion (in particular, relative to the boundaries of the crystal or the grain) at a velocity $\vec{v} = -\vec{v}_a$ [see Eq. (1.6)].

In the other limiting case, when vacancies are created and annihilated easily at the boundary between an inclusion (for example, a liquid drop) and its host, so that the presence of the inclusion does not distort the field of forces acting on the host atoms, we may assume that the flux of vacancies $\vec{I}_{v\infty}$ in the bulk of the host is not distorted by the inclusion so that $\vec{I}_v(\vec{r}) = \vec{I}_{v\infty}$ and $\vec{I}(\vec{r}) = \vec{I}_\infty$. It then follows from Eqs. (1.1) and (1.3) that in the system of coordinates linked to the lattice the velocity of an inclusion

$$\vec{v}' = \vec{I}_\infty \omega$$

is equal to the velocity of matter in the host crystal far from the inclusion, given by Eq. (1.5). In the system of coordinates linked to the matter of the host crystal far from the inclusion, the inclusion is immobile because $\vec{v} = 0$, in accordance with Eq. (1.6).

An inclusion practically always distorts the force field and the flux $\vec{I}(\vec{r})$ near an inclusion depends on the coordinates. Therefore, in general, the velocity of an inclusion does not vanish in either of the two systems of the coordinates we have considered. However, generally speaking, the velocities $\vec{v}'$ and $\vec{v}$ differ considerably and they can have different signs (this is discussed in detail in Secs. 2-4).

Even when a crystal contains a liquid or gaseous inclusion, the efficiency of the inclusion boundary as a vacancy source or sink depends strongly on the structure of the inclusion. A perfectly flat part of the interface may act as a sink or source of only a few vacancies because an excessive number of vacancies would increase the chemical potential of the flat surface, disturb the dynamic equilibrium conditions, and hinder the further transfer of vacancies from the bulk to the interface. The host—inclusion interface can be an efficient vacancy sink if it is sufficiently rough; for example, if it has a large number of steps of atomic height. Then, vacancies will travel from the bulk of the host crystal to the edges of such steps and will gradually "eat away" these steps without altering the concentration of the surface vacancies or their chemical potential. This process is accompanied by the motion of the interface but the efficiency of the interface as a vacancy source or sink will not be affected if the degree of roughness of the interface is maintained. This situation is in many respects analogous to that encountered in crystallization, which is also accelerated by the presence of a rough surface.

A sufficiently rough inclusion—host interface is often generated by natural phenomena. In particular, a rough interface is obtained if the energy of a single step is sufficiently small. Then, fluctuations will continuously generate (and destroy) atomic-size steps on the surface of a small inclusion, and these steps may extend over the whole cross section of an inclusion so that the surface is rough. If an inclusion with an equilibrium shape is faceted, transition regions appear near the edges and these regions are covered by a dense network of grooves typical of the transition between faces with different indices. In the case of small inclusions, the widths of these regions may be comparable with the distances between the edges of the faces and this may also produce a rough interface. Atomic-size steps appear also on faces which depart from the singular orientations corresponding to low indices. Finally, a screw dislocation can interact with an inclusion, producing steps of the type observed in the spiral growth of a crystal.

In all the situations described above, the inclusion—host interface is sufficiently rough to be regarded as an efficient vacancy source or sink. We shall concentrate our attention on rough inclusion—host interfaces and we shall assume that vacancies appear and disappear easily at the interface. We shall also postulate that the rate of motion of an inclusion is controlled by the diffusion processes and not by the boundary conditions. Cases can occur, at least in principle, when the appearance and disappearance of vacancies at the inclusion—host interface is difficult. However,

these cases have received little attention and they will not be discussed here.

Another important point is the ease with which the matter in
an inclusion can occupy the vacancies resulting from the diffusion-
induced loss of the host atoms in the front part of the inclusion.
In the case of liquid, gaseous, or sufficiently plastic solid inclusions, these vacancies are filled immediately by the matter in the
inclusion and their presence has little effect on subsequent diffusion. This is true also of amorphous or incoherent inclusion—host
interfaces and of situations in which the diffusion mobility within
the inclusion is high. However, when a crystalline inclusion is
bound coherently to the host lattice or when the diffusion mobility
of the atoms within the inclusion is low, clusters of vacancies
formed in the host are not filled easily with the matter in the inclusion and they give rise to back-diffusion fluxes, which cause at
least partial dispersal of these clusters. The rate of diffusional
migration of such coherently bound inclusions is controlled (irrespective of the mechanism by which the host atoms diffuse) by
the rate at which the vacancies in the host lattice are occupied by
the matter present in the inclusion, i.e., by the volume or surface
diffusion within the inclusion. Under these conditions, the rate of
migration may be quite low. These special cases of the migration
of coherently bound inclusions will not be considered in the present
book.

The second mechanism of diffusional migration of inclusions
is related to the surface diffusion at the micropore—host interface, or to the boundary diffusion in a thin amorphous layer at the
interface between a solid inclusion and the host crystal. Directional surface or boundary diffusion fluxes, which appear under
the action of external forces, transport the host atoms from the
front to the rear part of an inclusion, so that a micropore or a
solid particle is set in motion. This mechanism is particularly
important at moderate temperatures, when the ratio of the surface (or boundary) to the volume diffusion coefficient is large, and
for small inclusions, when the ratio of the surface area of the inclusion to its volume is fairly large.

By definition of the surface (or boundary) flux of host atoms
$\vec{I}_s$, the product $\vec{I}_s \vec{n}_s$ ($\vec{n}_s$ is a unit vector along the tangent to the
surface) defines the number of atoms traversing, per unit time, a
unit length of a line which is perpendicular to the vector $\vec{n}_s$ and

lies in the surface (boundary) layer. This flux depends on the co-ordinates of the points on the surface $\vec{r}_S$, so that the numbers of atoms entering and leaving any surface element will not be equal. The change in the number of host atoms arriving in unit time on a unit area of the surface $\partial N_S / \partial t$ is added to the surface divergence $\mathrm{div}_S \vec{I}_S$ (with its sign reversed) of the flux $\vec{I}_S$ to give a condition analogous to the equation of continuity;*

$$\frac{\partial N_S}{\partial t} + \mathrm{div}_S \vec{I}_S = 0. \tag{1.8}$$

The volume $-\omega \dfrac{\partial N_S}{\partial t} dS = \omega \, \mathrm{div}_S \vec{I}_S \, dS$ of the atoms which leave the element in question per unit time (as a result of surface or boundary diffusion) should be equal to the volume $\vec{v}(\vec{r}_S)\,\vec{n}(\vec{r}_S)\,dS$ occupied by the corresponding element of the moving inclusion. Hence, we find that the velocity of an inclusion migrating under the influence of surface diffusion is given by

$$\vec{v}\left(\vec{r}_S\right)\vec{n}\left(\vec{r}_S\right) = \omega \, \mathrm{div}_S \vec{I}_S. \tag{1.9}$$

If the dominant mechanism is the surface diffusion and the diffusion in the bulk of the host crystal is unimportant, we can ignore the motion of the host matter relative to the lattice, and we need not make any allowance for the difference between the velocities $\vec{v}(\vec{r}_S)$ and $\vec{v}'(\vec{r}_S)$ in the coordinate systems linked to the host matter and to the lattice. Moreover, the volume of the inclusion should remain constant in the absence of vacancy diffusion fluxes in the bulk of the host.

---

* We can consider a rectangular element of the surface with a local rectangular system of coordinates x' and y'. The reduction (per unit time) in the number of atoms in this element can be expressed in terms of the difference between the fluxes $-\dfrac{\partial N_S}{\partial t} = = \vec{I}_S(x'+dx') - \vec{I}_S(x') + \vec{I}_S(y'+dy') - \vec{I}_S(y')$. Equation (1.8) is obtained by expanding the right-hand side of this expression with respect to dx' and dy', and defining the surface divergence in this system of coordinates by means of the expression $\mathrm{div}_S \vec{I}_S = \dfrac{\partial I_{Sx'}}{\partial x'} + \dfrac{\partial I_{Sy'}}{\partial y'}$. Obviously, the surface (and volume) divergence are independent of the system of coordinates.

The shape of an inclusion, its orientation relative to the host crystal, and its velocity may vary during the initial period after the application of external forces. This period is defined by $t \lesssim \tau$, where

$$\tau \propto R^2/D_S \qquad (1.10)$$

(D_S is the surface or boundary diffusion coefficient). However, after the end of this initial stage (when $t \gg \tau$), steady-state conditions are attained and all parts of an inclusion move at the same velocity $\vec{v}$.

The velocity of migration of an inclusion under the influence of surface diffusion should also depend strongly on the conditions at the inclusion−host interface. This is primarily because of the strong dependence of the surface transport coefficients (specifically, the surface diffusion coefficient) on the state of the interface. Moreover, the ease with which vacancies are formed in the layer of the host crystal next to the inclusion and the ease with which vacancies in the host lattice are filled by host atoms are also important in the migration of inclusions under the influence of surface diffusion. As in the volume diffusion mechanism, we shall assume that the inclusion−host interface is sufficiently rough, that it can provide the necessary sources or sinks of vacancies, and that the atoms of the inclusion can fill immediately the vacancies formed in the host lattice.

The third migration mechanism is due to the transport of the host atoms from one end of an inclusion to the other, across the inclusion itself. In the case of solid or liquid inclusions, this mechanism is possible if the solubility of the host atoms in the inclusion and the value of the diffusion coefficient of these atoms inside the inclusion are sufficiently high. At inclusion moves as a result of the dissolution of the host atoms in the front part of the inclusion, the diffusion of these atoms across the inclusion, and their settling in the host lattice on the rear side of the inclusion.

We shall use $\vec{i}'$ to denote the flux of the host atoms inside an inclusion. The arrival of $\vec{i}'(\vec{r}_S)\vec{n}(\vec{r}_S)\,dS$ atoms (per unit time) at the inclusion−host interface alters the volume of the host crystal, bounded by a surface area element dS, by an amount $\omega \vec{i}'(\vec{r}_S)\vec{n}(\vec{r}_S)\,dS$. This change should be equal to the volume $-\vec{v}(\vec{r}_S)\vec{n}(\vec{r}_S)\,dS$, which is

liberated by the moving inclusion. Therefore, we obtain the following expression for the steady-state velocity of the inclusion

$$\vec{v}\left(\vec{r_s}\right)\vec{n}\left(\vec{r_s}\right) = -\omega\vec{I'}\left(\vec{r_s}\right)\vec{n}\left(\vec{r_s}\right).\qquad(1.11)$$

As in the case of the migration mechanism controlled by surface diffusion, the volume of an inclusion remains constant and the
velocities of all the parts of the inclusion—host interface are the
same (for $t \gg \tau$). Once again, no distinction need be made between
the velocities $\vec{v'}(r_s)$ and $\vec{v}(r_s)$.

The diffusion fluxes $\vec{I'}$ are obviously governed not only by the
diffusion mobility of the host atoms but also by the rate of their
dissolution in the inclusion. Therefore, the velocity of an inclusion
may depend strongly on the conditions at the inclusion—host interface (a detailed discussion will be given in Sec. 2).

Gas-filled inclusions (bubbles) may also move under the influence of volume diffusion across the inclusion if the rate of
evaporation of atoms from the front part of the interface is sufficiently high and a directional flux of the host atoms is established
across the gaseous phase or vacuum. In this case, the velocity of
a pore is again given by Eq. (1.11) where $\vec{I'}$ should be regarded as
the flux of the host atoms carried by evaporation.

Usually, comparable contributions to the velocity of an inclusion are made by several mechanisms. The contributions of different fluxes in the bulk of the host crystal and in the inclusion
and on the surface of the host crystal are obviously additive.
Therefore, it follows from Eqs. (1.7), (1.9), and (1.11) that the
velocity of an element of the surface of an inclusion $\vec{v}(r_s)$, expressed in the system of coordinates linked to the matter in the
host crystal (i.e., to the boundaries of this crystal or of grains)
is given by the formula

$$\vec{v}\left(\vec{r_s}\right)\vec{n}\left(\vec{r_s}\right) = \omega\left[\vec{I}\left(\vec{r_s}\right) - \vec{I_\infty} - \vec{I'}\left(\vec{r_s}\right)\right]\vec{n}\left(\vec{r_s}\right) + \omega\,\mathrm{div}_S\vec{I_S}.\qquad(1.12)$$

The directional diffusion fluxes resulting in the motion of an
inclusion may be due to various generalized thermodynamic forces
arising from the presence of a temperature gradient, an electric
field, an elastic stress gradient, an inhomogeneous magnetic field,
a vacancy concentration gradient, etc. The diffusional migration

of inclusions under the action of these forces will be considered more fully later. The present chapter will be restricted to the simplest case of one-component crystals, i.e., crystals consisting of atoms of one kind.

The theory of the diffusional migration of inclusions was first developed for motion in a temperature gradient as a result of diffusion inside the inclusion [1-5]. This was followed by the extension of the theory to the mechanism controlled by surface diffusion [4, 6-9]. The motion of inclusions due to diffusion in the host crystal was also considered in [7] but no allowance was made for the important difference between the velocities relative to the lattice and to the boundaries of the host crystal. A correct allowance for the motion of matter relative to the lattice was made in [10, 11]. The migration of inclusions in an inhomogeneous stress field, which can be due to various mechanisms, was considered in [10].

The migration of inclusions in metals subjected to an electric field was discussed in [12, 13] and the motion in an inhomogeneous magnetic field was considered in [14].

2. MIGRATION OF INCLUSIONS
IN A TEMPERATURE GRADIENT

Thermal Diffusion of Atoms

A temperature gradient in a crystal gives rise to a directional diffusion flux of vacancies. Since the motion of vacancies is accompanied by the interchange of sites with neighboring atoms, it must create a flux of atoms relative to the crystal lattice. In the linear theory, this flux $\vec{I}$, considered in a coordinate system linked to the lattice, is given by

$$\vec{I} = -N_0 \frac{\alpha}{T} D\nabla T, \qquad (2.1)$$

where D is the self-diffusion coefficient; α is the thermal diffusion ratio, divided by the number of atoms per unit volume N_0.

The product αD is, generally speaking, a tensor. However, in the case of cubic crystals, as well as in the case of liquids or gases, the quantities α and D are scalar. For simplicity, we shall restrict our treatment to cubic crystals and to liquids and gases.

An inhomogeneous temperature field gives rise also to surface or boundary fluxes $\vec{I}_S$ along the free surface of a crystal or along its interface with a second phase. The magnitudes of these fluxes are determined by the surface temperature gradient $\nabla_S T$ ($\nabla_S T$ is the projection of the vector ∇T onto the surface in question) and the expression for $\vec{I}_S$ is

$$\vec{I}_S = -N_0 \frac{\alpha_S}{T} D_S a \nabla_S T, \tag{2.2}$$

where $N_0 \alpha_S$ is the thermal diffusion ratio for surface diffusion; D_S is the surface diffusion coefficient for a monatomic layer of thickness a, i.e., this coefficient is defined as if the surface or boundary diffusion flux were localized entirely within a layer of thickness a.

Strictly speaking, the surface or boundary diffusion coefficient D_S and the product $\alpha_S D_S$ are tensors of the second rank (and not scalars) even in cubic crystals: they depend on the orientation of the surface element of the host crystal relative to its crystallographic axes. However, in order to simplify the problem, we shall assume that D_S and α_S are scalar quantities which are independent of the orientation of the surface of the host crystal. The effects associated with the anisotropy of the surface transport coefficients will be discussed separately in Sec. 7.

It is very difficult to carry out a microscopic calculation of the transport coefficients α and α_S. In particular, an allowance must be made for the fact that the diffusion in a temperature gradient is not only due to thermodynamic factors (the tendency for the entropy of the system to increase) but also due to the drag exerted on atoms by a directional flux of phonons (the phonon wind), which results from the presence of the temperature gradient. When phonons are scattered by an atom which is overcoming a potential barrier surrounding a vacancy, they transfer their momentum to this atom, producing the preferential motion of the atom along the direction of ∇T. Similarly, in metals with aspherical Fermi surfaces, the diffusing atoms are dragged by an electron wind which appears in a temperature gradient (the corresponding electric current is zero because oppositely directed electron fluxes corresponding to different parts of the Fermi surface are produced in a ∇T field but the drag forces are not generally balanced out and the resultant force acts on the diffusing atoms) [15].

The phonon and electron scattering cross sections of diffusing atoms cannot be calculated sufficiently accurately and all we can do is to estimate their orders of magnitude. Therefore, in the phenomenological theory used in the present book, the transport coefficient α will be regarded as a parameter which can be determined experimentally. We shall simply bear in mind the fact that α is dimensionless, that it depends weakly on the temperature (in contrast to D), and that its value is in the range 1-10.

Temperature Distribution Near a Spherical Inclusion

Far from an inclusion the temperature is a linear function of distance and the temperature gradient has the constant value ∇T_∞. Since the thermal conductivity of an inclusion, λ_0, generally differs from the thermal conductivity of the host crystal, λ, the temperature distribution $T(\vec{r})$ near an inclusion is distorted. This distorted steady-state temperature distribution can be found by solving the equation

$$\Delta T = 0 \qquad (2.3)$$

inside and outside the inclusion, subject to the boundary condition at infinity

$$T(\vec{r}) = T_0 + \vec{r}\,\nabla T_\infty \qquad (r \to \infty) \qquad (2.4)$$

(T$_0$ is a constant) and to the boundary conditions specifying continuity of the temperature and heat flow on the surface of the inclusion

$$T^+(\vec{r}) = T^-(\vec{r}), \quad \lambda\vec{n}\nabla T^+(\vec{r}) = \lambda_0\,\vec{n}\,\nabla T^-(\vec{r}) \qquad (\vec{r} = \vec{r}_s) \qquad (2.5)$$

[$T^+(\vec{r}_s)$ and $T^-(\vec{r}_s)$ represent, respectively, the temperature of the host crystal and of the inclusion at the inclusion—host interface].†

The solution of this boundary-value problem can be represented in an explicit form if the inclusion in question is of sufficiently simple shape. If we assume that the surface tension at the inclusion—host interface is isotropic, we may expect the inclusion to have a spherical shape. In this case, the solution of Eq. (2.3) subject to the boundary conditions (2.4) and (2.5) can be found quite easily:

$$\left.\begin{aligned} T\,(\vec{r}) &= T_0 + \vec{r}\,\nabla T_\infty + \varkappa\,\frac{R^3}{r^3}\,\vec{r}\,\nabla T_\infty \qquad (r \geqslant R), \\ T\,(\vec{r}) &= T_0 + (1 + \varkappa)\,\vec{r}\,\nabla T_\infty \qquad (r \leqslant R), \end{aligned}\right\} \tag{2.6}$$

where R is the radius of the spherical inclusion and the constant $\varkappa$ can be expressed in terms of the thermal conductivities λ and λ_0:

$$\varkappa = \frac{\lambda - \lambda_0}{2\lambda + \lambda_0}. \tag{2.7}$$

The validity of the solution given by Eq. (2.6) can be established quite easily by direct substitution. In particular, if the thermal conductivity of the inclusion is much lower than the thermal conductivity of the host crystal (this is true, for example, of a gas-filled probe in a metal), we find that

$$\varkappa \approx \frac{1}{2} \quad (\lambda_0 \ll \lambda). \tag{2.8}$$

The temperature gradient near an inclusion is not constant and, according to Eq. (2.6), it is given by the formula

$$\left.\begin{aligned} \nabla T\,(\vec{r}) &= \left(1 + \varkappa\,\frac{R^3}{r^3}\right)\nabla T_\infty - 3\varkappa\,\frac{R^3}{r^3}\,(\vec{n}\,\nabla T_\infty)\,\vec{n} \qquad (r > R), \\ \nabla T\,(\vec{r}) &- (1 + \varkappa)\,\nabla T_\infty \qquad (r < R), \end{aligned}\right\} \tag{2.9}$$

where $\vec{n} = \dfrac{\vec{r}}{r}$.

---

† Strictly speaking, an allowance should be made in Eq. (2.5) for the evolution or absorption of heat at the inclusion—host interface as a result of the dissolution of atoms or vacancies and for the change in the position of the inclusion with time. However, since the velocity of inclusions is low (because the diffusion coefficients are small compared with the thermal conductivity) such allowances need not be made (relevant estimates are given in [1, 2]).

In particular, on the surface of a sphere we have

$$T(\vec{r}) = T_0 + (1 + \varkappa)\vec{r}_S \nabla T_\infty = T_0 + (1 + \varkappa) R \,|\nabla T_\infty| \cos\theta \quad (r = R), \quad (2.10)$$

where θ is the angle between the radius vector of the surface point $\vec{r}_S$ under consideration and the direction of the temperature gradient ∇T_∞.

In the case of bubbles migrating at high temperatures, we must make allowance not only for the thermal conductivity of the gaseous phase but also for the radiative heat exchange. Then, the heat flux $-\lambda_0 \nabla T$, transported by the gas atoms in the inclusion, must be supplemented by the heat flux $-\frac{2}{3}\lambda' R \nabla T$ carried by photons ($\lambda' = \partial\Lambda/\partial T$ is the derivative, with respect to the temperature, of the rate of emission of radiation Λ by a unit surface of the crystal). Therefore, in this case, Eqs. (2.6)-(2.10) should be modified by the substitution

$$\lambda_0 \to \lambda_{0\,\text{eff}} = \lambda_0 + \frac{2}{3}\lambda' R. \quad (2.11)$$

The effective thermal conductivity of the inclusion is now $\lambda_{0\,\text{eff}}$ and the characteristic constant $\varkappa$, which determines the distortion of the temperature field near the inclusion, depends on the radius R.

However, even at high temperatures, the second term in Eq. (2.11) is usually small compared with the first. According to the Stefan−Boltzmann law, we have

$$\lambda' \propto 2.3 \cdot 10^{-7} ET^3, \; J \cdot m^{-2} \cdot sec^{-1} \cdot deg^{-1}$$

or

$$\lambda' \propto 5.5 \cdot 10^{-12} ET^3, \; cal \cdot cm^{-2} \cdot sec^{-1} \cdot deg^{-1},$$

where E is the relative emissivity of the surface. Hence, it follows that when $E \sim 1$, $T \sim 2 \times 10^3 \,°K$, and $R \sim 10^{-4}$ cm, $\lambda' R \approx 2 \times 10^{-2} \, J \cdot m^{-1} \cdot sec^{-1} \cdot deg^{-1}$ [5×10^{-6} cal $\cdot$ cm^{-1} $\cdot$ sec^{-1} $\cdot$ deg^{-1}], i.e., $\lambda' R \ll \lambda_0$ (with the exception of pores filled with a gas under very low pressure) and $\lambda_{0\,\text{eff}} \approx \lambda_0$.

The value of $\lambda_{0\,\text{eff}}$ may depend on R in the case of pores of very small radius (when R if comparable with or smaller than the mean free path of molecules in a gas l_g). If $R \ll l_g$, we find that

$$\lambda_{0\,\text{eff}} \propto \lambda_0 \frac{R}{l_g} \quad (R \ll l_g). \tag{2.12}$$

However, the influence of the dependence $\lambda_{0\,\text{eff}}(R)$ on $\varkappa$ and on the temperature distribution near an inclusion is important only in the case of pores in dielectric crystals with a low thermal conductivity, i.e., when λ_0 is comparable with λ.

Distribution of Vacancies and Steady-State Conditions

In considering diffusion fluxes resulting from the temperature gradient given by Eq. (2.9), we must bear in mind that the concentration of vacancies c_v in an inhomogeneous temperature field is a function of the coordinates and the expression for the flux $\vec{I}$ should include a term proportional to ∇c_v. However, if the distances l between the vacancy sources and sinks are small compared with the characteristic lengths in the problem — in particular, compared with the radius R — we find that $c_v(\vec{r})$ is equal to the equilibrium concentration c_v^0 and it depends on the temperature $T(\vec{r})$ at the point in question $\vec{r}\,(c_v(\vec{r}) = c_v^0\{T(\vec{r})\})$. In this case, the gradient $\nabla c_v(\vec{r})$ is proportional to $\nabla T(\vec{r})$ and the term proportional to $\nabla c_v(\vec{r})$ can be regarded as included in Eq. (2.1) if α is suitably renormalized. It is this renormalized value that is usually determined experimentally.

If the distance between vacancy sources and sinks is large compared with R, the vacancy distribution can also be regarded as being in equilibrium and Eq. (2.1) can be used provided vacancies appear and disappear freely on the inclusion—host interface and the distribution of vacancies is determined by the steady-state condition $\Delta c_v = 0$. In fact, $c_v(\vec{r})$ and $T(\vec{r})$ are defined by identical Laplace equations [because the thermal diffusivity is large, the steady-state distribution $T(\vec{r})$ with $\Delta T = 0$ is established relatively rapidly] and the value of $c_v(\vec{r})$ at the interface assumes the equilibrium value $c_v(\vec{r}) = c_v^0\{T(\vec{r})\}$. Therefore, in the linear theory, in which it is postulated that $\delta c_v \, (\vec{r})$ is small compared with $c_v \, (\vec{r})$, the concentration $c_v \, (\vec{r})$ at each point in the host crystal should also assume its equilibrium value and Eq. (2.1) can be used.

The condition under which vacancies are easily created and annihilated is always satisfied at the interface of a liquid or gaseous inclusion (micropore) and in the case of a crystalline inclusion separated from the host by a thin amorphous layer. If this layer is absent and the equilibrium concentration of vacancies is not established at the interface, we must consider separately the fluxes associated with the temperature gradient and with the vacancy distribution. Under these conditions, the results are naturally different. However, the absence of an amorphous layer at the interface between a crystalline inclusion and its host is relatively rare and it will not be considered here in detail.

A steady-state distribution of vacancies is obtained if the right-hand side of the diffusion equation

$$\frac{1}{D_v} \frac{\partial c_v}{\partial t} = \Delta c_v \, ,$$

i.e., the term $(1/D_v)(\partial c_v / \partial t)$, can be ignored compared with Δc_v. This can be done even if $l \gg R$ [for $t \gg \tau \propto R^2/D_v$, as indicated by Eq. (1.2)] and the distribution of vacancies is of the quasisteady-state type if the velocity of an inclusion v' is small. In this case, in a time $t_0 \sim R/v'$ the inclusion travels a distance which is of the order of its radius R and the vacancies diffuse over a distance $\sim \sqrt{D_v t_0} \sim \sqrt{D_v R/v'}$, which is large compared with R. This situation is obtained if

$$v' \ll \frac{D_v}{R} \, , \tag{2.13}$$

which is the condition of validity of the linear theory of the diffusional migration of inclusions in one-component crystals. If $l \ll R$, an equilibrium distribution of vacancies is established because of the presence of sources in the bulk of the host crystal and the condition imposed on the velocity of an inclusion is even less stringent ($v' \ll D_v R/l^2$).

Velocity of Inclusions Migrating

Under the Influence of Volume

Diffusion in the Host Crystal

Equations (2.1), (2.2), (2.9), and (2.10) can be used to find the diffusion fluxes of atoms and to determine [in the linear theory subject to the condition (2.13)] the velocity of an inclusion migrating under the influence of various diffusion mechanisms. For example, the volume thermal diffusion of the host atoms near the surface of a spherical inclusion, on which vacancies can be created and annihilated easily, is given by [Eqs. (2.1) and (2.9)]

$$\vec{I} = - N_0 \frac{\alpha}{T} D \left[(1 + \varkappa) \nabla T_\infty - 3\varkappa \left(\vec{n} \nabla T_\infty \right) \vec{n} \right] \quad (r = R). \tag{2.14}$$

If we substitute Eq. (2.14) into Eq. (1.1), we find that the velocities $\vec{v}'(\vec{r}_S)$ of elements of the surface of an inclusion, resulting from the diffusion of atoms in the bulk of the host crystal, are given by

$$\vec{n} \left(\vec{r}_S \right) \vec{v}' \left(\vec{r}_S \right) = - \frac{\alpha}{T} D (1 - 2\varkappa) \left(\vec{n} \left(\vec{r}_S \right) \nabla T_\infty \right). \tag{2.15}$$

This equation is derived on the assumption that $N_0 \omega = 1$. In this case, $\vec{v}'(\vec{r}_S)$ is independent of $\vec{r}_S$ and the inclusion travels relative to the lattice of the host crystal at a constant velocity $\vec{v}'$ given by

$$\vec{v}' = - \frac{\alpha}{T} D (1 - 2\varkappa) \nabla T_\infty. \tag{2.16}$$

It must be stressed that all parts of a spherical inclusion move at the same velocity at all times, including the initial period immediately after the application of a temperature gradient. This means that a spherical surface is not distorted and an inclusion (in particular, a micropore) retains its spherical shape also when it moves in a homogeneous field resulting from a temperature gradient. The invariance of the shape of an inclusion is due to the fact that the normal component $\vec{I}\vec{n}$ of the flux $\vec{I}$ and, consequently, the volume of atoms which leave the element of an inclusion under consideration as a result of diffusion are both proportional to $\cos \theta$, in accordance with Eq. (2.14). This angular dependence of the volume occupied by an element of the surface of an inclusion is possible only if the inclusion retains its spherical shape (Fig. 2).

If we substitute Eq. (2.15) into Eq. (1.4) and assume that $\vec{n}\,(\vec{r_S}) = \vec{r_S}/r_S,$, we can easily show that $d\Omega/dt = 0$, i.e., the thermal diffusion of vacancies does not alter the volume of a micropore. This happens because, according to the linear theory, a homogeneous field of external forces gives rise to a flux of vacancies traveling toward the front half of a pore which is exactly equal to the flux of the vacancies leaving the rear half. Consequently, the volume of the pore remains constant if there are no other factors which could change it.

We shall now go over to the natural system of coordinates which is linked to the matter in the host crystal far from an inclusion, or to the boundaries of the host crystal. It follows from Eqs. (1.5) and (2.1) that

$$\vec{v_a} = -\frac{\alpha}{T} D\nabla T_\infty$$

and, therefore, Eqs. (1.6) and (2.16) yield an expression for the velocity of an inclusion in this system of coordinates:

$$\vec{v} = 2\varkappa \frac{\alpha}{T} D\nabla T_\infty . \qquad (2.17)$$

If the thermal conductivity of an inclusion is much lower than the thermal conductivity of its host and if $\varkappa \approx 1/2$ in accordance with Eq. (2.8), we find that the velocity v' relative to the lattice of the host crystal is much lower than the velocity v relative to the matter in the host crystal far from the inclusion (the inclusion moves relative to the matter in the host crystal far from the inclusion at almost the same velocity $\vec{v} \approx \frac{\alpha}{T} D\nabla T_\infty = -\vec{v_a}$ as the mat-

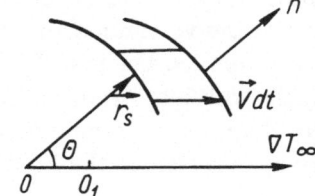

Fig. 2. Motion of a spherical inclusion.

ter in the host crystal moves relative to the lattice, but the directions of motion are opposite). On the other hand, if the thermal conductivities of an inclusion and its host are comparable, it follows from Eq. (2.7) that $\varkappa \ll 1$ and Eqs. (2.16), (2.17) yield $v \ll v'$, i.e., the inclusion moves comparatively slowly relative to the matter in the host crystal (this matter is considered at a large distance $r \gg R$ from the inclusion) but comparatively rapidly relative to the lattice. If $\lambda_0 < \lambda$, the velocities $\vec{v}$ and $\vec{v}'$ have opposite signs.

If the dependence of $\lambda_{0\,\text{eff}}$ (i.e., of $\varkappa$) on the radius R is unimportant, it follows from Eq. (2.17) that the radius of an inclusion does not affect its velocity. This is because, in the case of volume diffusion in the bulk of the host crystal, the diffusion fluxes are independent of R.

The influence of temperature on the velocity is mainly due to the exponential temperature dependence of the volume self-diffusion coefficient of the host atoms: when the temperature is increased, the velocity of an inclusion rises rapidly. The dependence of $\vec{v}$ on the temperature gradient is linear and the direction of motion of the inclusion is parallel to ∇T_∞.

We shall estimate possible values of the velocity in the case of volume diffusion in the host crystal by postulating that, for example, $2\varkappa \sim 1$, $\alpha \sim 1$, $T \sim 10^3 \,°K$, $|\nabla T_\infty| \sim 10^3$ deg /cm, $D \sim 10^{-8}$ cm^2/ sec (this self-diffusion coefficient has been obtained for uranium atoms at $0.8 T_{mp}$, where T_{mp} is the melting point). It then follows from Eq. (2.17) that $v \sim 1$ Å/sec. At lower temperatures, the value of v is much less. For example, if $D \sim 10^{-10}$ cm^2/sec, $v \sim 10^{-10}$ cm/sec.

We pointed out earlier that these results are applicable to the inclusion velocity if vacancies can appear or disappear easily at the inclusion—host interface or in the bulk of the host crystal. If the host crystal is sufficiently close to being perfect and vacancies cannot disappear easily on the inclusion—host interface, the lines of flow of the vacancies in a temperature gradient do not terminate on this interface but may bypass the inclusion (at least partially). Then, the velocity of an inclusion may be quite different. For example, there may be no vacancy sources and sinks in the bulk and on the boundaries of the host crystal and the inclusion may be impermeable to vacancies (this is the limiting case considered in Sec. 1). Under these conditions, the inclusion is rigidly bound to the lattice of the host crystal and it does not move relative to the lattice in a temperature gradient $(\vec{v}'=0)$. Then, the velocity of an inclusion relative to the host matter far from it is

$$\vec{v} = -\vec{v}_a = \frac{\alpha}{T} D \nabla T_\infty \qquad (2.17a)$$

and it differs in magnitude (and in direction, if $\varkappa < 0$) from the velocity of an inclusion with an amorphous boundary.

Velocity of Inclusions Migrating Under the Influence of Surface Diffusion

In order to determine the velocity of an inclusion moving under the influence of surface diffusion, we must consider the thermal diffusion flux of the host atoms at the inclusion—host interface, which is given by Eq. (2.2). The surface divergence of this flux

$$\text{div}_S \vec{I}_S = -N_0 \frac{\alpha_S}{T} D_S a \Delta_S T \left(\vec{r}_S\right) \qquad (2.18)$$

can be expressed in terms of the surface Laplace operator Δ_S of the temperature of the inclusion surface. This operator, which is defined on the surface of a sphere, acts only on the angle-dependent part of a given function. In particular, when applied to the function $\cos\theta$, the operator Δ_S simply multiplies this function by a constant:

$$\Delta_S \cos\theta = -\frac{2}{R^2}\cos\theta. \qquad (2.19)$$

If we use Eqs. (2.10) and (2.19), we can find the divergence $\text{div}_S \vec{I}_S$ from Eq. (2.18) and then apply Eq. (1.9) to calculate the velocity of an element of the surface of an inclusion:

$$\vec{n}\left(\vec{r}_S\right)\vec{v}\left(\vec{r}_S\right) = \frac{2(1+\varkappa)a}{R}\frac{\alpha_S}{T} D_S \left|\nabla T_\infty\right|\cos\theta$$

$$= \frac{2(1+\varkappa)a}{R}\frac{\alpha_S}{T} D_S \vec{n}\left(\vec{r}_S\right)\nabla T_\infty. \qquad (2.20)$$

It follows that a spherical inclusion moving under the influence of surface diffusion retains its shape and travels parallel to the temperature gradient at a velocity [8, 9]:

$$\vec{v} = \frac{2(1+\varkappa)a}{R} \frac{\alpha_S}{T} D_S \nabla T_\infty .$$ (2.21)

Under surface diffusion conditions, the velocity of an inclusion is inversely proportional to its radius. This dependence of $\vec{v}$ on R is due to the fact that the total diffusion flux along the surface of an inclusion is proportional to the perimeter $2\pi R$ and the flux per unit area (divided by πR^2) is inversely proportional to R.[†]

In this case, the temperature dependence of $\vec{v}$ is determined by the factor D_S. Since the activation energy of surface or boundary diffusion is always much lower than the corresponding energy of volume diffusion, it follows that the velocity $\vec{v}$ of inclusions moving under the influence of surface diffusion depends much less on the temperature than in the case when inclusions move under the influence of volume diffusion in the bulk of the host crystal.

Equation (2.21) differs from Eq. (2.17) by an additional small factor a/R and by the replacement of D with D_S. The orders of magnitude of all the other quantities in these two equations are the same. Therefore, the surface diffusion mechanism should dominate the motion of small-radius inclusions when the ratio a/R is not too small, and it should be important at relatively low temperatures when the ratio D_S/D is large.

For example, if $\varkappa \approx 1/2$, $\alpha_S \sim 1$, $T \sim 10^3$ °K, $|\nabla T_\infty| \sim 10^3$ deg/cm, $D_S \sim 10^{-6}$ cm^2/sec, we find that Eq. (2.21) yields v $\sim$ 10 Å/sec for R $\sim$ 100 Å and v $\sim$ 0.1 Å/sec for R $\sim 10^4$ Å.

[†] In principle, the diffusion mobility of the host atoms may be considerably higher along certain lines (for example, the edges of a polyhedral inclusion or along dislocation loops surrounding an inclusion). The corresponding activation energy may be considerably lower than that for other parts of the inclusion—host interface. If the temperature is sufficiently low, the migration of small-radius inclusions is then due to linear diffusion. In this case, the velocity must be a function of the radius of the inclusion: it should be proportional to $1/R^2$.

Velocity of Inclusions Migrating

Under the Influence of Internal

Diffusion

The diffusion flux across an inclusion depends strongly on the conditions at the inclusion—host interface, i.e., on how rapidly the host atoms cross this interface.[†] Therefore, the diffusion flux will be controlled either by the rate of diffusion in the bulk of the inclusion or by the processes occurring at the interface.

If the rate of exchange of the host atoms with a liquid or a crystalline inclusion is sufficiently high, an equilibrium concentration $c_A^0 \{T(\vec{r}_s)\}$ (corresponding to the temperature T of the region being considered) of the host atoms A, dissolved in the inclusion, is established at the inclusion—host interface. If we use Eq. (2.10) and expand c_A^0 with respect to the small change in the temperature associated with the gradient ∇T_∞, we can rewrite the concentration of the A atoms at the inclusion—host interface in the form

$$c_A(\vec{r}_S) = c_A^0 + \frac{\partial c_A^0}{\partial T}(1 + \varkappa)\,\vec{r}_S \nabla T_\infty \quad (r = R), \qquad (2.22)$$

where $c_A^0 = c_A^0(T_0)$. The solution of the equation

$$\Delta c_A = 0 \qquad (2.23)$$

for a steady-state distribution of the A atoms in a spherical inclusion, which satisfies the boundary conditions of Eq. (2.22) is of the form:

$$c_A(\vec{r}) = c_A^0 + \frac{\partial c_A^0}{\partial T}(1 + \varkappa)\,\vec{r}\,\nabla T_\infty. \qquad (2.24)$$

The inhomogeneous atomic distribution of Eq. (2.22) and the inhomogeneous temperature distribution of Eq. (2.6) are due to the ordinary diffusion and thermal diffusion fluxes of the A atoms in

---

[†] We shall consider only that case in which the atoms present in an inclusion are practically insoluble (and do not diffuse) in the host crystal at the temperatures under investigation.

an inclusion. It follows from Eqs. (2.9) and (2.24) that the total flux is[†]

$$\vec{I'} = - N_0' D_A \nabla c_A(\vec{r}) - N_0' \frac{\alpha_A}{T} c_A(\vec{r}) D_A \nabla T$$

$$= - N_0'(1 + \varkappa) D_A \left(\frac{\partial c_A^0}{\partial T} + \frac{\alpha_A c_A^0}{T} \right) \nabla T_\infty , \qquad (2.25)$$

where N_0' is the total number of atoms per unit volume of the inclusion; D_A is the diffusion coefficient of the A atoms across the inclusion; $N_0 c_A \alpha_A$ is the thermal diffusion ratio for these atoms.

In the case of cubic crystalline inclusions or liquid and gaseous inclusions, the quantities D_A and α_A are scalar.

It is evident from Eqs. (1.11) and (2.25) that in the case of motion of spherical inclusions under the influence of volume diffusion inside the inclusion and under the influence of the mechanisms considered earlier, the inclusion is not distorted (the product $\vec{In}$ is proportional to cos θ) and it moves at a velocity

$$\vec{v} = (1 + \varkappa) \frac{\omega}{\omega'} D_A \left(\frac{\partial c_A^0}{\partial T} + \frac{\alpha_A c_A^0}{T} \right) \nabla T_\infty , \qquad (2.26)$$

where $\omega' = 1/N_0'$ is the atomic volume in the inclusion.

In this case, as in the case of volume diffusion in the host crystal, the diffusion flux per unit area and, therefore, the velocity $\vec{v}$ are independent of the radius of the inclusion (provided $\varkappa$ is independent of R). The influence of temperature on $\vec{v}$ is primarily due to the temperature dependence of the product $D_A c_A^0$. In the case of liquid inclusions, this dependence can be weaker than for other migration mechanisms.

A comparison of Eqs. (2.26) and (2.17) shows that the mechanism considered here is important if $c_A^0 D_A \gg D$. This condition is usually satisfied by liquid inclusions which appreciably dissolve the host substance.

[†] We shall initially assume that there are no convection currents in liquid inclusions. Such currents may enhance considerably the transport of the host atoms across an inclusion and raise its velocity (this point will be considered later).

If the diffusion coefficient D_A is sufficiently large, the diffusion fluxes are controlled not only by the rate of diffusion but also by the rate of dissolution of the host atoms in the inclusion, i.e., by the processes which occur at the inclusion–host interface. In this case, the concentration $c_A(\vec{r}_S)$ of the A atoms in the inclusion close to the interface will differ from the equilibrium concentration $c_A^0\{T(\vec{r}_S)\}$.

The concentration at the interface $c_A(\vec{r}_S)$ can be found by equating the normal projections of the fluxes of the A atoms crossing the interface and diffusing in the inclusion (continuity conditions). If we assume that the rate of dissolution is proportional to the difference between the equilibrium and the actual concentrations of the A atoms at the interface, we can rewrite this condition in the form

$$\vec{n}\vec{I'}(\vec{r}) = \beta \left[c_A(\vec{r}) - c_A^0 \{T(\vec{r})\} \right] - \beta_T c_A^0 \, \vec{n} \nabla T^-(\vec{r})$$

$$= -N_0' D_A \vec{n} \nabla c_A(\vec{r}) - N_0' \frac{\alpha_A}{T} c_A(\vec{r}) D_A \vec{n} \nabla T^-(\vec{r}) \qquad (2.27)$$

$$(\vec{r} = \vec{r}_S)$$

or

$$\beta \left[c_A(\vec{r}) - c_A^0 \right] + N_0' D_A \vec{n} \nabla c_A(\vec{r}) =$$

$$= (1 + \varkappa) \left[\beta \frac{\partial c_A^0}{\partial T} R + \beta_T c_A^0 - N_0' \frac{\alpha_A}{T} c_A^0 D_A \right] (\vec{n} \nabla T_\infty) \quad (\vec{r} = \vec{r}_S), \qquad (2.28)$$

where β and β_T are the transport coefficients representing, respectively, the rate of dissolution of the A atoms under isothermal conditions and their rate of dissolution under the influence of phonon and electron fluxes which appear in a temperature gradient.

The solution of Eq. (2.23) for the interior of a spherical inclusion satisfying the boundary conditions of Eq. (2.28) is of the form

$$c_A(\vec{r}) = c_A^0 + (1 + \varkappa) \frac{\beta \dfrac{\partial c_A^0}{\partial T} R + \beta_T c_A^0 - N_0' \dfrac{\alpha_A}{T} c_A^0 D_A}{\beta R + N_0' D_A} \vec{r} \nabla T_\infty. \qquad (2.29)$$

In the case of macroscopic inclusions with $R \gg a$, the term $\beta_T c_A^0$

can be ignored because the order of magnitude of the first factor is $\beta_T \sim \beta a\alpha/T$. Substituting Eq. (2.29) into Eq. (2.27), we can find the normal component of the flux $\vec{i}'$ and then the velocity of an inclusion can be derived from Eq. (1.11):

$$\vec{v} = (1 + \varkappa)\frac{\omega}{\omega'}\ D_A \frac{\beta R}{\beta R + N_0' D_A}\left(\frac{\partial c_A^0}{\partial T} + \frac{\alpha_A}{T}\ c_A^0\right)\nabla T_\infty. \qquad (2.30)$$

If the inclusion is sufficiently large but the diffusion coefficient of the A atom is not too great, the condition

$$\beta R \gg N_0' D_A \qquad (2.31)$$

is satisfied, which means that the diffusion fluexes are controlled by the rate of diffusion (the diffusion resistance of the interface is small compared with the diffusion resistance of the bulk of the inclusion). In this case, Eq. (2.30) for the velocity of the inclusion reduces to Eq. (2.26).

However, if the inclusion is fairly small but the diffusion coefficient D_A is sufficiently large (this applies to small liquid inclusions), we find that Eq. (2.31) is replaced by the opposite inequality

$$\beta R \ll N_0' D_A \qquad (2.32)$$

(in this case, the diffusion resistance of the interface is large compared with the diffusion resistance of the bulk of the inclusion) and the diffusion fluxes are controlled by the rate of dissolution. We now find that Eq. (2.30) yields the following expression for the velocity of an inclusion:

$$\vec{v} = (1 + \varkappa)\omega\beta\left(\frac{\partial c_A^0}{\partial T} + \frac{\alpha_A}{T}\ c_A^0\right)R\nabla T_\infty. \qquad (2.33)$$

The diffusion coefficient does not appear in the above formula and the temperature dependence of $\vec{v}$ is governed by the temperature dependences of the solubility c_A^0 and of the transport coefficient β (the rate of dissolution).

If β and $\varkappa$ are independent of R, the velocity of an inclusion is proportional to its radius. The velocity depends on the radius because the diffusion fluxes are proportional to the difference between the temperatures at the opposite ends of the inclusion (and not to the temperature gradient), the difference itself being proportional to the radius. However, the dependence $\vec{v} \propto R$ applies only in a certain range of relatively small values of R which satisfy Eq. (2.32); when the value of R is large, this dependence tends to saturation in accordance with Eq. (2.30).

We should bear in mind that the transport coefficient β depends on the structure of the inclusion—host interface and, therefore, its value may be affected by the conditions under which the inclusion is formed and by its radius. A linear dependence of the rate of dissolution on the degree of supersaturation applies only if the interface is characterized by a sufficient "natural" roughness independent of the degree of saturation. The coefficient β can be regarded as constant if the degree of roughness is independent of the radius (or if the dissolution mechanism is "normal"). These conditions are satisfied if the inclusion is smaller than the growth steps under given supersaturation conditions and if the distances between the edges of the inclusion (or micropore) are comparable with the widths of the transition regions near the edges that are densely covered with grooves and are very inhomogeneous. Faces with different but low indices meet in these regions and there the surface orientation changes rapidly.

If an inclusion has regular facets (a polyhedron) and the dissolution processes are fairly rapid only at the edges of these facets, we find that the effective transport coefficient is $\beta \propto 1/R$. In the case of large-radius inclusions, with surface growth steps whose dimensions depend on the degree of supersaturation, the rate of dissolution is not a linear but a more complex function of $c_A(\vec{r}s) - c_A^0 \{T(\vec{r}s)\}$ and the results obtained are modified significantly.[†] However, inclusions of moderate radius should have no growth steps but the roughness of the inclusion boundary should be

---

[†] If, for example, the rate of dissolution is proportional to $[c_A(\vec{r}s) - c_A^0 \{T(\vec{r}s)\}]^n$ and the diffusion fluxes are controlled by the conditions at the interface, it follows from [7, 2] that

$$\vec{v} \propto R^n.$$

The velocity of large-radius inclusions may also depend strongly on their previous history and on the degree of imperfection of the host crystal.

sufficient (or the dissolution mechanism should be "normal") to ignore the dependence of β on the radius.

The formulas (2.26), (2.30), and (2.33) for the velocity of an inclusion migrating under the influence of the diffusion fluxes across it apply not only to liquid and crystalline inclusions but also to gas-filled micropores (bubbles) if the evaporation of the host atoms is sufficiently rapid in the range of temperatures in question. These formulas are applicable to gas-filled pores if the mean free path of gas atoms l_g is short compared with the pore radius R (in this case, the A atoms are transported across the pore by diffusion). The coefficient β now represents the rate of evaporation of the A atoms into the gaseous phase which fills the pore.

A characteristic feature of the diffusion coefficient D_A applicable to gas-filled pores (bubbles) is its dependence on the gas pressure: D_A is proportional to the mean free path of the gas molecules l_g, which is itself inversely proportional to the density of the gas or its pressure. Under these conditions, we have

$$l_g = \frac{kT}{\sqrt{2}\, P_0 \sigma_g}, \qquad (2.33a)$$

where σ_g is the scattering cross section of the gas molecules; P_0 is the total pressure of all the gases in the pore (and not just the partial pressure P_A of the A atoms). The partial pressure P_A is proportional to the equilibrium concentration of the A atoms in the gas $N'c_A^0 = c_A^0 / \omega'$ and is independent of the pore radius (if the terms $\sim a/R$ are ignored). The total pressure P_0 of all the gases in the pore (these may be inert gases) is frequently equal to the sum of the Laplace pressure P_L and the external pressure P^0 applied to the crystal:†

$$P_0 = P_L + P^0, \quad P_L = \frac{2\gamma}{R} \qquad (2.34)$$

---

† In this case, the pore is under steady-state conditions and the diffusion fluxes of vacancies which alter its radius are no longer active. Such a steady-state gas pressure is usually established in bubbles formed in fissionable materials. The inert gas atoms which are produced as a result of fission collect in the bubbles and produce the pressure defined by Eq. (2.34).

(γ is the surface tension). In small-radius pores, the total pressure P_0 is inversely proportional to the radius if $P_L \gg P^0$. In this case, $l_g \propto R$ and

$$D_A = \frac{1}{3} w_g l_g = \frac{1}{3} \frac{(kT)^{3/2}}{\sqrt{\pi m_A}\, \sigma_g \gamma} R \propto R, \qquad (2.35)$$

where $w_g = (8kT/\pi m)^{1/2}$ is the average thermal velocity of the A atoms in the pore and m_A is the mass of one A atom.

In this case, the velocity of a pore is proportional to R although the condition (2.31) is satisfied (this point will be discussed later). Obviously, the dependence given in Eq. (2.35) may apply in a wide range of values of R only if $P_A \ll P_0$. If the pressure in a pore is independent of the radius, as is the case in empty pores which contain only the vapor of the host substance (the pressure of this vapor is not related to the Laplace pressure), the diffusion coefficient D_A is independent of R and the velocity of the pore is also independent of R if the condition (2.31) is satisfied. The mean free path and the diffusion coefficient D_A are also independent of R for large-radius pores when the Laplace pressure is much smaller than the external pressure. For example, if $\gamma \sim 1$ J/m^2 (10^3 ergs/cm^2), these two pressures become comparable when $R \sim 10^{-3}$ cm if the external pressure is equal to atmospheric, and when $R_0 \sim 10^{-5}$ cm if the external pressure is $P^0 \sim 10^2$ bar (10^2 atm).

If the gas pressure in a pore is given by Eq. (2.34), the concentration of the gas is usually sufficiently high to satisfy the condition $l_g \ll R$. For example, if $\gamma \sim 1$ J/m^2 (10^3 ergs/cm^2), $\sigma_g \sim 10^{-15}$ cm^2, and $P^0 \ll P_L$, we find from Eqs. (2.33a) and (2.34) that $l_g \sim 0.1R$.

According to the kinetic theory of gases, the number of the A atoms which evaporate from unit area of the inclusion—host interface per unit time (this number determines the coefficient β) is given by

$$n_S = \chi \frac{c_A^0 N_0' (kT)^{1/2}}{(2\pi m_A)^{1/2}} = \frac{1}{4} \chi w_g c_A^0 N_0' = \beta c_A^0, \qquad (2.35a)$$

where χ is the evaporation coefficient. Therefore, the transport coefficient β in Eq. (2.27) is $\beta = \frac{1}{4}\chi w_g N_0^!$ and the condition (2.31) should always be satisfied if $\chi \sim 1$, $R \gg l_g$. [According to Eq. (2.35), $D_A = \frac{1}{3}w_g l_g$]. In this case, the velocity of a gas-filled pore, given by Eq. (2.26), can be expressed in terms of the pressure P_A of the saturated vapor of the A atoms:

$$\vec{v} = (1 + \varkappa)\frac{\omega D_A}{kT}\left(\frac{\partial P_A}{\partial T} + \frac{\alpha_A}{T}P_A\right)\nabla T_\infty. \tag{2.35b}$$

However, if $\chi \ll 1$, the condition (2.31) may not be satisfied. Then, it follows from Eqs. (2.30) and (2.35a) that

$$\vec{v} = (1 + \varkappa)\frac{\omega}{kT}\frac{D_A\chi w_g R}{4D_A + \chi w_g R}\left(\frac{\partial P_A}{\partial T} + \frac{\alpha_A}{T}P_A\right)\nabla T_\infty. \tag{2.35c}$$

In this case, the velocity of a gas-filled pore depends on its radius. The influence of temperature on this velocity is primarily due to the exponential temperature dependence of the saturated vapor pressure P_A.

In small empty pores which do not contain impurity gases and which satisfy the condition

$$l_g \gg R, \tag{2.36}$$

matter is transported not by diffusion in the pore but by evaporation and the evaporated atoms are free to travel across the pore because they do not encounter any gas atoms on the way. Once again, the flux of the A atoms inside the pore is independent of the coordinates and proportional to ∇T_∞.

If the flux of the evaporated atoms is independent of direction relative to the inclusion–host interface, i.e., if this flux is completely random, and if we can ignore the reflection of atoms from this interface, we find [12] that the flux $\vec{I'}$ of atoms in a pore is

$$\vec{I'} = -\frac{2}{3}M(1 + \varkappa)R\nabla T_\infty. \tag{2.37}$$

Here $M = \partial n_S/\partial T$, where n_S is the number of atoms that have evaporated from unit area of the interface per unit time [this number is given by Eq. (2.35a)].

It follows from Eqs. (1.11) and (2.37) that the velocity of pores migrating due to internal evaporation is given by the expression

$$\vec{v} = \frac{2}{3} (1 + \varkappa) \omega M R_\nabla T_\infty. \tag{2.38}$$

The temperature dependence of $\vec{v}$ is again practically exponential. The activation energy associated with this dependence is equal to the heat of evaporation. The migration due to evaporation is important in the case of substances which can be sublimated easily at high temperatures.

The flux of evaporating atoms given by Eq. (2.37) depends, for $R \ll l_g$, on the difference between the temperatures of the opposite ends of the pore (and not on the temperature drop over a distance l_g, which is the case when $R \gg l_g$) and, therefore, this flux is proportional to the pore radius. Consequently, the velocity of a pore given by Eq. (2.38) is also proportional to the pore radius R [this is true only if the condition (2.36) is satisfied].

In the case of relatively large liquid inclusions the transport of matter inside an inclusion may be affected considerably by convective currents in the liquid, which give rise to convective diffusion. Such currents appear whenever ∇T is directed at some angle with respect to the vertical (i.e., with respect to the force of gravity) and has a nonzero component $\nabla_\perp T$, which is perpendicular to the force of gravity vector. Convective current in typical inclusions are too weak to exert any significant influence on the transport of heat so that the temperature distribution given by Eq. (2.6) remains valid. However, since the diffusion coefficient is considerably smaller than the thermal diffusivity, convective transport of the host atoms across an inclusion may be considerably faster than the diffusional migration [174].

The importance of convection in the transport of the host atoms is given by the parameter

$$A_c = \frac{1}{20} (\varkappa + 1) \frac{g \alpha_l R^4 |\nabla_\perp T|}{\gamma_l D_A},$$

where g is the acceleration due to gravity; α_l is the expansion coefficient of the liquid; γ_l is the kinematic viscosity.

In the case of small-radius inclusions and weak temperature gradients, we find that $A_c \ll 1$ and convective currents modify only slightly the expressions for the velocity of an inclusion moving under the action of internal diffusion across the inclusion. However, if the inclusion radius and $|\nabla_\perp T|$ are sufficiently large or if the diffusion coefficient D_A is small, we find that $A_c \gg 1$ and the convective transport of the host atoms is much faster than the diffusional migration. In this case the expressions for the velocity of an inclusion are modified considerably. For example, if $\beta R \gg N_0' D_A$ and $\beta R \gg N_0' D_A A_c$, it follows from [174] that when $A_c \gg 1$ the velocity of an inclusion is almost normal to ∇T and its value is

$$|v| = \frac{8}{5}(1+\varkappa)\frac{\omega}{\omega}\left(\frac{\partial c_A^0}{\partial T} + \frac{\alpha_A c_A^0}{T}\right) D_A A_c |\nabla T_\infty|.$$

In this case v depends strongly on the radius (as R^4) and is a quadratic function of the temperature gradient, i.e., the migration of inclusions becomes essentially nonlinear.

Convection currents affect less the velocity of inclusions if $\beta R \ll N_0' D_A A_c$ ($A_c \gg 1$). In this case v differs from the value given by Eq. (2.33) by a factor which is of the order of unity.

Dependence of the Velocity of an Inclusion on Its Radius

In some cases, several migration mechanisms are of comparable importance. The resultant velocity can be found by adding Eqs. (2.17), (2.21), and (2.30) [or Eq. (2.38)]. The dependence of $\vec{v}$, on R is then found to be much more complex than the dependences given by these equations.

Thus, if the solubility of the host atoms in the inclusion is negligible, the dependence of v on R is determined by the sum of two terms, one of which is constant and the other decreases hyperbolically with increasing R (Fig. 3). At low temperatures, this dependence approaches an asymptotic line. If only the volume diffusion of atoms in the host crystal and in the inclusion is important, it follows from Eqs. (2.17) and (2.30) [or Eq. (2.38)] that initially v rises linearly with increasing R and then

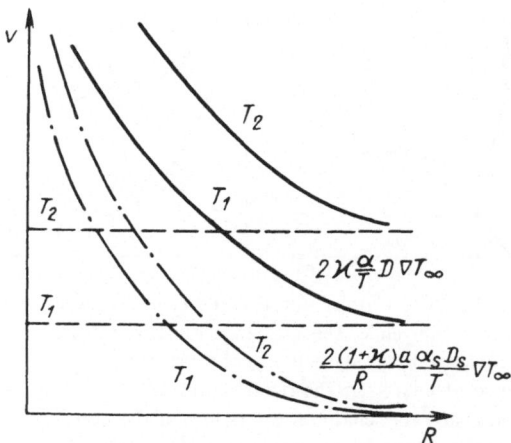

Fig. 3. Dependences of the velocity of an inclusion on its radius in the case of migration controlled by volume and surface diffusion in the host crystal at two temperatures T_1 and T_2 ($T_1 < T_2$). The dashed and the chain curves represent the dependences of v on R for, respectively, the volume and surface diffusion mechanisms; the continuous curves represent the resultant effect of both mechanisms.

tends to a constant limit (Fig. 4). Finally, when all three mechanisms are of comparable importance, the dependence v(R) can have a minimum (Fig. 5) or can resemble the dependences plotted in Fig. 3.

An additional dependence of $\vec{v}$ on R may arise from the dependence of the effective thermal conductivity (and of $\varkappa$) on the radius of the inclusion. For example, in the case of pores in a host crystal of low thermal conductivity, an appreciable dependence of $\varkappa$ on R is possible in the range where $R \sim l_g$, when $\lambda_{0\,eff}$ is given by Eq. (2.12). In the case of very small inclusions, when R is less than or comparable with the mean free path of electrons or pho-

Fig. 4. Dependence of the velocity of an inclusion on its radius in the case of migration controlled by volume diffusion in the host crystal and across the inclusion.

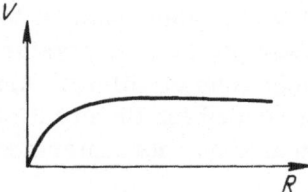

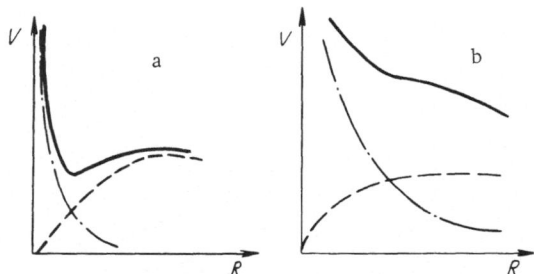

Fig. 5. Dependences of the velocity of an inclusion on its radius under the influence of all the diffusion mechanisms (the dashed curves represent the dependence of v on R for the two volume diffusion mechanisms, and the chain curves represent the corresponding dependence in the case of surface diffusion). The continuous curves are the resultant velocities of inclusions in the case of: a) a relatively small surface diffusion coefficient; b) a large surface diffusion coefficient).

nons, the macroscopic equations used to derive the diffusion fluxes cease to apply (for example, α_S depends on R) and an additional dependence of $\varkappa$ (and, consequently, of $\vec{v}$) on R may arise.

We must bear in mind that inclusions of the same radius but different in nature (for example, crystalline inclusions and micropores) may migrate at quite different velocities in the same host. This difference between the velocities applies not only to those cases when the diffusion fluxes in the interior of an inclusion are important, but also in the case when surface diffusion fluxes are significant (small inclusions). This happens because the coefficient of surface diffusion on a boundary of a crystal depends strongly on the nature of the phase outside this boundary: for example, the diffusion coefficient for a free boundary may be considerably greater than that for a boundary with another crystal. In this case, gas-filled micropores will travel much faster than crystalline inclusions. If the dominant mechanism is the volume diffusion in the host crystal, the difference between the velocities of inclusions of different kinds may be due to the different conditions needed for the creation and the annihilation of vacancies at the inclusion—host interface.

We mentioned in Sec. 1 that when the inclusion—host interface is coherent, the velocity of diffusional migration may be quite low if the atoms of the inclusion do not manage to fill the vacated sites in the host. We also mentioned in Sec. 1 that the velocity may also be quite low if the interface is not sufficiently rough and vacancies are not easily created or annihilated on this interface (see also [145]).

The motion of cylindrical inclusions in the field of a gradient perpendicular to the cylinder axis can be considered in the same way as the motion of spherical inclusions. The distribution of the temperature around a cylinder of radius R is of the form

$$\left. \begin{array}{c} T\,(\vec{r}) = T_0 + \vec{r}\,\nabla T_\infty + \varkappa_c\,\dfrac{R^2}{r^2}\,\vec{r}\,\nabla T_\infty \quad (r \geqslant R); \\[2mm] T\,(\vec{r}) = T_0 + \vec{r}\,\nabla T_\infty \quad (r \leqslant R), \end{array} \right\} \qquad (2.39)$$

where

$$\varkappa_c = \frac{\lambda - \lambda_0}{\lambda + \lambda_0}. \qquad (2.40)$$

We can easily show that all the expressions for the velocities of spherical inclusions apply also to cylindrical inclusions if Eqs. (2.26), (2.30), (2.33), and (2.38) are modified by the substitution

$$\varkappa \to \varkappa_c, \qquad (2.41)$$

and Eqs. (2.16), (2.17), and (2.21) are altered by substituting

$$2\varkappa \to \varkappa_c. \qquad (2.42)$$

Experimental Investigations of the Migration of Inclusions in a Temperature Gradient

We shall now consider the results of experimental studies in which the migration of inclusions in an inhomogeneous temperature field was observed and investigated. They include the data obtained in studies of one-component crystals, as well as those consisting of two or more types of different atom. However, we

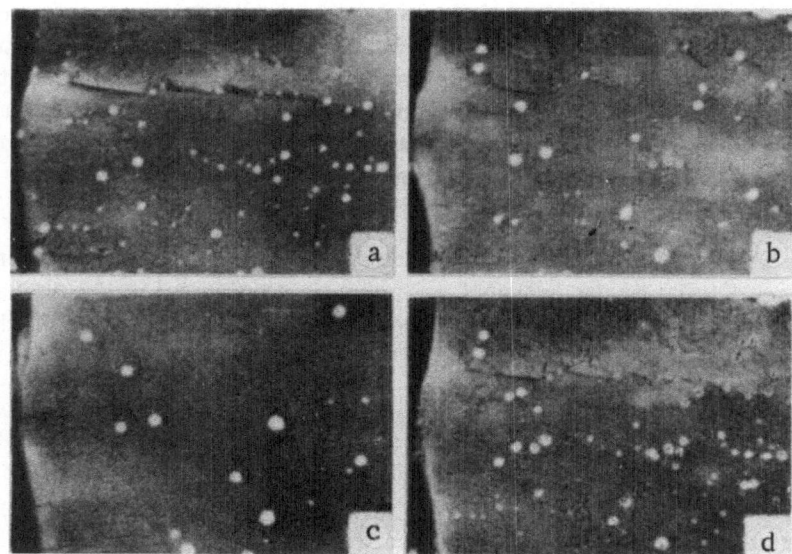

Fig. 6. a) Helium bubbles (some lying on dislocation lines) in a thin
copper film after several heating pulses (reference points are the grain
boundaries and surface impurities); b) the same area after successive
electron-beam pulses (many bubbles have moved, others have left the
film, and some have coalesced; the smaller bubbles have moved further than the larger ones); c) the same region after two more pulses; d)
composite micrograph made by superposition of Figs. 6a and 6b.

shall consider only those properties of multicomponent crystals
which do not depend on the number of components. The theory of
the relevant effects and the experimental data relating to multicomponent crystals will be considered in Chap. II.

A clear experimental demonstration of the dominant influence
of surface diffusion on the motion of small gas-filled bubbles in
a temperature gradient is reported in [16, 17].

The motion of helium-filled bubbles in copper was investigated by Barnes and Mazey [16]. These bubbles were generated
by the bombardment of copper plates with α particles. A total
dose of $\sim 1.7 \times 10^{17}$ cm^{-2} produced He^4 atoms in amounts corresponding to a relative concentration of $\sim 10^{-3}$ in copper. Since
helium is practically insoluble in copper, it collected in isolated
bubbles of ~ 300-400 Å radii.

In these experiments, the temperature gradient was established by the localized heating of a sample with an electron beam under a microscope. A copper plate was heated uniformly by an external source to 700°C and the electron bombardment raised the temperature locally to 800°C. The resultant temperature gradient was very high, of the order of 10^5 deg/cm. The motion of individual bubbles was studied by comparing the electron micrographs of the same part of the sample subjected to cycles of local pulsed heating by electron-beam bombardment (Fig. 6).

A similar experimental procedure was followed by Williamson and Cornell [17], who studied the motion of krypton-filled bubbles in uranium dioxide. Figure 7 shows schematically the contours of these bubbles during the successive stages of their migration. All the bubbles moved along practically the same direction, which was governed by the direction of the temperature gradient ∇T. The slight changes in the direction of motion of individual bubbles

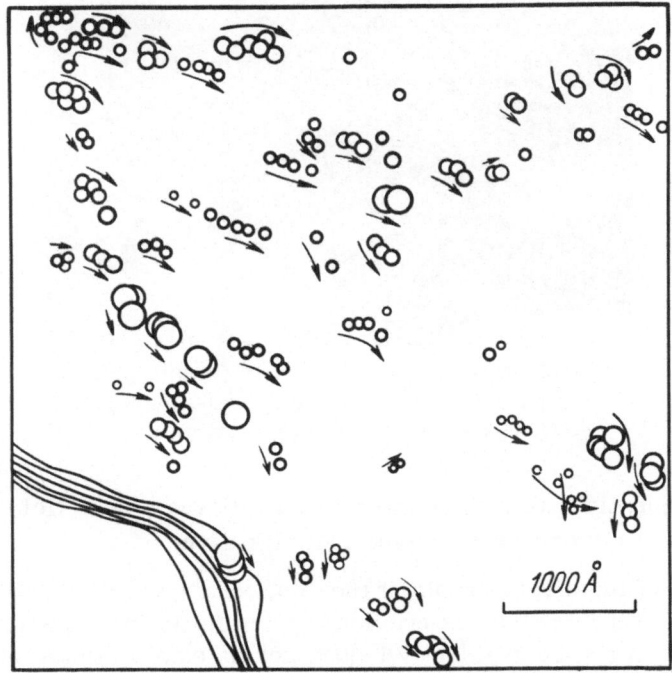

Fig. 7. Successive positions of gas-filled bubbles in uranium dioxide, moving under a temperature gradient.

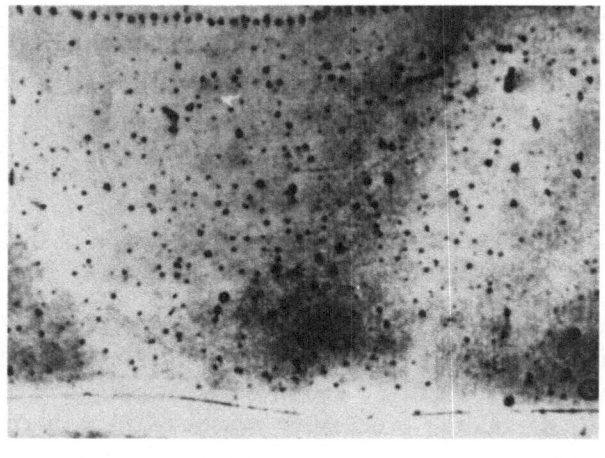

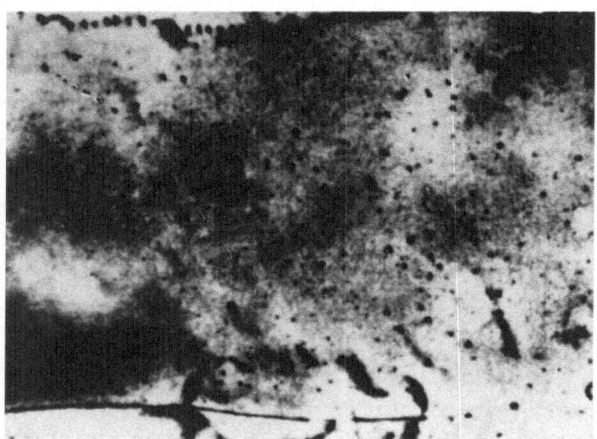

Fig. 8. Escape of gas-filled bubbles from the surface of a plate. It can be seen that the number of bubbles decreases strongly after the pulse heating.

were evidently due to their interaction with the stress field of dislocations encountered during the migration.

One of the direct proofs of the motion of gas-filled bubbles in a temperature gradient is provided by the following experiment. If the temperature gradient ∇T does not lie exactly in the plane of a film investigated under an electron microscope, the gas-filled bubbles may travel at an angle with respect to the film's surface. Consequently, if a bubble at a depth λ travels a distance $l >$

$\lambda/\sin\varphi$, it reaches the surface of the film and bursts open. The gas contained in this bubble escapes and the crater produced in this way on the surface is healed by diffusion. Consequently, the number of gas bubbles in a thin plate annealed in a temperature gradient should decrease with time. This was demonstrated in experiments on uranium dioxide films containing krypton [18]. Figure 8 shows two electron micrographs of the same part of the film at different stages of annealing. These micrographs show clearly the pronounced reduction in the number of gas-filled bubbles.

It follows from Eq. (2.21) that in migration under the influence of surface diffusion the velocity of a bubble should depend on its radius in accordance with the law $v \propto 1/R$. This law is obeyed approximately in the experiments described in the preceding paragraphs. A rigorous check of the law $v \propto 1/R$ can be made by ensuring that the value of ∇T along the path of a moving bubble is maintained more precisely than is possible when a temperature gradient is generated by short-duration localized heating with an electron beam.

Studies of the migration of helium-filled bubbles in copper revealed that small bubbles ($R \approx 3 \times 10^{-6}$ cm) migrated at a velocity of $\sim 10^{-5}$ cm/sec, and that they traveled a path which was many times greater than their radius.† This high bubble velocity was in agreement with Eq. (2.21) if it were assumed that the surface self-diffusion coefficient was $D_S \sim 10^{-6}$ cm^2/sec. This value of D_S was close to that obtained from the experimental data on the kinetics of the healing of scratches on copper at 800°C [19, 20].

It is worth noting that, like the healing of scratches or the development of thermal-etching grooves, the motion of gas-filled bubbles is a variant of the mass transfer method which is used in experimental studies of the surface diffusion parameters [21].

The mechanism of diffusional migration of inclusions cannot be deduced simply from their velocity. The value of the volume diffusion coefficient D at the temperature of the experiments, which can be found from Eqs. (2.16) and (2.17), does not provide sufficient grounds for assuming that volume diffusion is the dom-

---

† The motion of a given bubble could be followed over a path considerably longer than its radius because a helium-filled bubble retained its shape (helium is practically insoluble in copper).

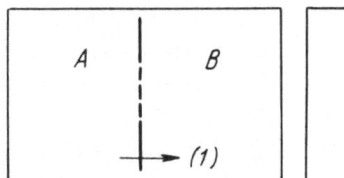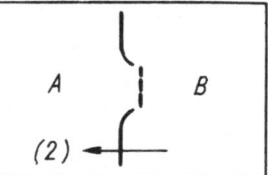

Fig. 9. Schematic representation of the positions of tungsten foils and wires moving in the field of a temperature gradient [22].

inant mechanism in the migration of inclusions. If we compare Eqs. (2.17) and (2.21), we can readily demonstrate that the effective value of D in the studies of inclusions of radius $R \sim a D_S /D$ is of the same order of magnitude as that in the volume diffusion case, even when the actual mechanism responsible for the migration of inclusions is surface diffusion. In this situation errors are very likely especially as the high-temperature value of the inclusion radius, estimated in accordance with the formula just given, is $R \sim 10^{-3}\text{-}10^{-4}$ cm, i.e., it is close to the dimensions of the inclusions and bubbles which are usually encountered in experiments.

A convincing proof of the predominance of the volume diffusion mechanism can be provided by showing experimentally that the velocity of an inclusion is independent of its size, or by finding the activation energy of the migration. In many experimental investigations, particularly those reported in [22], of the thermal diffusion of vacancies, studies are made of the motion of inert markers relative to the external boundaries of a crystal, or relative to reference markers within the crystal, which remain fixed with respect to the lattice. Adda et al. [22] used a reference marker in the form of tungsten foil which was embedded in copper at right angles to the direction of ∇T_∞, and inert markers, which acted as second-phase inclusions, in the form of tungsten wires of $\sim 15 \mu$ radius. Before the diffusion annealing the embedded tungsten foil and wires were in the same plane. The motion of the wires could be followed by studying their displacement relative to the foil (Fig. 9). In order to ensure that the displacements were as large as possible, the tungsten foil and the wires were initially located in the region where the value of $(D\alpha/T)\nabla T_\infty$ was greatest [see Eq. (2.17)]. Adda et al. [22] showed that at 880-870°C tung-

sten wires migrated in iron at a velocity of $v \approx 6 \times 10^{-10}$ cm/sec.
The corresponding displacements of tungsten wires in silver, copper, and gold, obtained under the same experimental conditions,
were too small to be measured. The thermal diffusion ratio of
iron, deduced from these experiments, was $\alpha \approx 7$. This estimate
of α was valid provided the motion of the tungsten wires was described correctly by Eq. (2.17), i.e., if the migration was governed
by the volume diffusion in the host crystal.

Many workers have investigated experimentally [23-29] the migration of inclusions (foreign particles) and gas-filled bubbles as a
result of diffusion fluxes across the inclusions and the bubbles,
which were subjected to a temperature gradient. The most con-
clusive results were obtained for liquid inclusions and for bubbles
filled with inert gases.

We shall consider first the investigations in which the motion
of liquid inclusions in crystalline hosts was studied.

Lemmlein [23] conducted the first experimental investigation
involving motion of an inclusion in a temperature gradient. He
studied the motion of inclusions consisting of an aqueous solution
of sodium nitrate which migrated in a single crystal of sodium
nitrate toward a source of heat.

The spontaneous transformation of the shape of an elongated
nonisomeric inclusion consisting of a saturated solution of the
host crystal, observed for $h \gg d$ (h and d were, respectively, the
length and the transverse size of the inclusions), was accompanied by the appearance of narrow "necks" [29]. The breakup of
these necks produced large as well as small (ranging from 5×10^{-4}
to 10^{-3} cm) inclusions. The formation of these inclusions was
similar to the formation of tiny satellite droplets in the separa-
tion of a large drop from a faucet or a pipette. These satellite
droplets are formed because of the loss of stability of the elongated
neck joining the main drop to the source of the liquid, or to another drop which is breaking away from the main one. This process occurs in many real situations encountered in the motion and
deformation of inclusions in solids. The formation of satellite
droplets is demonstrated clearly by the sequence of cineframes
shown in Fig. 10. The moment at which the water drop breaks
away from the pipette can be seen clearly in these frames.

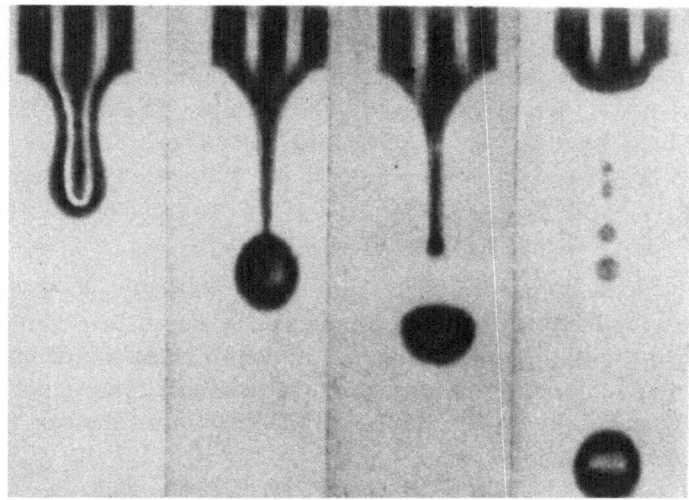

Fig. 10. Consecutive frames showing generation of satellite
droplets in a "neck" formed during breakaway of a large drop.

In Lemmlein's experiments, the surface area of a liquid sodium nitrate inclusion decreased after the inclusion had broken
away and the excess surface energy was dissipated in the form
of heat. Thus, a temperature gradient was established around the
inclusion. This gradient acted on a small satellite inclusion
which had acquired an equilibrium shape a little earlier. Sodium
nitrate was dissolved in that part of the surface of the small inclusion which faced the larger inclusion responsible for the temperature gradient. Sodium nitrate was transported across the
small inclusion and deposited on the opposite part of its surface.
This process resulted in the migration of the small inclusion toward the larger one. Such migration stopped when the larger
inclusion acquired its equilibrium shape so that it no longer emitted heat. In the experiments discussed here, the smaller inclusion sometimes merged with the larger heat-emitting inclusion
(Fig. 11).

It was difficult to estimate the value of $|\nabla T|$ from the re-
sults of Lemmlein [23] but it was known that it was not very large
(< 10 deg/cm). In spite of the low value of the temperature gradient, the motion of inclusions could be observed clearly in sodium nitrate crystals because of the strong temperature dependence of the solubility of this compound in water [see Eq. (2.26)].

The motion of liquid inclusions in crystals under the influence of a temperature gradient has been studied quantitatively by other workers [24-28]. These studies were concerned with ice crystals containing brine inclusions.

The motion of liquid inclusions of saturated brine in ice has been studied by many workers because it is one of the stages in the process of the desalination of seawater, which is of great practical importance. The basis of the partial desalination of sea-water is that the crystallization results in the formation of highly concentrated brine inclusions on grain boundaries and in internal defects, in accordance with the water −salt equilibrium phase diagram. These brine inclusions may drain from ice under gravity, flowing along channels between ice grains. Completely inclosed brine inclusions may migrate under the influence of a temperature gradient. Such gradients may arise because of the difference between the temperature of the cold air above ice and that of the warmer water below it. Consequently, the brine inclusions will move downward in the direction of the water. This drainage of liquid brine inclusions increases the strength of the ice, which can then be used as a structural material.

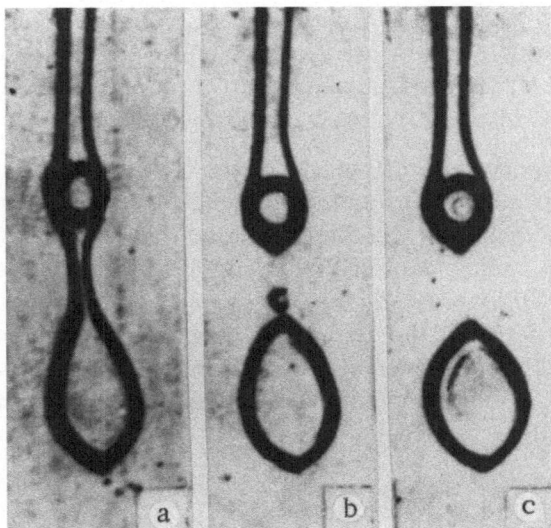

Fig. 11. Formation and disappearance of a small inclusion in a sodium nitrate crystal [23] (a, b, and c show successive stages).

When the motion of an inclusion is due to the diffusion of matter across it, i.e., the inequality $c_A^0 D_A \gg D$ is satisfied, and when the diffusion fluxes are not limited by the rate of dissolution of the host substance in the liquid inclusion [see Eq. (2.31)], the velocity of the inclusion should be independent of its radius but linearly dependent on ∇T_∞ and it should vary with temperature in the same way as $c_A^0 D_A$. These features of the motion of inclusions follow from Eq. (2.26), deduced for one-component crystals, and from the more general formula (12.51a), which applies to solid solutions and approximately to the motion of pockets of aqueous solutions of salts. It also follows from Eq. (12.51a) that when the product of the concentration and the diffusion coefficient in the inclusion is considerably smaller than that for one of the components of the solution (salt) than for the other component (water), the velocity of inclusions can still be described by Eq. (2.26) if c_A^0, D_A, and α_A are understood to apply to the salt.

In the case of inclusions of aqueous solutions of NaCl and KCl in ice, the second term in Eq. (2.26) — which describes the transport of matter as a result of thermal diffusion — is negligible compared with the term which describes the diffusion flux proportional to ∇c. This is due to the small value of the Soret coefficient $S = \alpha_A/T$ and to the strong temperature dependences of the solubilities of these salts in water. The relative contributions of the two diffusion fluxes are given by the dimensionless ratio $S c_A^0 / (\partial c_A^0 / \partial T)$. Estimates show that in the case of inclusions consisting of dissolved KCl, this ratio is less than 10^{-3} at $-5°C$.

These theoretical predictions were tested experimentally in [24-26]. The results of the experiments of Kingery and Goodnow on ice single crystals [24] demonstrated that the velocity of inclusions was independent of their radii, in full agreement with Eq. (2.26). The experiments described in [25, 26] showed that $v \propto \nabla T$: in these experiments the velocity of brine inclusions was determined for different values of $|\nabla T|$ at a practically constant temperature (from -8.6 to $-8.9°C$). The results obtained are plotted in Fig. 12.

The most convincing evidence of the dominant effect of the diffusion in the interior of inclusions on the motion of liquid pockets in ice crystals was provided by the temperature dependences of the velocity of these pockets. The results obtained [27] are plotted in Fig. 13, which shows not only the experimental dependences $v(T)$

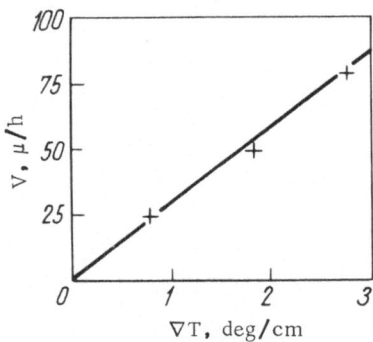

Fig. 12. Dependence of the velocity of brine inclusions in ice single crystals on the temperature gradient (T = −8.75°C) [25].

but also the corresponding theoretical dependences v (T) deduced from Eq. (2.26). Throughout the investigated range of temperatures, the experimentally determined velocities were found to be 2–4 times smaller than the calculated values. However, in agreement with $(D_A c_A^0)_{KCl} > (D_A c_A^0)_{NaCl}$, which was true at all temperatures, the KCl solution inclusions traveled faster. The rise of the velocity of in-

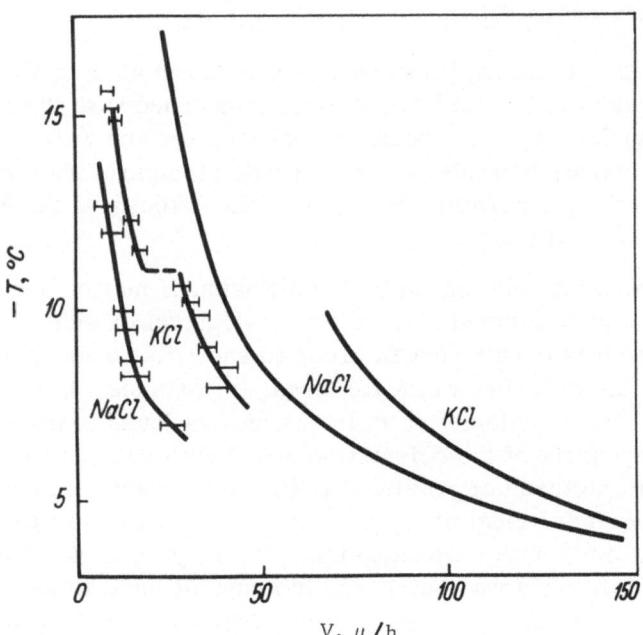

Fig. 13. Temperature dependence of the velocity of brine inclusions in ice single crystals subjected to a temperature gradient [27].

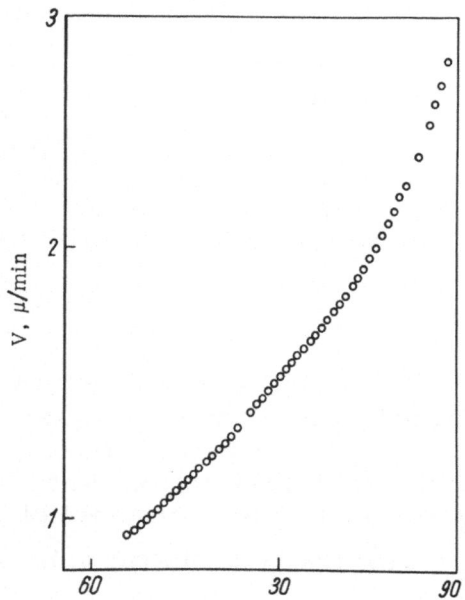

Fig. 14. Increase in the velocity of brine inclusions in an ice
single crystal of 24 μ diameter
as they approach the boundary
between the ice and water (temperature gradient 15 deg/cm)
[28].

clusions with increasing temperature was shown also by the experiments reported in [28], which were concerned with the migration of inclusions near the boundary between ice and water. When
inclusions approached this boundary (whose temperature was
higher than the temperature of the ice), the velocity of the inclusions increased (Fig. 14).

The dominant role played by the diffusion of matter across
inclusions was confirmed also by several control experiments.
One of these was carried out in order to show that convective mass
transfer in an inclusion was not of great importance. In this experiment [27], the velocity of an inclusion was found to be practically independent of the orientation of ∇T with respect to the
direction of the force of gravity. A different experiment showed
that the velocity of most of the inclusions was independent of the
orientation of ∇T with respect to the symmetry axes of an ice
single crystal. This was due to the isotropy of the diffusion coefficient in the liquid inclusions. The migration of inclusions
would have been anisotropic had the dominant process been the
melting – crystallization or the dissolution – precipitation mechanism.

A study [28] of the motion of brine droplets near the boundary between ice and water demonstrated that these droplets increased in size on approaching the boundary. This was expected because of the reduction in the equilibrium concentration of salt in water at the temperature close to the melting point of ice: a constant amount of salt was present in a liquid droplet whose volume increased on approach to the melting point of ice.

A study of the motion of KCl solution inclusions in a temperature gradient [28] demonstrated that, at temperatures below the eutectic point T_{eu} of the KCl$-$H$_2$O system ($T_{eu} = -10.9°C$), the inclusions moved in the form of solidified crystalline particles. Mass transfer in the solid host and inclusions was unlikely at the experimentally observed migration velocities of inclusions whose size was $\sim 3 \times 10^{-3}$ cm (Fig. 13). These solid inclusions probably migrated as a result of the following mechanism. A liquid film of the host substance (ice) may appear at the inclusion$-$host interface because of the tendency of the interfacial surface energy to decrease. The condition for the appearance of such a film can be written in the form

$$\delta\gamma = \gamma_{12} - (\gamma_{1l} + \gamma_{2l}) > 0, \tag{2.43}$$

where γ_{12} is the interfacial surface energy; γ_{1l} and γ_{2l} are, respectively, the surface energies at the interface between the liquid film and the inclusion and between this film and the host.

The upper limit of the thickness of the liquid film L^* is given by [30]:

$$L^* \simeq \delta\gamma/\delta F ,$$

where δF is the difference between the free energy (per unit volume) of the liquid film and the host crystal.

Since $\delta F = q\delta T/T_{mp}$ (T_{mp} and q are, respectively, the melting point and the heat of melting per unit volume of the host substance; $\delta T = T_{mp} - T \ll T_{mp}$), it follows that

$$L^* \sim \frac{\delta\gamma}{q} \frac{T_{mp}}{\delta T} . \tag{2.44}$$

If we assume that $L^* \sim (2-3)a$, a liquid film should form when

$$\delta T \sim \frac{\Delta\gamma}{aa} T_{mp} . \tag{2.45}$$

Even in the case of ice, which has an exceptionally high heat of melting (~336 J/cm^3 or ~ 80 cal/cm^3), we find that $\delta T \approx 1$ deg for $\delta \gamma \approx 1$ erg/cm^2. An inclusion separated from its host substance by a liquid film can travel as a result of the melting and the crystallization in the liquid film. In this case, matter is transported from the front to the rear of an inclusion by viscous flow in the liquid film. This mechanism of the migration of solid inclusions in a temperature gradient may explain also the motion of glass particles on ice [31].

Geguzin, Simeonov, and Mostovoi [173] studied the motion of lithium inclusions in LiF single crystals subjected to a temperature gradient. These inclusions were introduced by electric breakdown near a pointed electrode. The resultant dendrites consisted of branched "needles" oriented along the [100] axes. These needles were unstable and split into separate inclusions at high temperatures.

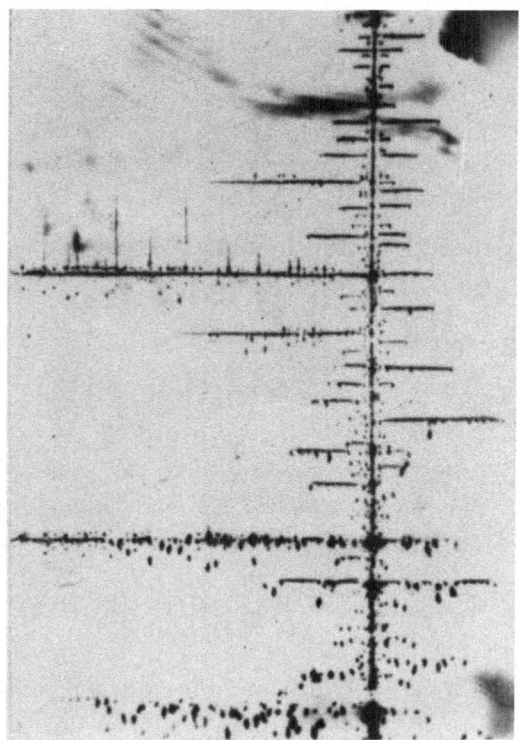

Fig. 15. Lithium dendrite in an LiF single crystal after annealing in ∇T = 1100 deg/cm [173].

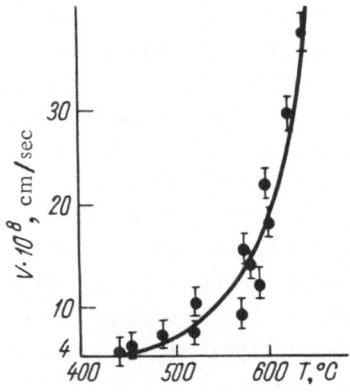

Fig. 16. Temperature dependence of the velocity of lithium inclusions in LiF subjected to a temperature gradient ∇T = 1100 deg/cm [173].

The splitting and spheroidization of the inclusions were used to generate isomeric inclusions of lithium in LiF single crystals. These inclusions migrated under a temperature gradient, which was directed along the main "stem" of a dendrite. The temperature within the dendrite ranged from 20 to 700°C, which corresponded to $|\nabla T|$ = 1100 deg/cm. The parts of the dendrite located in the cold end of the crystal were practically immobile, and they could be used as reference points for the measurement of the displacements of the inclusions in the hot end of the crystal (Fig. 15). It was found experimentally that spherical inclusions of 30-70 μ diameter migrated at a velocity which was independent of their size. The velocity varied exponentially with temperature and, at a fixed temperature was proportional to the temperature gradient (Figs. 16 and 17). The activation energy of the process, which determined the kinetics of the migration of lithium in LiF single crystals, was $(1.0 \pm 0.10) \times 10^{-7}$ pJ $(0.5 \pm 0.05$ eV).

Fig. 17. Dependence of the velocity of a lithium inclusion in an LiF single crystal on ∇T at 700°C [173].

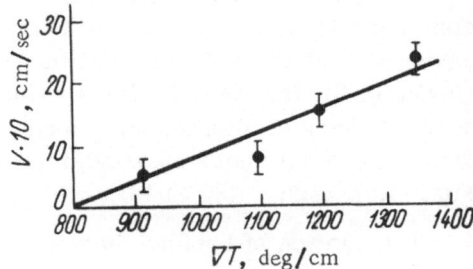

The velocity of the inclusions of liquid lithium in LiF was governed by the dissolution—precipitation mechanism and matter was transported across these inclusions by internal diffusion. The diffusion mechanism in the Li—LiF system could be described as follows. Near the front of an inclusion, where the temperature was higher, the concentration of fluorine in the inclusion was greater than that near the rear surface where the temperature was lower, i.e., the solution of fluorine in lithium was characterized by $\partial c / \partial T > 0$. The difference between the concentrations of fluorine at the front and the rear of the inclusion was responsible for the diffusion of this element toward the rear. This diffusion resulted in the appearance of new atomic planes (crystallization) near the rear surface and in the disappearance of atomic planes (dissolution) at the front of the inclusion—host interface, i.e., the inclusion migrated relative to the lattice of the host crystal.

The velocity of the inclusions was independent of their size, in accordance with Eq. (2.30). Consequently, this velocity was limited by the rate of diffusion of fluorine across liquid lithium and not by the rates of dissolution and precipitation occurring at the inclusion—host interface.

The results of these experiments were used to determine C_A, $\partial C_A / \partial T$, and φ_1 (the last quantity is the heat of solution of fluorine in liquid lithium). It was assumed that at low concentrations $C_A \propto \exp(-\varphi_1 / kT)$, $D \propto \exp(-\varphi / kT)$, and, therefore, $\varphi = \varphi_1 + \varphi_2$ in accordance with Eq. (2.30). Assuming that $\varphi_2 \approx 0.05$ eV [142], it was found that $\varphi_1 \approx 0.45$ eV. Since $\partial C_A / \partial T = (C_A / T) \varphi_1 / kT \gg \alpha C_A / T$ ($\alpha \simeq 1$), it was found from Eq. (2.30) that $C_A \simeq kT^2 v \times (\varphi_1 D \nabla T_\infty)^{-1}$. It was thus found that at $T = 700°C$, $C_A \approx 10^{-4}$ and $\partial C_A / \partial T \simeq 10^{-6}$ deg^{-1}.

The experimental results indicated that cylindrical lithium inclusions migrated more slowly than spherical inclusions of the same radii as the cylinders. This effect was a natural consequence of the fact that the local value of the temperature gradient near a sphere of high thermal conductivity is 1.5 times greater than the gradient near a cylinder made of the same material [see Eqs. (2.6), (2.39), and (2.40)].

The motion of bubbles in a temperature gradient, governed by the transport of matter across the occluded gas, was inves-

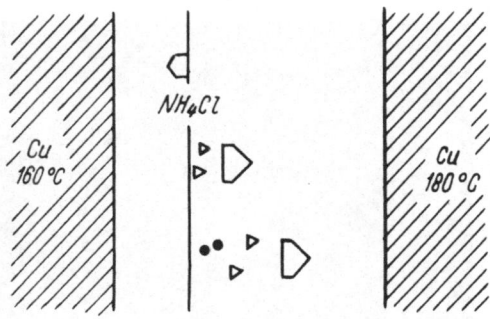

Fig. 18. Formation and migration of bubbles in NH_4Cl subjected to a temperature gradient [5].

tigated experimentally by Diez and Biersack [5]. They used NH_4Cl single crystals because the vapor pressure of this compound was sufficiently high for the transport of matter across the bubbles to govern the kinetics of motion of the bubbles. The bubbles were generated artificially. Small holes were drilled in two identical single crystals. These single crystals were bonded together along their drilled faces. In this way, the drilled holes and accidental defects formed closed cavities whose migration was then investigated. A definite temperature gradient was maintained by placing the bonded NH_4Cl single crystals between two copper blocks whose temperatures were 160 and 180°C, respectively (Fig. 18).

Diez and Biersack first demonstrated [5] that, other conditions being equal, the velocity of bubbles increased with their size. This dependence was clearly demonstrated in the successive photographs shown in Fig. 19 and by the curves plotted in Fig. 20.

The observed dependence of the velocity was due to the fact that, as a result of the dissociation of the NH_4Cl vapor, the evaporation coefficient χ was small (it was estimated to be $\chi \sim 10^{-3}$). Therefore, the coefficient β, calculated from Eq. (2.35a), was small and the condition (2.32) was satisfied. Thus, the motion of the bubbles was controlled not by the diffusion across them but by the rate of evaporation at the interface with the host crystal. The velocity of the bubbles v was given by Eq. (2.35c) which predicted, for $\chi < 4D_A/\omega_g R \simeq l_g/R$, that v should increase with the radius of the bubble (this point will be discussed later).

According to Eq. (2.35c), the velocity of an inclusion should depend strongly on the evaporation coefficient χ. This coefficient is affected by any contamination on the surface of the bubble. If impurities located on this surface impede the evaporation of the

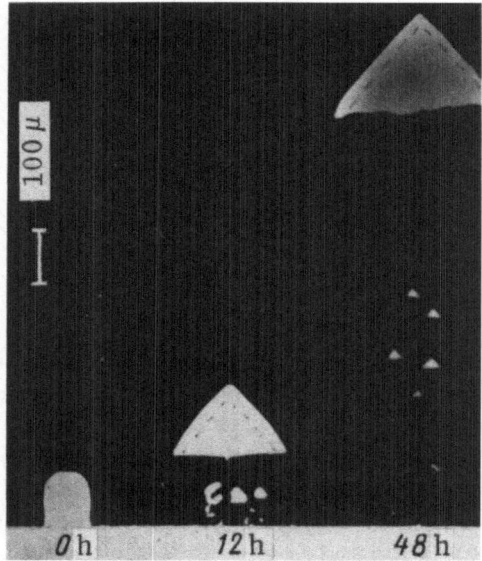

Fig. 19. Increase in the linear size of a bubble migrating in an NH_4Cl single crystal subjected to a temperature gradient $\nabla T = 225$ deg/cm (T = 160°C) [5].

host atoms, the velocity of the bubble may be reduced quite considerably. The experiments on NH_4Cl single crystals confirm that the introduction of active impurities of the type that hinder the evaporation can completely stop the migration of bubbles. Diez and Biersack [5] used zirconium oxychloride. They found that the bubbles in crystals containing this impurity remained immobile.

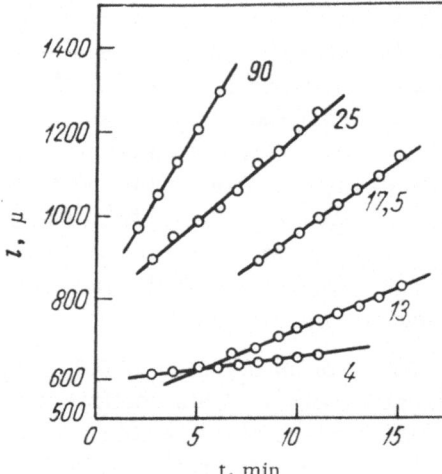

Fig. 20. Displacement–time dependences for bubbles of different sizes in an NH_4Cl single crystal subjected to a temperature gradient $\nabla T = 124$ deg/cm (T = 175°C) [5]. The size of the bubbles is given (in μ) alongside each line.

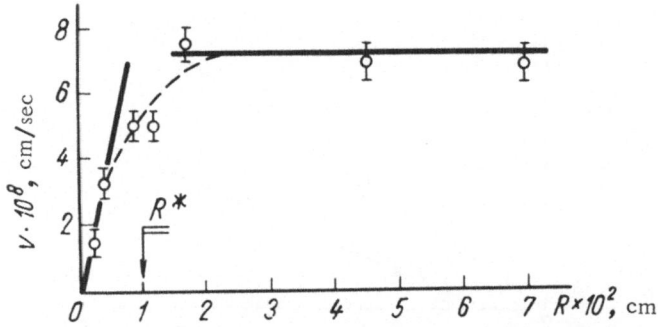

Fig. 21. Dependence of the velocity of air-filled bubbles in a KBr single crystal subjected to a temperature gradient $\nabla T =$ 500 deg/cm (T = 650°C) [170].

The experimental observations made on NH_4Cl single crystals could be explained in a natural manner by the dominant influence of the transport of matter across the occluded gas, and they could be described formally by Eqs. (2.30) and (2.35).

The motion of gas-filled bubbles and of empty cavities in alkali halide crystals was investigated by Geguzin and Simeonov [170]. These experiments were carried out on KBr, KCl, and NaCl crystals containing large numbers of gas-filled bubbles. The dimensions of these bubbles ranged from 10^{-3} to 7×10^{-2} cm. The dependence v(R) obtained in this study is plotted in Fig. 21.

The mean free path of gas molecules in the bubbles was $l_g \approx$ $(1-3) \times 10^{-4}$ cm, i.e., it was less than the linear size of the smallest bubbles. Consequently, the transport of matter across these bubbles took place by diffusion in the occluded gas. However, if $\beta \ll N_0' D_A/R$, the velocity of such bubbles should be limited not by the diffusion in the occluded gas but by the evaporation and condensation of atoms [see Eq. (2.32)]. In this case, the motion of a cavity should obey the law $v \propto R$, as in the experiments performed on NH_4Cl [5]. This dependence was obtained in [173] for $R < 10^{-2}$ cm.

The transport coefficient β was estimated by means of Eq. (2.33):

$$\beta \cong \frac{vT}{\omega R \alpha |\nabla T|}. \tag{2.46}$$

It was found that $\beta \sim 10^{18}$ $cm^{-2} \cdot sec^{-1}$ for $\alpha \sim 1$, $R \approx 10^{-2}$ cm, $v \approx 10^{-7}$ cm/sec.

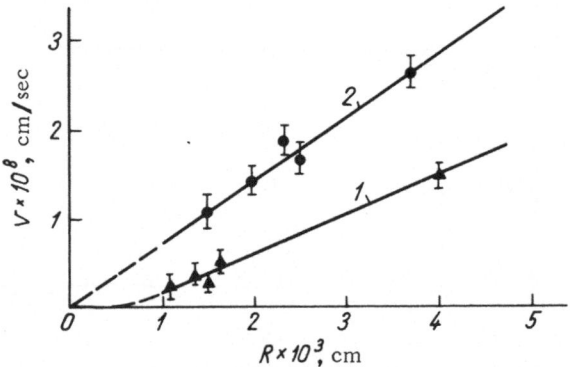

Fig. 22. Dependence of the velocity of cavities in KBr single crystal subjected to a temperature gradient ∇T = 500 deg/cm (T = 650°C): 1) gas-filled bubbles; 2) empty cavities.

In spite of the fact that $\beta \ll N_0' D_A / R$, the kinetics of the diffusion of atoms across the occluded gas affected slightly the migration of gas-filled bubbles, as demonstrated in special experiments in which the migration of very small (R < 5 × 10^{-3} cm) empty and gas-filled cavities was investigated in a KBr single crystal. The results of the experiments (Fig. 22) indicated that a reduction in the pressure from P = 100 kN/m^2 (10^6 dyn/cm^2) to P = 1 kN/m^2 (10^4 dyn/cm^2) increased the velocity only slightly. If this velocity were governed by the diffusion in the gas, such a change in the pressure would have altered the velocity of five orders of magnitude because $D \propto 1/P$.

In larger bubbles, when $\beta \gg N_0' D_A / R$, the process that controls the migration can be the diffusion in the occluded gas [see Eq. (2.31)]. In this case [see Eq. (2.30)], the velocity should be independent of the bubble radius. This was indeed observed experimentally by Geguzin et al. [173].

Obviously, in the range $R \approx R^*$ (the meaning of R^* is explained in Fig. 21), we should have

$$\beta \cong \frac{N_0' D_A}{R^*}. \qquad (2.47)$$

This estimate gave a reasonable value $\beta \sim 10^{18}$ $cm^{-2} \cdot sec^{-1}$.

A dependence $v(R)$, similar to that shown in Fig. 21, was obtained also in a study of the migration of gas-filled bubbles in camphor single crystals [175]. The value of $R^* \approx 1.4 \times 10^{-3}$ cm was obtained for $T = 37.4°C$ and $\nabla T = 6.1$ deg/cm . Application of Eq. (2.46) yielded $\beta \approx 10^{20}$ cm$^{-2} \cdot$ sec^{-1}. According to Eq. (2.47), this value of β corresponds to $N_0' D_A \approx 10^{17}$ cm$^{-1} \cdot$ sec^{-1}. If $D_A \approx 1$ cm^2/ sec, it follows that $N_0' \approx 10^{17}$ cm^{-3}, which corresponds to a reasonable value of the vapor pressure of camphor in the bubbles $P = N_0' kT \approx 10^3$ dyn/cm^2. The temperature dependences reported in [173, 175] yield activation energies close to the heats of evaporation of the crystals under investigation, in agreement with the hypothesis that the "evaporation—condensation" mechanism is the dominant process.

The motion of inclusions, gas-filled bubbles, and pores under a temperature gradient is accompanied by the deformation of the migrating inclusions irrespective of the nature of mass transport. During their motion, the inclusions assume nonequilibrium dynamic shapes.

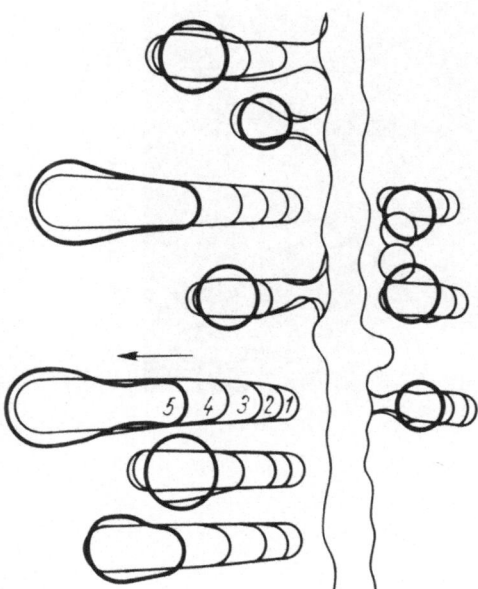

Fig. 23. Consecutive positions and shapes of lithium inclusions migrating in LiF single crystals subjected to a temperature gradient (× 700) [173]. Migration is accompanied by spheroidization. The direction of the temperature gradient is indicated by the arrow. 1) Initial shape; 2) after 3 h; 3) after 9 h; 4) after 18 h; 5) after 33 h.

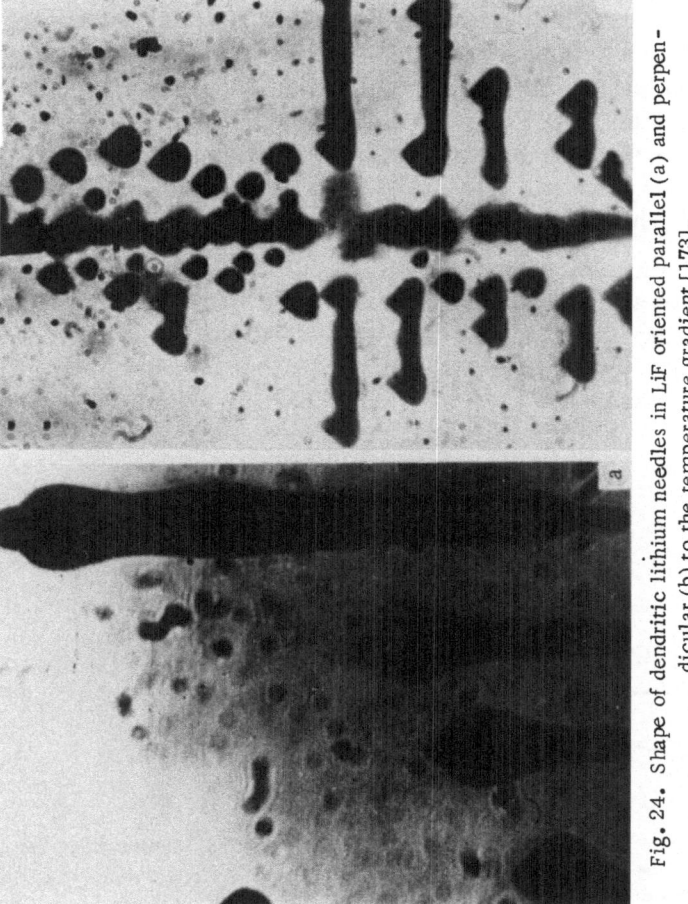

Fig. 24. Shape of dendritic lithium needles in LiF oriented parallel (a) and perpendicular (b) to the temperature gradient [173].

   The experimentally observed deformation of inclusions and gas-filled bubbles can be due to various causes. One of them is the tendency for the area separating a nonisomeric inclusion from the host crystal to decrease; another is the presence of local inhomogeneities of the fluxes associated with the temperature distribution within the inclusion. These two effects are in competition in the sense that they tend to produce different shapes. The first effect gives rise to fluxes tending to establish the equilibrium shape of the inclusion for which the surface energy of the inclusion–host interface has its minimum value, whereas the second effect produces fluxes which tend to disturb the equilibrium shape.

   The consequences of the first effect were observed in studies of the migration of lithium inclusions elongated along the [100] axis in LiF single crystals [173]. Figure 23 shows the sequence of shapes observed for moving inclusions. These inclusions tended to become spherical during their motion. Consequently, different parts of the inclusion surface traveled at different velocities. In this situation, the front of an inclusion could move in the opposite direction to the motion of the center of gravity. This effect was the consequence of the superposition of two diffusion fluxes of atoms. One of them controlled the kinetics of the motion of an inclusion under the influence of a temperature gradient and it was directed from the rear to the front of the inclusion. The other, which determined the kinetics of spheroidization of the inclusion, which was elongated along the direction of motion, appeared under the influence of a gradient of the chemical potential of the atoms located on the end and lateral surfaces of the inclusion (this gradient was due to the different curvatures of these surfaces). This flux was directed from the front and rear ends of the elongated inclusion toward its lateral surface. The combined effect of the fluxes traveling toward the front part of the inclusion and away from it could be responsible for the tendency of the front of the inclusion to approach more closely the center of gravity.

   The deformation of inclusions under the influence of a temperature gradient is illustrated in Figs. 24 and 25. The temperature can vary over the surface of the inclusion and, therefore, different parts of the surface may attract fluxes of different intensity. This may result in the deformation of an inclusion, which can intensify the local inhomogeneities of the temperature field.

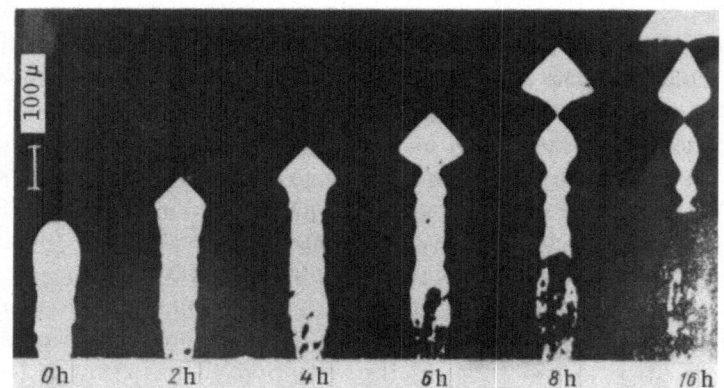

Fig. 25. Motion and breakup of an elongated bubble in NH$_4$Cl subjected
to a temperature gradient ∇T = 320 deg/cm, T = 160°C) [5].

Figure 26 shows schematically successive temperature fields
near a moving deformable inclusion and it explains qualitatively
the dynamic shapes of the inclusions shown in Figs. 24 and 25 [5].

When the dominant mechanism is the transport of matter
across the gas occluded in a bubble, the shape of a migrating bubble may be influenced strongly by the dependence of the coefficient β on the crystallographic orientation. This dependence was

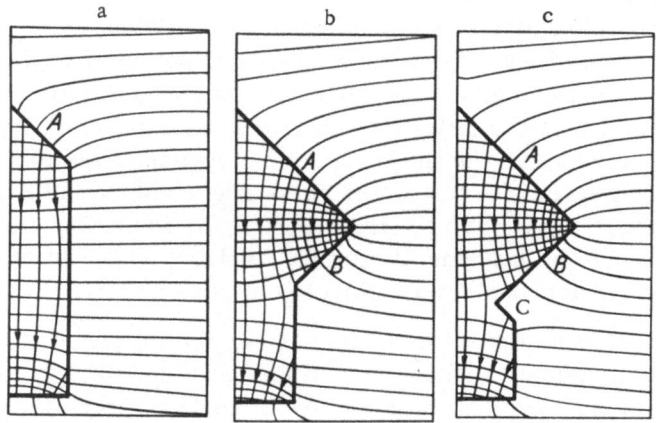

Fig. 26. Schematic explanation of the results shown in Fig. 25 in
terms of the temperature fields near an inclusion which deforms
during its migration (a, b, and c are the three consecutive stages).
The arrowed curves represent the temperature gradients and the
other curves are isotherms [5].

observed clearly in the experiments on NH_4Cl single crystals [5], which indicated that the shape of a bubble depended strongly on the orientation of ∇T relative to the crystallographic axes.

3. MIGRATION OF INCLUSIONS IN METALS SUBJECTED
TO ELECTRIC FIELDS

Volume and Surface Electrotransport
of Ions in a Metal

An external electric field gives rise to directional diffusion fluxes of ions in ionic crystals and metals. These fluxes are responsible for the motion of inclusions. In ionic crystals, there are several complications associated with the presence of ions of opposite signs and with the need to observe the conditions of electrical neutrality: these crystals will be considered later (Sec. 15).

The present selection will be restricted to the diffusional migration of inclusions in metals under the influence of electric fields [12, 13].

The influence of an electric field on the diffusion jumps of ions in a crystal is related to the forces which act on a diffusing ion. The probability of ion jumps in the direction of an applied force is higher than the probability of jumps in the opposite direction and this gives rise to the directional electrotransport of ions in a metal.

Two forces act on a diffusing ion in a metal. One is directly related to the external electric field $\vec{E}$, which exerts a force $z_0 e\vec{E}$ on an ion whose charge is $z_0 e$ (e is the electronic charge but with the positive sign). The other force is related to the flux of electrons or holes which move in the electric field and collide with the diffusing ion. The electrons (or holes) transfer their momentum to the ion and drag it in the direction of their motion (see, for example, [33-35, 15]). This drag force is proportional to the electric current and, therefore, to the external electric field, i.e., it can be written in the form $z^* e\vec{E}$, where z* is some effective charge.

The resultant force $\vec{F}_i$, which acts on the ion is equal to the sum of the two aforementioned forces:

$$\vec{F}_i = z_0 e\vec{E} + z^* e\vec{E} = ze\vec{E}, \quad z = z_0 + z^*. \tag{3.1}$$

The magnitude of the effective charge z* is governed by the values of the electron (σ_e) and the hole (σ_h) scattering cross sections of the ion measured at the top of the potential barrier (more exactly, the charge is governed by the cross sections averaged out over the barrier region), by the densities of the electrons and the holes n_e and n_h, and by their mean free paths l_e and l_h. For example, if the electron and the hole energies are quadratic functions of the crystal momentum, the value of z* is given by the formula [15]

$$z^* = \sigma_h n_h l_h - \sigma_e n_e l_e.$$

(3.2)

Depending on the values of the parameters in this formula, the effective charge can be negative (this applies to pure n-type conduction) or positive (for example, in the case of pure p-type conduction). It is important to note that the experimental results (see, for example, [35]) show that the effective charge of an ion z* can sometimes be an order of magnitude greater than the "true" charge z_0.

When the force $\vec{F}_i$ [Eq. (3.1)] is applied to a monatomic crystal, it gives rise to a directional diffusion flux of ions $\vec{I}$, equal to the product of $\vec{F}_i$, the mobility of ions, and their number per unit volume N_0:

$$\vec{I} = N_0 \frac{D}{fkT} ez \vec{E}.$$

(3.3)

In this formula, f is the correlation factor. This factor represents the observation that, in the vacancy diffusion mechanism, the probability that an ion returns — after a diffusion jump — to its former position (occupied temporarily by a vacancy with which it changes places) is higher than the probability of its transition to another position. This correlation exists only between the probabilities of consecutive ion jumps but not between the probabilities of vacancy jumps (the vacancies are surrounded on all sides by identical neighbors). Therefore, this correlation affects the self-diffusion coefficient D but not the flux of vacancies (or ions) in the field of external forces. Therefore, the correlation gives rise to an additional factor 1/f in the Einstein relationship between the mobility and the diffusion coefficient [36, 37]. In the case of monatomic crystals, the correlation factor f differs little from unity (for example, in the case of fcc crystals, we find that f = 0.781 [37]).

The application of an electric field parallel to the boundary of a metal crystal should also give rise to the surface electrotransport of ions. As in the case of volume electrotransport, ions near the surface are subject to two forces $z_{0s}e\vec{E}$ and $z_s^{\bullet}e\vec{E}$, which represent, respectively, the direct effect of the field on an ion and the drag exerted by carriers. The "true" surface charge z_{0S} and, particularly, the effective surface charge z_S^* may differ considerably from the corresponding values in the bulk of a metal.

A calculation of the drag force exerted on the surface ions by electrons or holes in a metal with a planar interface with a dielectric (or free space) was carried out in [12] on the assumption that the energies of the electrons and holes are quadratic functions of their crystal momenta. In this case, the effective charge z_S^*, which represents the drag force exerted on an ion diffusing near the surface, is given by a formula similar to Eq. (3.2):

$$z_S^{\bullet} = \frac{1}{2} \left(1 + p_h \right) \sigma_h^S n_h \, l_h - \frac{1}{2} \left(1 + p_e \right) \sigma_e^S n_e \, l_e, \qquad (3.4)$$

where σ_e^S and σ_h^S are, respectively, the electron and the hole scattering cross sections of the surface ion, averaged out over the potential barrier region (σ_e^S and σ_h^S can differ considerably from σ_e and σ_h); p_e and p_h are, respectively, the specular reflectivities of electrons and holes from the surface.

Usually, the inflection of electrons and holes from the surface of a metal is diffuse and $p_e \ll 1$, $p_h \ll 1$; however, in the case of "poor" metals with low carrier densities, the values of p_e and p_h can be close to unity. If electrons and holes are reflected from an interface between a metal and a second conducting phase, the explicit expression for z_S^* and its magnitude are different: z_S^* depends on the nature of the second phase but retains its order of magnitude.

The application of a force

$$\vec{F_i} = z_S e \vec{E}_S \qquad \left(z_S = z_{0S} + z_S^{\bullet} \right) \qquad (3.5)$$

($\vec{E}_S$ is the surface component of the external field) gives rise to a surface diffusion flux

$$\vec{I}_S = N_0 \frac{D_S \, a}{f_S \, kT} \, e z_S \, \vec{E}_S, \qquad (3.6)$$

where f_S is the correlation factor for surface diffusion (this factor is also close to unity).

Similar forces and diffusion fluxes should also appear on the surface of a spherical inclusion of sufficiently large radius:

$$R \gg l_e, \quad R \gg l_h. \tag{3.7}$$

In this case, a small section of the inclusion–metal interface can be regarded as planar and the transport equation for electrons and holes is the same as that for the surface of a metal. However, if the condition (3.7) is not obeyed and the mean free paths of the carriers are comparable with the radius of an inclusion, the solution of the transport equation will be different from that in the planar case and the effective charge z_S^* will depend on the inclusion radius. In the case of very small inclusions when $R \ll l_e$, $R \ll l_h$, the value of z_S^* again tends to a constant (but different) limiting value.

Electric fields near an inclusion can be found from the equations

$$\left. \begin{array}{l} \operatorname{div}\vec{E} = 0, \ \vec{E}_S^+ = \vec{E}_S^-, \ \lambda_e \, \vec{n}\vec{E}^+ = \lambda_{0e} \, \vec{n}\vec{E}^- \quad (\vec{r} = \vec{r}_S); \\[2mm] \vec{E} = \vec{E}_\infty \quad (r \to \infty), \end{array} \right\} \tag{3.8}$$

where $\vec{E}_\infty$ is the field far from the inclusion; λ_e and λ_{0e} are, respectively, the electrical conductivities of the host crystal and the inclusion (they are scalar for cubic crystals and liquids); $\vec{E}^+$ and $\vec{E}^-$ are the fields in the host crystal and in the inclusion near the inclusion–host interface.

In the case of spherical inclusions, the electric fields are given by formulas of the same type as those which apply to a temperature gradient [Eq. (2.9)]:

$$\left. \begin{array}{l} \vec{E}\,(\vec{r}) = \left(1 + \varkappa_e \dfrac{R^3}{r^3}\right)\vec{E}_\infty - 3\varkappa_e \dfrac{R^3}{r^3}\,(\vec{n}\vec{E}_\infty)\,\vec{n} \quad (r > R); \\[3mm] \vec{E}\,(\vec{r}) = \left(1 + \varkappa_e\right)\vec{E}_\infty \quad (r < R), \end{array} \right\} \tag{3.9}$$

where

$$\varkappa_e = \frac{\lambda_e - \lambda_{0e}}{2\lambda_e + \lambda_{0e}}.$$

In the case of dielectric inclusions, $\lambda_{0e} = 0$ and $\varkappa_e = 1/2$. On the surface of a nonconducting pore, the tangential component of the field is

$$\vec{E}_s(\vec{r}_s) = \frac{3}{2}\left[\vec{E}_\infty - \left(\vec{E}_\infty \vec{n}\right)\vec{n}\right].$$

(3.10)

Velocities of Inclusions for Different Electrotransport Mechanisms

Vacancies in metals are electrically neutral (when considered in association with the screening electron charges) and their energy is independent of the potential. If vacancies are generated relatively easily at the inclusion–host interface or in the bulk of the host crystal and if vacancies do not accumulate at the interface, we may assume that the electric field does not affect the vacancy concentration c_v. Therefore, in calculations of the diffusion fluxes in an electric field, we have to allow only for the directional diffusion under the action of the external forces but not for the diffusion flux proportional to ∇c_v.

It is evident from Eqs. (3.3) and (3.9) that the normal component of the volume diffusion flux which reaches the inclusion–host interface from the host crystal is given by the following formula (in the system of coordinates linked to the host lattice):

$$\vec{n}\,\vec{I} = (1 - 2\varkappa_e)\,N_0\frac{Dez}{fkT}\,(\vec{n}\vec{E}_\infty).$$

(3.11)

This quantity is proportional to $\cos\theta$ (Fig. 2) and, therefore, the shape of the inclusion does not change during migration in a homogeneous electric field (this is analogous to the situation in a temperature gradient).

It follows from Eqs. (1.1) and (3.11) that the velocity of migration of an inclusion as a result of volume diffusion in the host crystal is given by the formula (in the system of coordinates linked to the host lattice):

$$\vec{v}' = (1 - 2\varkappa_e)\frac{Dez}{fkT}\vec{E}_\infty.$$

(3.12)

Subtracting from this expression the velocity of electrotransport of the host atoms relative to their lattice,

$$\vec{v}_a = \omega \, \vec{I}_\infty = \frac{Dez}{fkT} \, \dot{E}_\infty, \qquad (3.13)$$

we can determine the velocity of an inclusion relative to the matter in the host crystal far from the inclusion, i.e., relative to the boundaries of the crystal (this applies to homogeneous conditions):

$$\vec{v} = -2\varkappa_e \frac{Dez}{fkT} \, \vec{E}_\infty. \qquad (3.14)$$

This velocity depends strongly on the ratio of the electrical conductivity of the host crystal to that of the inclusion. For example, in the case of nonconducting (or poorly conducting) inclusions, we have $\varkappa_e \approx 1/2$ and

$$\vec{v} = -\frac{Dez}{fkT} \, \vec{E}_\infty. \qquad (3.15)$$

However, if the conductivity of the inclusion is considerably higher than that of the host crystal, we find that $\varkappa_e \approx -1$ and the velocity

$$\vec{v} = 2 \frac{Dez}{fkT} \, \vec{E}_\infty \qquad (3.16)$$

has the opposite direction.

As in the case of a temperature gradient, the velocity of inclusions is different if the crystal is perfect ($l \gg R$) and free of vacancy sources and sinks in its bulk and at the inclusion–host interface (the interface is assumed to be impermeable to vacancies). In this case, $\vec{v}' = 0$ irrespective of the values of the conductivities and

$$\vec{v} = -\vec{v}_a = -\frac{Dez}{fkT} \, \vec{E}_\infty. \qquad (3.17)$$

It is evident from Eqs. (3.12)-(3.17) that in the electrotransport mechanism the velocity of inclusions $\vec{v}'$ or $\vec{v}$ differs only by a numerical factor from the electrotransport velocity of the host atoms relative to their lattice. This point is important in the interpretation of the experimental data on electrotransport because the velocity $\vec{v}_a$ is frequently deduced from the motion of markers which are at rest relative to the lattice. Our analysis shows that, in fact, such markers are at rest relative to the lattice only if their electrical conductivity is considerably less than that of the host crystal or if the host crystal contains very few defects and has a coherent interface with the inclusion. In the general case, the electrotransport results should be interpreted making allowance for the motion of inclusions which has been considered above [13, 176].

The surface divergence of the flux $\vec{I}_S$ at the interface between a metal and a nonconducting inclusion follows from Eq. (3.6):

$$\text{div } \vec{I}_S = - N_0 \frac{D_S\,a}{f_S\,kT}\, ez_S\, \Delta_S\, \varphi = - 3N_0 \frac{D_S\,aez_S}{Rf_S\,kT}\, E_\infty \cos\theta. \qquad (3.18)$$

The expression given above is derived making allowance for Eq. (2.19) and for the fact that, according to Eq. (3.9), the potential φ on the surface of a nonconducting inclusion ($\varkappa_e = 1/2$) is $\varphi = -\frac{3}{2}\vec{E}_\infty \vec{r} = -\frac{3}{2}E_\infty R\cos\theta$. Since $\text{div } \vec{I}_S$ is proportional to $\cos\theta$, once again the shape of the inclusion does not change during its migration and its velocity, governed by surface diffusion fluxes, follows from Eqs. (1.9) and (3.18):

$$\vec{v} = - 3 \frac{D_S\,aez_S}{Rf_S\,kT}\, \vec{E}_\infty. \qquad (3.19)$$

The velocity given by Eq. (3.19) may be greater than the velocity associated with the volume fluxes in the host crystal, given by Eq. (3.14), which applies to small inclusions at moderate temperatures when

$$\frac{a}{R} D_S \gg D. \qquad (3.20)$$

The motion of an inclusion in an electric field resulting from the flux of host atoms across the inclusion may be quite rapid if the inclusion is conducting. The rate of dissolution $\beta\,(c_A - c_A^0)$ of the host atoms A in the inclusion is practically independent of the field[†] and is determined solely by the difference between the equilibrium concentration $c_A^0(T)$, corresponding to a given temperature, and the actual concentration c_A of the A atoms in the inclusion. The number of the A atoms dissolved per unit time can be equated to the number of the A atoms carried away by the electrodiffusion flux into the interior of the inclusion away from the inclusion—host interface:

$$\vec{I}' = N_0' \frac{D_A c_A^0}{f_A kT} ez_A (1 + \varkappa_e) \vec{E}_{\infty} - N_0' D_A \nabla c_A. \tag{3.21}$$

Here, f_A and $z_A = z_{A0} + z_A^*$ are, respectively, the correlation factor and the total charge of the A atoms or ions in the inclusion.[‡] In this way, we obtain the boundary condition for the determination of the actual concentration of the A atoms in the inclusion:

$$\vec{n}\,\vec{I}' = \beta\,(c_A - c_A^0) = N_0'(1 + \varkappa_e) \frac{D_A c_A^0 ez_A}{f_A kT} \vec{n}\,\vec{E}_{\infty} - N_0' D_A \vec{n}\,\nabla c_A. \tag{3.22}$$

If the above boundary condition is applied to the solution of Eq. (2.23), which determines the steady-state concentration of the A atoms in the inclusion, we find that

$$c_A = c_A^0 + N_0'(1 + \varkappa_e) \frac{D_A c_A^0 ez_A}{f_A kT} \frac{1}{R\beta + N_0' D_A} \vec{r}\,\vec{E}_{\infty}. \tag{3.23}$$

If we use this solution in Eq. (3.22) for $\vec{n}\cdot\vec{I}'$, and in Eq. (1.11), we find that the velocity of migration controlled by diffusion across an inclusion given by

$$\vec{v} = -\frac{\omega}{\omega'}(1 + \varkappa_e) \frac{D_A c_A^0 ez_A}{f_A kT} \frac{R\beta}{R\beta + N_0' D_A} \vec{E}_{\infty}. \tag{3.24}$$

[†] In view of the drag effects, an ion crossing an interface between phases and located near the top of a potential barrier is subject to a force $ez'^* \vec{En}$ ($z'^* \sim 1$-10 is an effective charge). Consequently, the expression for the flux of the dissolved host atoms acquires the term $\sim \beta \frac{c_A^0 aez'^*}{kT} \vec{En}$. However, an allowance for this term gives rise to a small correction $\lesssim a/R$ to the velocity and, therefore, it can be ignored.

[‡] If the A atoms dissolved in the inclusion are neutral, the charge $z_A = z_A^*$ is also the effective charge, related solely to the drag forces.

If the condition (2.31) is satisfied and the diffusion resistance at the inclusion—host interface can be ignored, we obtain

$$\vec{v} = -\frac{\omega}{\omega'}\,(1 + \varkappa_e)\,\frac{D_A\,c_A^0\,ez_A}{f_A\,kT}\,\vec{E}_\infty,\qquad(3.25)$$

i.e., the velocity is independent of the radius of the inclusion. However, if the opposite condition (2.32) is satisfied and the diffusion flux is limited by the rate of dissolution, the velocity

$$\vec{v} = -\omega\,(1 + \varkappa_e)\,\frac{c_A^0\,ez_A\,\beta R}{f_A\,kT}\,\vec{E}_\infty\qquad(3.26)$$

is proportional to the radius of the inclusion.

In the case of nonconducting inclusions, in which the A atoms are not ionized, there are no electron fluxes which can exert drag on the A atoms and, therefore, the charge z_A is zero. Consequently, the velocity of electrotransport controlled by volume diffusion in the host crystal vanishes, in accordance with Eqs. (3.24)–(3.26).[†]

The dependences of the electrotransport velocities of the temperature and radius are of the same type, for the same migration mechanisms, as in the case of a temperature gradient (Sec. 2, Figs. 3–5).

We shall estimate the velocity of an inclusion in an electric field by assuming that the density of the electric current is $j_\infty = 10^5$ A/cm^2. We then find that $E_\infty \sim 1$ V/cm if $\lambda_e \sim 10^5\ \Omega^{-1} \cdot$ cm^{-1}.

[†] In this case, the finite velocity is due to the drag forces acting on the host atoms during dissolution (see Footnote 1 in the present section). However, the normal component of the electric field in the host lattice at the interface with a nonconducting inclusion is zero. Therefore, the effective charge z'* is of the order of l_e/R if $l_e \lesssim R$ and z'* $\approx$ 1–10 if $l_e \gg R$, whereas the velocity of an inclusion is of the order of

$$v \sim \frac{\beta R}{\beta R + N_0' D_A}\,\frac{\omega}{kT\omega'}\,D_A\,c_A^0\,\frac{a}{R}\,\frac{l_e}{R}\,eE_\infty$$

and is usually less than the velocity given by Eq. (3.19), which is due to surface diffusion fluxes (because in practically all cases $D_S \gg D_A c_A^0$). These comments also apply to the migration of micropores associated with the evaporation of the host matter at the pore—host interface.

TABLE 1. Results of a Study of Electrotransport of
Inclusions in Copper and Silver

Material	T, °C	i, A/mm^2	D, cm^2/sec	z
Copper	820	84	$1 \cdot 10^{-10}$	-24
	830	110	$1 \cdot 10^{-10}$	-26
	860	133	$1.65 \cdot 10^{-10}$	-15
Silver	800	100	$3.3 \cdot 10^{-10}$	-17
	820	100	$5.6 \cdot 10^{-10}$	-13
	850	180	$1 \cdot 10^{-9}$	-4

We shall also assume that $\varkappa_e \approx 1/2$, $z \sim 10$, and $D \sim 10^{-9}$ cm$^2 \cdot$ sec^{-1}. It follows from Eq. (3.14) that $v \sim 10^{-7}$ cm/sec. For the same parameters, except for $D_S \sim 10^{-5}$ cm^2/sec, Eq. (3.19) gives $v \sim 10^{-6}$ cm/sec for $R = 10^{-4}$ cm and $v \sim 10^{-4}$ cm/sec for $R = 10^{-6}$ cm. The very high current densities assumed in this example can be achieved conveniently in thin wires or plates (efficient cooling would have to be provided).

Similar formulas apply also to the motion of cylindrical inclusions in a field perpendicular to the cylinder axis. We have to modify Eqs. (3.24)-(3.26) by replacing $\varkappa_e$ with $\varkappa_{ec}$, where

$$\varkappa_{ec} = \frac{\lambda_e - \lambda_{0e}}{\lambda_e + \lambda_{0e}}, \qquad (3.27)$$

and by replacing $2\varkappa_e$ with $\varkappa_{ec}$ in Eqs. (3.12) and (3.14). In Eq. (3.19), the factor 3 should be replaced with 2.

In the case of pores located near the boundary of a crystal or in a thin plate the electric field distribution may differ somewhat from that given by Eq. (3.9) because of the image forces at the boundary. Consequently, the expressions for the velocity of the diffusional migration of pores in an electric field will be modified somewhat. The necessary corrections are given in [176].

Experimental investigations of the motion of inclusions in an electric field were reported in [22]. These investigations were carried out on copper and silver. The procedure adopted in the case of copper was as follows. Two copper rods were welded along a plane in which were embedded a tungsten foil and a series of thin ($d = 15\ \mu$) tungsten wires. The foil acted as a reference plane. The motion of the tungsten wires (Fig. 9) was measured relative to this plane (the electrical conductivity of tungsten was considerably less than that of copper).

In the experiments on silver, the moving inclusions were deep scratches on the surface which were displaced by the passage of a current along wire-shaped samples.

The experimentally determined displacements were used, in conjunction with Eq. (3.15), to determine the effective charge of ions, z [22].

The results of the experiments reported in [22] are summarized in Table 1.

Similar values of z were obtained in the experiments on the electrotransport of host atoms [35, 38].

4. MIGRATION OF INCLUSIONS IN A VACANCY

CONCENTRATION GRADIENT

The distribution of vacancies in a crystal $c_v\ (\vec{r})$ can be inhomogeneous. An inhomogeneity of $c_x\ (\vec{r})$ may result, for example, when vacancies emerge on the surface of a crystal or on grain surfaces when a crystal supersaturated with vacancies is quenched or irradiated with fast particles. The vacancy fluxes reaching different parts of a pore in such a crystal are not equal and should result in the motion of the pore as a whole [12]. Since a vacancy concentration gradient is related to the gradient of the chemical potential of vacancies, i.e., it is due to a generalized thermodynamic force, the motion of a pore in the field of a vacancy concentration gradient can be regarded as a special case of the migration under the action of external forces.

We shall consider such migration in the simplest case when the vacancy concentration gradient far from a pore is constant and the concentration $c_r\ (\vec{r}) = c_{v\infty}\ (\vec{r})$ varies linearly with distance:

$$c_{v\infty}\ (\vec{r}) = c_{v0} + \vec{r}\, \nabla c_{v\infty}.$$
(4.1)

Near the surface of a spherical pore of radius R, the vacancy concentration has a constant value c_{vR}^0. This value depends on the gas pressure in the pore P_0 and on the Laplace pressure of

Eq. (2.34), i.e., it depends on the radius R:

$$c_{vR}^0 = c_v^0(T)\left[1 + \left(\frac{2\gamma}{R} + P^0 - P_0\right)\frac{\omega}{kT}\right].$$

(4.2)

Since practically always $R|\nabla c_{v\infty}| \ll 1$ (because $c_v \ll 1$), the velocity of a pore, $v' \propto D_v|\nabla c_{v\infty}|$ [see Eq. (4.7)], satisfies the condition (2.13), under which the distribution of vacancies assumes a steady-state configuration after a time interval equal to the relaxation time of Eq. (1.2) and is then described by the equation $\Delta c_v = 0$. The solution of this equation, subjected to the boundary conditions (4.1) and (4.2) at infinity and on the surface of the pore, is of the form

$$c_v(\vec{r}) = c_{v0} + \vec{r}_p\nabla c_{v\infty} + (c_{vR}^0 - c_{v0} - \vec{r}_p\nabla c_{v\infty})\frac{R}{r'} + \left(1 - \frac{R^3}{r'^3}\right)\vec{r}'\nabla c_{v\infty}.$$

(4.3)

Here, $\vec{r}_p$ is the radius vector of the center of the pore and $\vec{r}' = \vec{r} - \vec{r}_p$.

When the vacancy concentration $c_v(\vec{r})$ is distributed nonuniformly, a flux of vacancies $\vec{I}_v$ and a flux of atoms $\vec{I} = -\vec{I}_v$ relative to the lattice appears in the bulk of the host crystal. At the pore surface, the flux of atoms is

$$\vec{I} = -\vec{I}_v = N_0D_v\nabla c_v = -N_0D_v\left(c_{vR}^0 - c_{v0} - \vec{r}_p\nabla c_{v\infty}\right)\frac{\vec{n}}{R} + 3N_0D_v(\vec{n}\nabla c_v)\vec{n}.$$

(4.4)

The velocity of each part of the pore surface due to this flux can be divided, in accordance with Eq. (1.1), into a component which determines the radius of the pore [the first term in Eq. (4.4)] and the component which determines the velocity of the translational migration of the pore as a whole [the second term in Eq. (4.4)]. Since $\vec{n}\vec{I} = \text{const} + \text{const}\cdot\cos\theta$, the pore retains its spherical shape.

It follows from Eqs. (1.1), (1.4), and (4.4) that the rate of change of the pore radius is

$$\frac{dR}{dt} = -D_v\frac{c_{vR}^0 - c_{v0} - \vec{r}_p\nabla c_{v\infty}}{R},$$

(4.5)

and the velocity of the translational migration of the pore relative to the lattice is

$$\vec{v}' = 3D_v \nabla c_v. \tag{4.6}$$

Since the flux of vacancies results in the motion of matter in the host crystal far from the inclusion and the velocity of such motion is

$$\vec{v}_a = \omega \vec{I}_\infty = D_v \nabla c_{v\infty},$$

we find that (in a system of coordinates linked to the matter far from the pore) the pore has the velocity

$$\vec{v} = 2D_v \nabla c_v, \tag{4.7}$$

which is determined by the vacancy diffusion coefficient and by the vacancy concentration gradient.

The change in the radius of a pore as a result of diffusion and of diffusion-viscous processes is considered in [39-43] for a homogeneous distribution of the vacancies (the behavior of an ensemble of pores is analyzed in [44, 45]). It follows from Eq. (4.5) that the time dependence of the pore radius can have several special features if the vacancy distribution is inhomogeneous. These features are due to the fact that the right-hand side of Eq. (4.5) must include the dependence of the concentration c_{vR}^0 on R, in accordance with Eq. (4.2), as well as the linear dependence of $\vec{r}_p = \vec{v}t$ on t. Consequently, when the distribution of vacancies is inhomogeneous, the time dependence of the pore radius may be very complex. In particular, when $2\gamma/R \approx P_0 - P^0$, the dependence R(t) may have a minimum.

A vacancy concentration gradient appears whenever the vacancy concentration in the interior of a crystal c_v' differs from the equilibrium value c_v^0. Such an enhanced vacancy concentration may be the result of quenching from high temperatures or irradiation with fast particles. Since the equilibrium vacancy concentration c_v^0 is established rapidly at the boundary of a crystal, a vacancy concentration gradient may be established and this gradient will result in the flow of vacancies from the interior to the boundaries of the crystal (and to vacancy sinks inside the crystal).

Under the action of such vacancy fluxes, a pore may move at a velocity given by Eq. (4.7). We can readily show that a pore located at the boundary of a region, which is free of vacancy sinks and has the shape of a plate of thickness d, is displaced by an amount δx in a time interval equal to the relaxation time of the excess vacancy concentration $\delta c_V = c_V' - c_V^0$. This displacement is directed away from the boundary of a crystal (for $\delta c_V > 0$) and is given by the following simple relationship:*

$$\delta x = \delta c_v d. \tag{4.8}$$

If a crystal has a sufficiently low concentration of defects and the region free of vacancy sinks is sufficiently thick, the displacement δx can be considerable even when the supersaturation δc_V is low. The displacement of a pore will be considerably larger in the case of a crystal which contains vacancy sources in its interior (for example, other pores), which can maintain a constant value of δc_V and a constant gradient ∇c_V at the boundary for a long time τ_0. Then $\delta x \sim D_V \delta c_V \tau_0 / x'$, where x' is the distance from the boundary of a crystal to the region where the concentration of vacancy sources is high.

Equations (4.6) and (4.7) can both be used to determine the velocity of inclusions in the field of a vacancy concentration gradient if vacancies can be created and annihilated easily on the surfaces of these inclusions. It should be noted that inclusions

---

*This can be shown quite easily because integration of the equation

$$\frac{\partial c_v}{\partial t} = D_v \frac{\partial^2 c_v}{\partial x^2}$$

with respect to time between the limits $t=0$ $(c_v'(\vec{r})=\text{const}=c_v')$ and $t=\infty$ $(c_v'(\vec{r})=c_v^0)$ yields

$$\delta c_v = -D_v \frac{\partial}{\partial x} Y(x), \quad Y(x) = \int_0^\infty \frac{\partial c_v}{\partial x} dt.$$

Since the initial value $c_v'(\vec{r})-c_v^0=\delta c_v$ is independent of x and since $\partial c_V / \partial x = 0$ at x = 0 (x = 0 is the midpoint of the plate), integration of the above equation with respect to x yields

$$Y(x) = -\frac{\delta c_v}{D_v} x, \quad Y\left(-\frac{d}{2}\right) = \frac{\delta c_v d}{2D_v}.$$

Hence, we obtain Eq. (4.8) for a rectangular plate.

differ from pores in the following respect: a rearrangement of stresses in a crystal containing inclusions makes c_{vR}^0 equal to c_v^0 and the radius of an inclusion ceases to vary after a certain relaxation period (Sec. 5).

5. MIGRATION IN AN INHOMOGENEOUS STRESS FIELD

AND THE RELAXATION OF STRESSES

AROUND INCLUSIONS

A homogeneous stress field simply orients inclusions in a crystal but does not give rise to migration. However, if the stress field is inhomogeneous, diffusion fluxes give rise to the translational motion of inclusions at a velocity proportional to the stress gradient [10].

Inhomogeneous stress fields are produced not only by the application of concentrated loads to the surface but also when the average strain in a polycrystalline crystal is homogeneous. This is due to the following reasons.

First, the stresses in grains oriented in different directions relative to one another are different because of the elastic anisotropy. Therefore, the stress distribution must be inhomogeneous in order to satisfy the stress-matching condition at the grain boundaries, which are usually quite complex. The stress inhomogeneity due to this cause is proportional to the elastic anisotropy parameter ζ, and the gradient σ' is of the order of $\zeta \sigma / L_g$, where L_g is the average size of a grain.

Second, even when a polycrystalline sample consists of elastically isotropic grains, diffusion fluxes of vacancies appear between grain boundaries at high temperatures even if the stress field is initially homogeneous. These fluxes give rise to diffusion–viscous flow in the polycrystalline sample (this flow is considered in [46, 47] and detailed treatments are given in [48, 49]). When such steady-state flow is established, some stress relaxation takes place and the stresses become inhomogeneous in the individual grains. The resultant gradient is of the order of σ / L_g.

Third, strong local stress gradients may appear near certain defects in crystals, for example, near dislocations or dislocation loops.

Diffusion Fluxes in an Elastic

Stress Field

Phonon and electron fluxes do not appear in the absence of temperature gradients and electric fields. Therefore, diffusing atoms are not acted upon by the drag forces. Vacancy diffusion fluxes, resulting from stress or vacancy concentration gradients, are simply due to the tendency of all the atomic subsystems to approach thermodynamic equilibrium and to minimize the Gibbs free energy. The thermodynamics of irreversible processes allows us to express the vacancy flux $\vec{I}_v$ in terms of the gradient of the chemical potential of vacancies μ_V:

$$\vec{I}_v = -K_v \nabla \mu_v. \tag{5.1}$$

The chemical potential of vacancies μ_V is defined as the work which must be done in the reversible replacement of a vacancy with an atom [47]. This potential is related to the vacancy concentration c_V by the standard expression applicable to weak solutions:

$$\mu_v = \mu_v^0 + kT \ln c_v. \tag{5.2}$$

Here, the quantity μ_V^0 is independent of c_V.

The transport coefficient K_V can be expressed in terms of the vacancy diffusion coefficient D_V. This can be done by considering an unstressed crystal, in which μ_V changes because of an inhomogeneous distribution of vacancies. Then, in accordance with Eq. (5.2), we find that

$$\nabla \mu_v = \frac{kT}{c_v} \nabla c_v. \tag{5.3}$$

Substituting Eq. (5.3) into Eq. (5.1) and comparing the resultant expression for $\vec{I}_v$ with Eq. (4.4), we obtain

$$K_v = \frac{N_0 D_v c_v}{kT}. \tag{5.4}$$

The diffusion coefficient of vacancies D_V and the self-diffusion coefficient of atoms D are related by

$$D = f D_v c_v, \tag{5.5}$$

where f is the correlation factor introduced in Eq. (3.3). Therefore, it follows from Eqs. (5.1), (5.4), and (5.5) that the expression for the diffusion flux of atoms $\vec{i} = -\vec{i}_v$ relative to the lattice of the host crystal can be written in the form

$$\vec{i} = \frac{N_0 D}{fkT} \nabla \mu_v. \tag{5.6}$$

The diffusion flux defined by Eq. (5.6) depends strongly on the ease with which the vacancies are created and annihilated in the interior of the host crystal. If the concentration of defects in the crystal or in its grains is sufficiently low so that vacancies can appear and disappear only on the grain boundaries and on the surfaces of inclusions, it follows that, under steady-state conditions div $\vec{i} = 0$ inside a grain and, in accordance with Eq. (5.6), the chemical potential obeys the Laplace equation

$$\Delta \mu_v = 0. \tag{5.7}$$

The distribution of the chemical potential and the diffusion fluxes can be found by solving Eq. (5.7) subject to the boundary conditions on the surfaces of grains and near vacancy sources and sinks (if these are present within a grain).

In specifying the boundary conditions for the chemical potential on the surface of a grain, we must bear in mind that the appearance of an atom on the surface of a crystal in the presence of external stresses requires a certain amount of work which must be done against the forces responsible for the stresses. Since the establishment of equilibrium is accompanied by the motion of a part of the surface along the normal $\vec{n}$, this work is equal to the product of the atomic volume ω and the normal stress σ_{nn} :

$$\omega \sigma_{nn} = \omega \sigma_{ij} n_i n_j,$$

where σ_{ij} are the components of the stress tensor (the summation over the repeated indices i and j is carried out between 1 and 3). This work must be included in the expression for the chemical potential μ_v. Therefore, the boundary conditions for μ_v near an external surface of a crystal are of the form [47, 48]

$$\mu_v = \mu_{v0} + \omega \sigma_{nn}, \tag{5.8}$$

where μ_{v0} is the chemical potential in the absence of stresses.

The solution of Eq. (5.7) subject to the boundary conditions of Eq. (5.8) allows us to express the chemical potential μ_v and, consequently, the diffusion fluxes in a perfect crystal in terms of the normal stresses on the surface of this crystal.

We shall now consider the opposite case when the concentration of vacancy sources and sinks is so high that the distances between them are short compared with the characteristic dimensions of the problem (in particular, compared with R). In this case, we may assume that a quasiequilibrium value of the chemical potential μ_v is established at any given point in a crystal and its value corresponds to the stresses at that point. If the distribution of vacancy sources and sinks is, on the average, of cubic symmetry (no texture), this quasiequilibrium value of μ_v can be found by averaging over the various directions the value given by Eq. (5.8):

$$\mu_v = \mu_{v0} + \frac{1}{3}\,\omega\sigma_{ii}. \tag{5.9}$$

In accordance with Eq. (5.6), the diffusion flux $\vec{I}$, associated with a stress gradient is given by

$$\vec{I} = \frac{D}{3\,fkT}\,\nabla\sigma_{ii}. \tag{5.10}$$

Similar expressions for the diffusion fluxes of atoms in inhomogeneous stress fields in solid solutions are considered in [50-52].

The concentration of vacancies or other point defects at the surface of a micropore, or an inclusion surrounded by an amorphous boundary, can always be regarded as the equilibrium concentration. Therefore, in the isothermal case (and in the absence of a field), the surface diffusion flux $\vec{I}_s$ can be expressed directly in terms of the chemical potential gradient μ_S of the host atoms on the surface and the resultant equation resembles Eq. (5.6):

$$\vec{I}_S = -\frac{N_0 D_S a}{f_S kT}\,\nabla_S \mu_S, \tag{5.11}$$

in which the surface diffusion coefficient D_S and the surface correlation factor f_S replace the corresponding bulk parameters.

Under surface diffusion conditions, equilibrium is established not as a result of the injection of vacancies from the surface of a pore, as in the case of volume diffusion in a perfect crystal, but as a result of the motion of atoms along the surface. Therefore, the dependence of the chemical potential μ_S on the stresses is determined, as in Eq. (5.9), not only by the normal stress σ_{nn} but also by the sum σ_{ii} of the diagonal elements of the stress tensor:

$$\mu_S = \mu_0 - \frac{1}{3}\omega\sigma_{ii}. \tag{5.12}$$

Here, μ_0 is the chemical potential of atoms in the absence of stresses.

Substituting Eq. (5.12) into Eq. (5.11), we obtain an expression for the surface flux in a stress field:

$$\vec{I}_S = \frac{D_S a}{3 f_S kT}\nabla_S\sigma_{ii}. \tag{5.13}$$

The diffusion fluxes can be determined from Eqs. (5.6), (5.10), and (5.13), if we know the stresses at the inclusion–host interface. In the case considered here, when the radius of a spherical inclusion in an elastically isotropic crystal is small compared with the grain dimensions, the stress field around an inclusion in the presence of a stress gradient can be determined by applying the standard methods of the theory of elasticity [53]. It is convenient to expand the stresses near a spherical inclusion in terms of spherical harmonics and their derivatives. Only the first harmonic is used in the determination of the migration velocity of inclusions because the higher and zeroth harmonics of the stresses simply distort the surface and alter the volume of an inclusion (Sec. 7).

Far from an inclusion, the first harmonic of the stresses is a linear function of the distance and it corresponds to a constant stress gradient σ_1 along some axis z. If the stresses have an axial symmetry (no torsion), the asymptotic expressions for stresses far from an inclusion can be written in terms of the spherical coordinates r, θ, and φ in the form

$$\left.\begin{array}{l} \sigma_{rr} = \sigma_1 r\cos\theta, \quad \sigma_{\theta\theta} = \sigma_{\varphi\varphi} = 2\sigma_1 r\cos\theta; \\[2mm] \sigma_{r\theta} = \frac{1}{2}\sigma_1 r\sin\theta, \quad \sigma_{r\varphi} = 0, \quad \sigma_{\theta\varphi} = 0 \quad (r\to\infty). \end{array}\right\} \tag{5.14}$$

Near an inclusion, the stress field represented by Eq. (5.14) is distorted and the nature of the distortion depends on the boundary conditions applicable to the stresses on the inclusion−host interface. In the case of micropores or liquid inclusions, the tangential stresses on the surface should vanish. Moreover, in the absence of an external force acting on an inclusion, the radial component of the first harmonic of the stresses should also vanish (the total force $\int \sigma_{rr} \cos \theta dS$, acting on a particle, is zero). Therefore, we find that †

$$\sigma_{rr} = 0, \ \sigma_{r\theta} = 0, \ \sigma_{r\varphi} = 0 \qquad (r=R). \tag{5.15}$$

The same boundary conditions apply also after a certain stress-relaxation time in the case of easy slip along an amorphous interface between a crystalline inclusion and its host.

If we solve the elasticity theory equations subject to the boundary conditions of Eqs. (5.14) and (5.15), we find that [10, 12]:

$$\left.\begin{array}{l} \sigma_{rr} = \sigma_1 R \left[\dfrac{r}{R} - \left(\dfrac{R}{r} \right)^4 \right] \cos \theta, \\[3mm] \sigma_{\theta\theta} = \sigma_{\varphi\varphi} = \sigma_1 R \left[2\dfrac{r}{R} + \dfrac{1}{2} \left(\dfrac{R}{r} \right)^4 \right] \cos \theta, \\[3mm] \sigma_{r\theta} = \dfrac{1}{2} \sigma_1 R \left[\dfrac{r}{R} - \left(\dfrac{R}{r} \right)^4 \right] \sin \theta, \ \sigma_{r\varphi} = 0; \ \sigma_{\theta\varphi} = 0. \end{array}\right\} \tag{5.16}$$

In this case, the stresses vanish inside an inclusion. It follows from Eq. (5.16) that the sum of the diagonal matrix elements σ_{ii} is

$$\sigma_{ii} = 5\sigma_1 r \cos \theta \qquad (r \gg R). \tag{5.17}$$

Velocity of Inclusions in an Inhomogeneous Stress Field under Different Migration Conditions

The velocity of an inclusion in the field of a stress gradient is determined by the diffusion fluxes in the interior of the host crystal and in the inclusion and the fluxes on the inclusion−host

---

† We shall ignore here the internal stresses associated with the Laplace pressure and the pressure of the gas in a pore. Their role will be discussed later (Sec. 7).

interface. If the flux of the host atoms in the interior of a perfect crystal (free of vacancy sources and sinks) is defined by Eq.
(5.6), an allowance must be made for the fact that near an inclusion the distribution of the chemical potential is distorted. This
distortion is due to the following circumstance. In the absence of
the zeroth stress harmonic (which alters the volume of an inclusion, as discussed later), it follows from Eq. (5.16) that $\sigma_{nn} =$
$\sigma_{rr} = \sigma_{rr}^0$ assumes a constant value on the surface of the inclusion
and this value is independent of the angular orientation [σ_{rr}^0 is the
zeroth harmonic of σ_{rr} corresponding to r = R, which has not been
included in Eq. (5.16)]. Therefore, the boundary value of the chemical potential μ_v given by Eq. (5.8) is also constant:

$$\mu_v = \mu_{v0} + \omega\sigma_{rr}^0 = \text{const} \tag{5.18}$$

(provided that the vacancies can appear freely on this boundary).

The solution of Eq. (5.7) subject to the condition (5.18) on the
surface of a sphere and to the condition $\mu_v = \mu'_{v0} + r\nabla\mu_{v\infty}$, which corresponds to a constant value of the chemical potential gradient
$\nabla\mu_{v\infty}$ far from an inclusion (μ'_{v0} is some constant which is unimportant in the present case), is of the form

$$\mu_v = \mu'_{v0} + \left(\mu_{v0} + \omega\sigma_{rr}^0 - \mu'_{v0}\right)\frac{R}{r} + \left(1 - \frac{R^3}{r^3}\right)r\nabla\mu_{v\infty} \tag{5.19}$$

(the origin of the coordinate system is assumed to be at the center of the sphere).

If we use Eqs. (5.6) and (5.19), we can find the normal component of the flux of the host atoms reaching the surface of an inclusion and the velocity of this inclusion (relative to the matter
in the host crystal far from the inclusion:

$$v = 2\frac{D}{fkT}\nabla\mu_{v\infty} = 2\omega\vec{I}_\infty, \tag{5.20}$$

where $\vec{I}_\infty = -\vec{I}_{v\infty} = \frac{N_0 D}{fkT}\nabla\mu_{v\infty}$ is the flux of the host atoms far
from the inclusion (measured relative to the host lattice).

If vacancies cannot appear at the inclusion–host interface,
we find that an inclusion which is rigidly bound to the host lattice
moves − relative to the matter in the host crystal far from the

inclusion – at a velocity $\vec{v} = -\omega \vec{I}_\infty = -\dfrac{D}{fkT}\nabla\mu_{v\infty}$, directed opposite to the velocity defined by Eq. (5.20).

If the flux of vacancies or atoms is due to discontinuities in the normal stresses at the grain boundaries, $\delta\sigma_{nn}$, it follows from Eq. (5.8) that $|\nabla\mu| \sim \omega\delta\sigma_{nn}/L_g$ and the order of magnitude of the velocity is given by

$$v \sim \frac{D\omega}{kT}\frac{\delta\sigma_{nn}}{L_g} .$$
(5.21)

If the distances between the vacancy sources and sinks are small compared with R, the fluxes near the surface and in the interior of the host crystal are given by Eqs. (5.10) and (5.17). It follows from Eqs. (1.1), (1.5), and (1.6) that the velocities $\vec{v}'$ and $\vec{v}$ of an inclusion are given by

$$\vec{v}' = \frac{5}{3}\frac{D\omega}{fkT}\sigma_1\vec{e}_z, \ \vec{v} = 0,$$
(5.22)

where the unit vector $\vec{e}_z$ is parallel to the stress gradient. † If the stress gradient results from the rearrangement of stresses at the boundaries, we find that $\sigma_1 \sim \delta\sigma_{nn}/L_g$, where $\delta\sigma_{nn}$ can be of the same order of magnitude as the average stresses in the polycrystalline sample. Then, Eq. (5.22) for $\vec{v}'$ yields the same order of magnitude of the velocity (5.21) as Eq. (5.20). In this case, the velocity $\vec{v}$ relative to matter far from the inclusion is zero.

For example, if the stresses $\delta\sigma_{nn} \sim 10$ J/cm^3 ($\sim 10^8$ ergs/cm^3) do not exceed the flow stress, we find that v ~ 1 Å/sec for D $\sim 10^{-10}$ cm^2/sec, $\omega \sim 10^{-23}$ cm^3, T $\sim 10^3$ °K, L$_g \sim 10^{-4}$ cm. Under such stresses an inclusion travels a distance of the order of L$_g$ in a time $\sim 10^4$ sec.

It follows from Eqs. (5.13) and (5.17) that the diffusion flux of the host atoms, which is generated by the stress field (5.16) and

---

†Equation (5.22) is derived on the assumption that the distribution of vacancy sources and sinks throughout the region surrounding an inclusion has a cubic symmetry. If the vacancy sources and sinks have a preferential orientation, in particular, if they are located mainly behind a moving inclusion, Eq. (5.22) is, strictly speaking, no longer valid and can be used only in rough estimates.

which reaches the inclusion−host interface, is given by

$$\vec{I}_S = \frac{5}{3}\frac{D_S a}{f_S kT}\sigma_1 R \nabla_S \cos\theta. \tag{5.23}$$

It follows from Eqs. (1.9), (2.19), and (5.23) that the expression for the velocity of an inclusion migrating as a result of surface diffusion fluxes in a stress field is

$$\vec{v} = -\frac{10}{3}\frac{\omega D_S}{f_S kT}\frac{a}{R}\sigma_1 \vec{e}_z. \tag{5.24}$$

For example, if $\sigma_1 \sim \delta\sigma_{nn}/L_g \sim 10^{13}$ J/m^4 (10^{12} ergs/cm^4), $D_S \sim 10^{-7}$ cm^2/sec, $\omega \sim 10^{-23}$ cm^3, $T \sim 10^3$ °K, it follows from Eq. (5.24) that v $\sim 10^2$ Å/sec for R $\sim 10^2$ Å and v ~ 1 Å/sec for R $\sim 10^4$ Å.

There are no stresses inside an inclusion whose surface provides favorable conditions for slip. Moreover, inside this inclusion there is no diffusion transport of atoms under the action of elastic forces. Nevertheless, a concentration gradient gives rise to diffusion fluxes of the host (A) atoms within an inclusion in an elastically stressed host crystal because the concentrations of atoms dissolved in the inclusion depend on the stresses in the host crystal near the inclusion−host interface. Consequently, they depend on the coordinates of a given point on this interface.

The equilibrium concentration $c_A^0(\vec{r}_S)$ of the A atoms in an inclusion (near the inclusion−host interface) can be found by equating the chemical potential $\mu_A(c_A)$ of these atoms and the chemical potential μ_S of the host atoms near the inclusion−host interface. It follows from Eq. (5.12) that the latter potential includes a term which depends on the stresses. Therefore, the stresses of Eq. (5.17) alter the equilibrium concentration of the A atoms in the inclusion near the inclusion−host interface and the change in the concentration is

$$\delta c_A^0(\vec{r}_S) = -\frac{1}{3}\sigma_{ii}\omega\left(\frac{\partial\mu_A}{\partial c_A}\right)^{-1} = -\frac{5}{3}\omega\sigma_1 R\left(\frac{\partial\mu_A}{\partial c_A}\right)^{-1}\cos\theta. \tag{5.25}$$

If the concentration c_A^0 of the A atoms in the inclusion is low, we find that $\mu_A = kT \ln c_A^0 + \text{const}$ and

$$\delta c_A^0(\vec{r}_S) = -\frac{5}{3}\frac{\omega c_A^0}{kT}\sigma_1 R \cos\theta. \tag{5.26}$$

Because diffusion occurs inside the inclusion, the actual change in the concentration $\delta c_A(\vec{r}_s)$ at the inclusion–host interface differs somewhat from the equilibrium value $\delta c_A^0(\vec{r}_s)$, and this gives rise to a flux of the A atoms dissolved in the inclusion, which is equal to the diffusion flux inside the inclusion:

$$\vec{n}\vec{I}' = \beta\left[\delta c_A(\vec{r}_s) - \delta c_A^0(\vec{r}_s)\right] = -N_0' D_A \vec{n} \nabla c_A\big|_{\vec{r}=\vec{r}_s}. \tag{5.27}$$

The above expression is the boundary condition for Eq. (2.23), which gives the concentration of the A atoms inside an inclusion. The solution of this equation is of the form

$$\delta c_A(\vec{r}) = -\frac{5}{3}\frac{\beta R}{\beta R + N_0' D_A}\frac{\omega c_A^0}{kT}\sigma_1 \vec{r}\,\vec{e}_z. \tag{5.28}$$

Substituting Eq. (5.28) into Eq. (5.27), we can determine the velocity of the inclusion (for $c_A^0 \ll 1$) from Eq. (1.11):

$$\vec{v} = -\frac{5}{3}\frac{\beta R}{\beta R + N_0' D_A}\frac{\omega}{\omega'}\frac{\omega c_A^0 D_A}{kT}\sigma_1 \vec{e}_z. \tag{5.29}$$

If the diffusion fluxes are limited by the processes occurring inside an inclusion or on its surface, we find that, respectively,

$$\vec{v} = -\frac{5}{3}\cdot\frac{\omega}{\omega'}\cdot\frac{\omega c_A^0 D_A}{kT}\sigma_1 \vec{e}_z \tag{5.30}$$

or

$$\vec{v} = -\frac{5}{3}\,\omega\beta R\frac{\omega c_A^0}{kT}\sigma_1 \vec{e}_z. \tag{5.31}$$

The migration of small-radius micropores, which satisfy the condition (2.36), in an elastic stress field is due to evaporation processes. If we carry out calculations similar to those employed in the derivation of Eq. (2.38), we can show that in the present case the velocity is

$$\vec{v} = -\frac{10}{9}\,M\,\frac{\omega^2 T}{u_S}\,\sigma_1 R \vec{e}_z, \tag{5.32}$$

where u_S is the activation energy of the rate of evaporation. We have assumed that the number of atoms n_S which are evaporated per unit time from unit surface is proportional to $\exp\left(-\dfrac{u_S}{kT} - \dfrac{1}{3}\dfrac{\sigma_{ii}\omega}{kT}\right)$ and, therefore,

$$\frac{\partial n_S}{\partial \sigma_{ii}} = -\frac{1}{3}\frac{\omega T}{u_S}\frac{\partial n_S}{\partial T} = -\frac{1}{3}\frac{\omega T}{u_S}M.$$

The dependences of the velocity $\vec{v}$ on the temperature and the radius of an inclusion, which are obtained for different migration mechanisms, are the same as those which apply to migration in a temperature gradient or in an electric field.

Diffusional Stress Relaxation
Around Particles

The stresses which depend on the angle (the first harmonic of the stress tensor) are responsible for the differences between the vacancy fluxes reaching different parts of an inclusion. They are also responsible for the motion of inclusions. However, some inhomogeneous stresses are independent of the angle (the zeroth harmonic of the stress tensor). They give rise to radial vacancy fluxes which do not disturb an inclusion but result in a relaxation of the stresses surrounding a solid or liquid inclusion.

These angle-independent internal stresses appear, for example, as a result of phase transitions when spherical particles of a new phase are precipitated in an elastically isotropic crystal. At the boundary of a particle, these stresses are σ_{rr}^0 and they decrease rapidly with distance from the particle:

$$\sigma_{rr} = \sigma_{rr}^0 \frac{R^3}{r^3}. \tag{5.33}$$

The presence of internal stresses around such particles has a strong influence on different properties of a two-component mixture, particularly the plastic properties (Chap. III). The relaxation of stresses may alter considerably the properties of such a two-component disperse system.

We shall investigate the relaxation of stresses as a result of vacancy diffusion [10] by considering just one spherical inclusion in a host crystal in the special case when the distances between vacancy sources and sinks l are large compared with the radius

of the inclusion. If the changes in the concentrations can be ignored, the chemical potential is given by Eq. (5.7) and the boundary conditions are deduced from Eq. (5.8):

$$\mu_v = \mu_{v0} + \omega \sigma_{rr}^0 \quad (r = R), \qquad \mu_v = \mu_{v0} \quad (r \to \infty). \qquad (5.34)$$

It is assumed that there are no external stresses.

The solution of Eq. (5.7), subject to the boundary conditions
of Eq. (5.34), is of the form

$$\mu = \mu_{v0} + \omega \sigma_{rr}^0 \frac{R}{r} . \qquad (5.35)$$

If we substitute this expression into Eq. (5.6), we can find the
flux of atoms $\vec{I}$ and the total number of atoms $d\delta n/dt$ reaching
the surface of an inclusion as a result of vacancy diffusion (it is
assumed that the exchange of atoms between the host crystal and
the inclusion is impossible):

$$\frac{d\delta n}{dt} = -\int \vec{I}\,\vec{n}\,dS = 4\pi \frac{D}{fkT} \sigma_{rr}^0 R. \qquad (5.36)$$

When the number of atoms in the host crystal changes by δn
near the inclusion–host interface, the radial stresses at this interface change by the amount [54]

$$\delta \sigma_{rr}^0 = -\frac{\Gamma G \omega}{\pi R^3} \delta n, \qquad \Gamma = \frac{1}{3} \frac{1+v}{1-v} , \qquad (5.37)$$

where G is the shear modulus; v is the Poisson ratio.

Therefore, it follows from Eq. (5.46) that the radial stresses
obey the equation

$$\frac{d\sigma_{rr}^0 (t)}{dt} = -\frac{1}{\tau} \sigma_{rr}^0 (t), \qquad (5.38)$$

where

$$\tau = \frac{fkT}{4\Gamma \omega G} \cdot \frac{R^2}{D} . \qquad (5.39)$$

It follows from this equation that the stresses relax in accordance with the exponential law

$$\sigma_{rr}^{0}(t) = \sigma_{rr}^{0}(0) \exp\left(-\frac{t}{\tau}\right).$$ (5.40)

The factor $fkT/4\Gamma\omega G$ in Eq. (5.39) is usually $\sim 10^{-2}$, i.e., an order-of-magnitude estimate of the relaxation time yields

$$\tau \sim 10^{-2} \frac{R^2}{D}.$$ (5.41)

It follows that the relaxation time τ is short even if the diffusion coefficient is relatively small. For example, if $R \sim 10^{-6}$ cm, $D \sim 10^{-16}$ cm^2/sec, the relaxation time is $\tau \sim 10^2$ sec.

The quasisteady-state solution just derived can be used if the diffusion paths of the vacancies corresponding to the relaxation time τ are much longer than the radius of the particle R, i.e., if

$$D_v\tau \gg R^2.$$ (5.42)

An examination of Eqs. (5.41) and (5.5) shows that the above condition reduces to the requirement $c_v \ll 10^{-2}$, which is practically always satisfied.

If the number of particles is large, the relaxation of stresses around these particles as a result of the diffusion of vacancies which are annihilated at the particle boundaries may reduce the total concentration of vacancies by δc_v. The order of magnitude of δc_v is $\delta c_v = \delta n N'/N_0 \sim \delta\rho/\rho$, where N' is the number of particles per unit volume and $\delta\rho$ is the change in the density ρ of a two-component system as a result of relaxation. This change in c_v does not reduce significantly the self-diffusion coefficient D and does not alter greatly Eq. (5.6) for the flux of atoms if δc_v is small compared with the equilibrium vacancy concentration c_v^0, i.e., if

$$\frac{\delta\rho}{\rho} \ll c_v^0,$$ (5.43)

or if the vacancy concentration relaxes as a result of diffusion from the grain boundaries or from sources within the grains.

Such relaxation is determined by the condition $L'^2 \ll \tau D_v$ (L' is the smaller of the two quantities L_g and l) or by the condition

$$c_v^0 \ll 10^{-2} R^2 / L'^2,$$ (5.44)

which follows from Eq. (5.41) for τ.

Thus, Eq. (5.41) for the relaxation time is valid only when the vacancy concentration is sufficiently low to satisfy Eq. (5.44), or when the vacancy concentration is high but changes in the density as a result of relaxation are small, in accordance with the condition (5.43).

If neither of these two conditions is satisfied, the relaxation time τ' of the stresses surrounding most of the particles in a grain is limited by the process of establishment of an equilibrium vacancy concentration, i.e., the relaxation time becomes equal to the time necessary for the diffusion of vacancies across a distance L':

$$\tau' \sim L'^2 / D_v.$$ (5.45)

In time intervals shorter than τ', the stresses relax only around those particles which are near grain boundaries or other vacancy sources and the number of such particles increases with time.

The relaxation of stresses around particles [particularly when these particles are large and the relaxation time of Eq. (5.41) is large] may also be due to the motion of dislocations, i.e., due to the usual mechanism of plastic deformation.

6. MIGRATION OF INCLUSIONS UNDER THE INFLUENCE OF EXTERNAL FORCES AND THE BROWNIAN MOTION OF INCLUSIONS

Theory

Sometimes, external forces act not only on diffusing atoms in the host crystal or in an inclusion but also on the inclusion as a whole. We can then assume that such forces are applied directly to the inclusion. For example, external forces act on charged particles which are subjected to an electric field, or on ferromagnetic or ferroelectric particles (which have dipole moments) when these are subjected to an inhomogeneous magnetic or electric field. Other sources of external forces acting on inclusions are grain boundaries, particles in contact with inclusions, dislocations and other extended defects which surround or intersect inclusions.

The force acting on an inclusion may give rise to inhomogeneous elastic stresses in the host crystal, and these can generate directional fluxes of atoms and may give rise to diffusional migration of the inclusion [55], resembling the migration in the field of external elastic stresses, considered in Sec. 5.

The stress fields generated by a force $\vec{F}$ applied to a particle in an elastically isotropic crystal can be determined by the standard methods of the theory of elasticity (see, for example, Chap. VI in [53]). The boundary conditions at the interface between the host crystal and a gaseous or liquid inclusion and at the interface with a solid inclusion in which slip is easy are of the form

$$\sigma_{rr} = -\frac{3}{4\pi}\frac{F}{R^2}\cos\theta, \quad \sigma_{r\theta} = 0, \quad \sigma_{r\varphi} = 0 \quad (r = R), \tag{6.1}$$

where θ is the angle between the normal to the surface and the direction of the force.

Such stresses produce a force $\vec{F}' = \int \sigma_{rr}\cos\theta dS \dfrac{\vec{F}}{F} = -\vec{F}$, which compensates the external force $\vec{F}$.

The sum of σ_{ii} for fields which vanish at infinity and which satisfy the boundary conditions of Eq. (6.1) is

$$\sigma_{ii} = -\frac{3\Gamma}{4\pi}\frac{F}{r^2}\cos\theta, \quad \Gamma = \frac{1}{3}\frac{1+\nu}{1-\nu} \quad (r > R). \tag{6.2}$$

Inside an inclusion, the sum σ_{ii} is

$$\sigma_{ii} = -\frac{9}{4\pi R^3}\vec{F}\vec{r}. \tag{6.3}$$

The above expression applies if the density of the external forces is uniform but the sum σ_{ii} is a complex function of the coordinates if the forces are concentrated in some parts of the inclusion.

If a crystal has a low concentration of defects so that the distances l between vacancy sources and sinks are large compared with R, the chemical potential of the vacancies at the inclusion–host interface is given by Eqs. (5.8) and (6.1). The solution of the Laplace equation (5.7), satisfying the boundary conditions of Eqs. (5.8) and (6.1), is of the form

$$\mu_v(\vec{r}) = \mu_{v0} - \frac{3}{4\pi} \frac{\omega \vec{F} \, \vec{r}}{r^3} \,.$$ (6.4)

It follows from Eq. (5.6) that the diffusion flux of atoms in the interior of the host crystal corresponding to the above distribution $\mu_v(\vec{r})$ is

$$\vec{I} = -\frac{3}{4\pi} \frac{D}{fkT} \left[\frac{\vec{F}}{r^3} - \frac{3(\vec{F}\vec{r})\vec{r}}{r^5} \right].$$ (6.5)

If the density of the vacancy sources and sinks in the interior of the host crystal is high ($l \ll$ R) and their distribution is isotropic, the flux of atoms in the bulk of the host crystal resulting from the stresses of Eq. (6.2) can be found from Eq. (5.10):

$$\vec{I} = -\frac{\Gamma D}{4\pi fkT} \left[\frac{\vec{F}}{r^3} - 3 \frac{(\vec{F}\,\vec{r})\vec{r}}{r^5} \right].$$ (6.6)

The fluxes of Eqs. (6.5) and (6.6) decrease rapidly with increasing distance and the flow of the host matter relative to the lattice vanishes far from the inclusion ($\vec{v}_a = 0$). Therefore, the velocities $\vec{v}'$ and $\vec{v}$ are equal in the systems of coordinates which are linked to the host lattice and to the host matter far from the inclusion, respectively. It follows from Eqs. (1.1), (6.5), and (6.6) that the velocity of an inclusion under the action of an external force associated with vacancy fluxes in the interior of the host crystal is

$$\vec{v} = \frac{3\Gamma_1'}{2\pi} \frac{D\omega}{fkT} \frac{\vec{F}}{R^3} \,.$$ (6.7)

Here, $\Gamma_1' = 1$ for $l \gg$ R and $\Gamma_1' = \Gamma/3$ for $l \ll$ R.

If the external force $\vec{F}$ applied to an inclusion is independent of its radius, the velocity of the inclusion is inversely proportional to the cube of the radius. However, if – as is frequently the case – the force is proportional to the volume of the inclusion, the velocity of Eq. (6.7) is independent of the radius.

When the motion of an inclusion under the action of an external force is associated with surface diffusion fluxes, the velocity is deduced from Eqs. (1.9), (5.13), (6.2), and (2.19):

$$\vec{v} = \frac{\Gamma}{2\pi} \frac{D_S a \omega}{f_S kT} \frac{\vec{F}}{R^4} . \tag{6.8}$$

In this case, the dependence of the velocity of the inclusion on its radius is determined by the factor $\vec{F}/R^4$.

Diffusion fluxes of the host atoms A, associated with stress fields of the (6.3) type and with a gradient of the concentration c_A, may appear in an inclusion. If we assume that the density of the vacancy sources and sinks inside the inclusion is high $(R \gg l)$, we obtain the following expression for the flux of these atoms $\vec{l'}$ (if $c_A \ll 1$) in the form

$$\vec{l} = \frac{D_A c_A^0}{3 f_A kT} \nabla \sigma_{ii} - N_0' D_A \nabla c_A. \tag{6.9}$$

The equilibrium concentration of the A atoms in the inclusion at the inclusion–host boundary in a stress field given by Eqs. (6.2) and (6.3) changes by the amount

$$\delta c_A^0 = \frac{1}{3} c_A^0 \frac{1}{kT} \left[\omega_A \sigma_{ii} \big|_{r=R-0} - \omega \sigma_{ii} \big|_{r=R+0} \right] = \frac{1}{4\pi} \cdot c_A^0 \frac{\Gamma \omega - 3\omega_A}{kT} \frac{F}{R^2} \cos \theta. \tag{6.10}$$

If the diffusion resistance of the inclusion–host interface can be neglected, we find that

$$\delta c_A (\vec{r}) = \delta c_A^0 \frac{r}{R} \tag{6.11}$$

and it follows from Eqs. (1.11), (6.3), and (6.9)-(6.11) that, in the case of a constant density of applied forces, the velocity of an inclusion which migrates because of the diffusion fluxes across it is

$$\vec{v} = \frac{3}{4\pi} \frac{D_A c_A^0 \omega}{f_A kT} \left(1 - \frac{f_A \omega_A}{\omega'} + \frac{f_A \Gamma}{3} \frac{\omega}{\omega'} \right) \frac{\vec{F}}{R^3} . \tag{6.12}$$

An example of the motion of particles in the field of an external force is the migration of a spherical ferromagnetic particle in
a nonferromagnetic host under the action of an inhomogeneous external magnetic field $\vec{H}(r)$ [14]. In this case, the force $\vec{F}$ is

$$\vec{F} = \Omega \nabla (\vec{M}\vec{H}), \qquad (6.13)$$

where $\Omega\vec{M} = N_0'\Omega\vec{\mu_0}$ is the magnetic moment of the particle; Ω is the
volume of the particle; $\vec{\mu_0}$ is the atomic magnetic moment. It fol-
lows from Eqs. (6.7), (6.8), and (6.12) that under the action of this
force a particle should move at a velocity

$$\vec{v} = \frac{D\text{eff}\omega}{kT\omega'} \nabla (\vec{\mu_0}\vec{H}). \qquad (6.14)$$

Here, $D_{eff} = 2\Gamma a\, D_S / 3Rf_S$ if the diffusion fluxes at the boundary of
the particle predominate, and $D_{eff} = 2\Gamma_i' D/f$, if the diffusion in the
interior of the host is the dominant process.

For example, if $D_S \sim 10^{-6}$ cm^2/ sec, $2a/R \sim 10^{-2}$, $D_{eff} \sim 10^{-8}$
cm^2 /sec, $T \sim 10^3$ °K, $\omega \sim \omega'$, $|\nabla H| \sim 10^9$ At /m (10^5 Oe /cm), and
$\mu_0 \sim 5\,\mu_B$ (μ_B is the Bohr magneton), the velocity of such a ferromagnetic particle is of the order of ~3×10^{-2} Å/sec. If the annealing time is ~10^5 sec, this particle migrates over a distance
~10^3 Å. Obviously, in order to observe such migration, we must
select a host with a low melting point in which diffusion can pro-
ceed easily at relatively low temperatures because high temperatures would destroy the ferromagnetism of the inclusions.

A different migration mechanism applies to inclusions subjected to an inhomogeneous magnetic field, which gives rise to
forces acting on individual atoms rather than on the whole particle. This mechanism is manifested particularly clearly in the
case of inclusions in ferromagnetic hosts and it will be considered
in Sec. 7, where the motion of the pores in a ferromagnet is analyzed.

The special case of external forces acting on an inclusion are
the forces resulting from the presence of an electric current or a
flux of phonons (in a temperature gradient). These forces are due
to the transfer of momentum from electrons or phonons which are
scattered by the boundary of an inclusion. For example, an elec-

tric current gives rise to a force whose order of magnitude is

$$\vec{F} \sim \frac{\vec{j}}{e}\, p_0 R^2 \sim enl_e R^2 \vec{E}, \tag{6.15}$$

where p_0 is the momentum of an electron on the Fermi surface.
However, if we calculate the velocity of an inclusion by means of
the formulas given in the present section, we find that this velocity
is $R\sigma_e/\omega \sim R/a$ times lower than the velocity resulting from the
influence of the forces acting on individual atoms (see Sec. 3).
Therefore, the effects associated with the forces described by Eq.
(6.15) or with the analogous forces in a temperature gradient can
be ignored if the treatment is restricted to the macroscopic approximation ($R \gg a$).

In the absence of external forces, an inclusion can also migrate under the influence of fluctuations which can give rise to the
Brownian motion of inclusions. The mean-square value $\overline{x^2}$ of the
displacement of an inclusion along some direction x in a time t
is then given by the well-known Brownian-motion formula

$$\overline{x^2} = 2D_i t. \tag{6.16}$$

The value of $\overline{x^2}$ depends on the Brownian diffusion coefficient D_i .

The coefficient D_B can be found from the Einstein relationship

$$D_i = kTu, \tag{6.17}$$

if we determine the mobility u by means of the expressions for the
velocity of inclusions migrating under the action of external forces
(the mobility is defined by $\vec{v} = u\vec{F}$).

In this case, the migration of inclusions associated with the
flux of vacancies in the interior of the host crystal is described
by the diffusion coefficient, which is deduced from Eqs. (6.7) and
(6.17):

$$D_i = \frac{3\Gamma_1'}{2\pi f}\, \frac{D\omega}{R^3}. \tag{6.18}$$

The diffusion coefficient of the inclusions is inversely proportional
to the cube of the radius and it is $\sim (R/a)^3$ times smaller than the
diffusion coefficient of atoms.

It follows from Eqs. (6.8) and (6.17) that, in the case of migration associated with surface diffusion fluxes, we have

$$D_i = \frac{\Gamma}{2\pi f_S} \frac{D_S a \omega}{R^4} , \qquad (6.19)$$

i.e., D_i is inversely proportional to R^4 and it is $\sim (R/a)^4$ times smaller than the surface diffusion coefficient D_S.

Finally, it follows from Eqs. (6.12) and (6.17) that, in the case of migration associated with diffusion fluxes across an inclusion, the Brownian diffusion coefficient obtained subject to the condition (2.31) is

$$D_i = \frac{3}{4\pi f_A} \left(1 - \frac{f_A \omega_A}{\omega'} + \frac{\Gamma f_A \omega}{3\omega'} \right) \frac{D_A c_A^0 \omega}{R^3} . \qquad (6.20)$$

In this case, D_i is inversely proportional to R^3 and is governed by the quantities c_A^0 and D_A. In particular, the expression (6.20) for a gas-filled micropore ($l_g \ll R$) becomes

$$D_i = \frac{\Gamma}{4\pi} \frac{\omega^2 P_A D_A}{kTR^3} , \qquad (6.21)$$

which is obtained by assuming that $\omega_A = \omega'$, $f_A = 1$, and $c_A^0 = P_A \omega'(kT)^{-1}$.

The formulas for D_i, analogous to Eqs. (6.19) and (6.21) but differing by numerical factors, have also been derived by a microscopic analysis of the fluctuation-induced Brownian motion [6, 56].

In the case of small-radius empty pores, when the condition (2.36) is satisfied, Eq. (5.32) yields (see also [57])

$$D_i \sim \frac{n_S \omega^2}{R^2} . \qquad (6.22)$$

Although the diffusion coefficients of inclusions are considerably smaller than those of atoms, the distances traveled by small-radius inclusions under the Brownian motion conditions may be considerable and can often give rise to significant effects (Sec. 10).

Brownian Motion of Gas-Filled

Cavities

The spontaneous random motion of gas-filled cavities has been observed directly and indirectly. This information has been ob-

tained in studies of the kinetics of the high-temperature evolution
of gases from metal foils into which a gas, insoluble in the material of the foil, has been forced under pressure.

Direct observations of the Brownian motion of microscopic
gas-filled cavities are reported in [58]. These observations were
made on UO_2 plates irradiated with neutrons in an amount suffi-

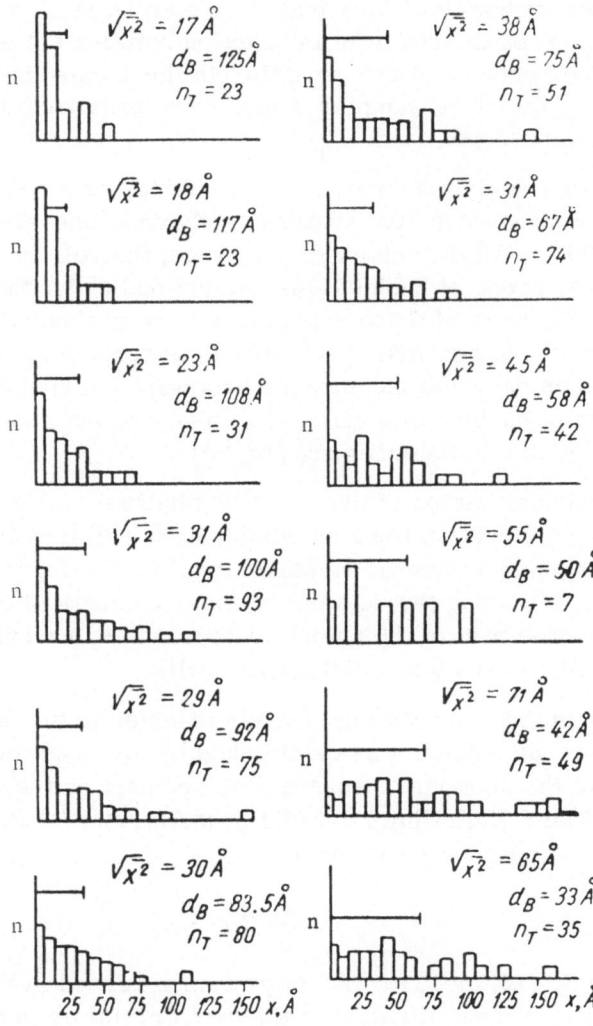

Fig. 27. Histograms describing the Brownian motion of gas-
filled bubbles in UO_2. T = 1500°C [58].

cient to produce fission at a rate of 10^9 cm^{-3} · sec^{-1}. The gaseous fission fragments, which were krypton and xenon atoms, coalesced during high-temperature annealing to form gas-filled bubbles. After the annealing at 1500°C, these bubbles acquired equilibrium properties in the sense that the regions surrounding them were free of stresses (the gas and Laplace pressures became equal).

The absence of stress fields around the bubbles was important because it showed clearly that the observed motion of the bubbles could not be the result of interaction between the stress fields and the free surfaces of the crystals. In these experiments, use was made of thin-film samples, which were examined under an electron microscope.

The relative displacements of the bubbles as a result of their Brownian motion were studied after isothermal annealing for 30 min at 1500°C. After each annealing stage, the relative position of a selected group of bubbles was determined under the microscope. The results of these experiments are presented in Fig. 27 in the form of histograms. Each of the histograms gives the size of the bubbles on which the observations were carried out $(d_B = R_0)$, the number of bubbles of a given size (n_T), and the root-mean-square value of the displacement $[(x^2)^{1/2}]$.

The random motion of the gas-filled bubbles could be due to the drag by dislocation lines and dislocation walls, which could move during high-temperature annealing. The histograms of Fig. 27 were plotted using the data on the displacements of the bubbles in a defect-free part of the sample, which was sufficiently far from the single dislocations and dislocation walls.

The principal mechanism of mass transfer in the Brownian migration of the bubbles was established by investigating the dependence of the root-mean-square displacement on the linear size of the cavities. Following Eqs. (6.16), (6.18), and (6.22), the dependence was written in the form

$$\log \sqrt{\overline{x^2}} = \text{const} + m \log (1/R), \qquad (6.23)$$

where m = 2, 3/2, or 1, for the migration governed by the surface diffusion, the volume diffusion in the host crystal or in the inclusion, or evaporation from the surfaces of "empty" bubbles, respectively. The dependences of $\log (\overline{x^2})^{1/2}$ on $\log (1/R)$, plotted in Fig.

28, were based on the results obtained at three different temperatures. It is evident from Fig. 28 that the displacements of the bubbles of radii exceeding 37 Å was governed by the volume diffusion
in the host crystal (the slope of the lines was 1.5). Some reduction in the slope, which was observed in the range R < 37 Å, was
probably due to some experimental error or a secondary effect.

The predominant role of the volume diffusion in the Brownian
migration of the gas-filled bubbles with R > 37 Å in UO_2 was supported also by the value of the activation energy, which was 546 ±
105 kJ/mole (130 ± 25 kcal/mole). The motion of these very small
bubbles was expected to be governed by surface diffusion. However, it was found that the surface diffusion was quite weak, probably due to the presence of surface contamination, which prevented
free mass transfer along the outer faces of the plates. This hap-
pened, in particular, due to the deceleration of the motion of steps.

The Brownian migration of helium-filled bubbles of 10^{-6}-10^{-7}
cm size was observed in experiments on gold and copper foils
[177]. These experiments showed that the velocity of such bubbles
was governed by the kinetics of the transport of mass along their

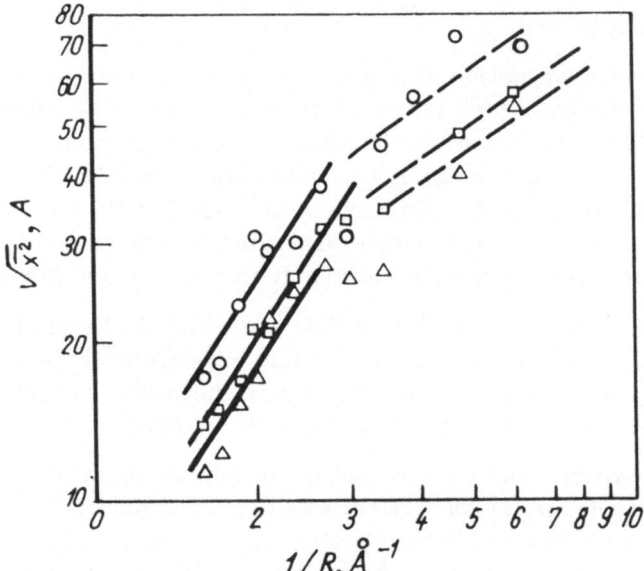

Fig. 28. Dependences of the rms Brownian displacements of
gas-filled bubbles on the reciprocal of their radii [58].

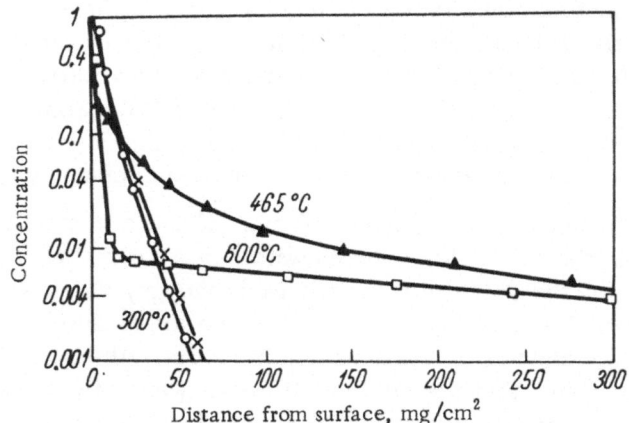

Fig. 29. Distribution of xenon with depth in an aluminum plate after its annealing at various temperatures. The xenon was introduced by bombarding the surface of the plate with 40 keV Xe ions. The duration of the annealing was t = 20 min. The unlabeled curve refers to an unannealed sample.

surfaces. However, such transport was limited not by the mobility of atoms but by the kinetics of nucleation of steps acting as the sources of atoms on that part of the surface which was at the front of the moving bubble.

The Brownian motion of gas-filled bubbles affects also the distribution of the gas which is not dissolved in a given substance and is distributed nonuniformly. The distribution of krypton in platinum and of xenon in aluminum was investigated in [57, 59]. The gases were introduced by bombarding the metals with the corresponding ions. High-temperature annealing produced gas-filled bubbles which were located mainly near the bombarded surface.

A study of the distribution of the gases with depth established that high-temperature annealing resulted not only in the expulsion of the gases but also in a redistribution in the metal samples as a result of the Brownian migration of gas-filled bubbles.

In the experiments carried out on aluminum containing xenon-filled bubbles (Fig. 29), this effect was observed particularly clearly because a surface oxide film prevented the escape of the gas from the sample and made the migration into the interior of the sample the only possible mechanism by which the inhomogeneity of the distribution of the gas could be reduced.

A study of the changes in the distribution of krypton in a surface layer in platinum [59] established that isothermal annealing
of prescribed duration altered the distribution of the gas because
of the Brownian migration of the krypton-filled bubbles. The migration of these bubbles was governed by the surface-self-diffu-
sion mechanism. The xenon-filled bubbles in aluminum, whose
surface was covered by an oxide film, could migrate into the interior of the sample by volume self-diffusion in the metal.

In these experiments, a redistribution of the gas could occur
solely as a result of the migration of gas-filled bubbles as a whole.
The normal diffusion of gas in the lattice was not significant because krypton is practically insoluble in platinum and the same is
true of xenon in aluminum.

The Brownian motion of gas-filled bubbles could exert a considerable influence on the high-temperature stages of the expulsion
of a gas from a solid sample. A gas-filled bubble located in the
direct vicinity of a free surface could break open and the gas contained therein might escape. The Brownian mechanism of the
evolution of inert gases was observed also in the experiments on
platinum.

7. DISTORTION OF A MOVING PORE

We have considered only the ideal cases of the migration of
inclusions whose shape is exactly spherical and whose surface and
bulk properties are perfectly isotropic. We have assumed that the
properties of the host crystal are also perfectly isotropic and that
the field of the external force responsible for the migration is
homogeneous. In all these ideal cases and in all the types of diffusion considered so far, the number of atoms leaving any given
element of the inclusion—host interface is proportional to the co-
sine of the angle θ formed by the direction of the external force
and the vector normal to a given region of the interface. The volume swept out by this part of the spherical interface is also proportional to $\cos \theta$, as shown in Fig. 2. Moreover, the normal component of the velocity is proportional to $\cos \theta$. Therefore, in this
ideal case, all the parts of the interface move at the same velocity
from the moment of application of the external forces and the inclusion migrates as one unit, retaining its spherical shape.

In real cases, the situation is more complex because the external forces are not necessarily homogeneous and the equilibrium and kinetic properties of the inclusion—host interface (and, in the case of noncubic crystals, the bulk properties of the inclusion and host crystal) are always at least slightly anisotropic. Even in the absence of external forces, the shape of an inclusion is usually aspherical. In view of these circumstances, the number of atoms carried away by diffusion fluxes from a given part of the inclusion—host interface is no longer proportional to cos θ and the normal component of the velocity of this region is also not proportional to cos θ. Therefore, during the initial period, different parts of the surface of an inclusion move at different velocities and the shape of the inclusion changes.

Such distortion of an initially spherical inclusion (or further changes in the shape of an aspherical inclusion) should give rise to additional diffusion fluxes which are not related to the external forces but to the curvature of the inclusion—host interface. This curvature changes the chemical potentials of atoms and vacancies. The additional diffusion fluxes are also due to internal stresses which appear near a distorted inclusion. The distortion of the inclusion surface and the appearance of internal stresses are observed until the additional diffusion fluxes, acting in combination with the fluxes generated by the external forces, equalize the velocities of all parts of the inclusion surface. The next stage is a steady-state motion during which an inclusion moves as one unit. The steady-state shape of a distorted moving inclusion and the steady-state internal stress fields can be deduced from the equality of the velocities of all parts of the inclusion surface. In the case of gas-filled pores, such translational motion of a pore as a whole is superimposed on the normal motion of the parts of its surface which alter the shape of the pore.

We shall consider changes in the shape of a gas- or liquid-filled pore during its motion. We shall assume that the velocity of the pore is low so that the linear theory may be applied and the distortion of the pore shape is not very great [63, 64, 14, 65]. The effects predicted by the nonlinear theory are observed more easily in solid solutions and will be discussed in Sec. 14.

We shall start by considering the distortion of an initially spherical pore with isotropic surface properties, subjected to inhomogeneous external forces. We shall then discuss the distortion associated with the anisotropy of the surface properties.

Diffusion Fluxes Associated with the Distortion of the Surface of a Pore

The distortion of the surface of a pore can be described by a curvature K defined by

$$K = \frac{1}{2}\left(\frac{1}{R_1} + \frac{1}{R_2}\right),\tag{7.1}$$

where R_1 and R_2 are the principal radii of curvature.

Nonzero curvature gives rise to two effects which affect the diffusion fluxes in the bulk of the host crystal and on the surface of the pore.

One of these effects is the distortion-induced Laplace pressure P_L' which is due to the surface tension γ:

$$P_L' = -2\gamma K.\tag{7.2}$$

This pressure is additional to the pressure P_0 of the gas (or liquid) in the pore and it alters the boundary conditions for the stress tensor of the host crystal surrounding the pore. The pressure on the surface of the pore, corresponding to the stresses in the host, should be balanced by the sum of the gas and Laplace pressures. The combined pressure is directed along the normal to the surface $\vec{n}(\vec{r}_S)$ and, in the general case of an aspherical pore, the boundary conditions on the surface are of the form

$$\sigma_{ij}n_j = -\left(P_0 - 2\gamma K\right)n_i \quad (\vec{r} = \vec{r}_S).\tag{7.3}$$

If the curvature varies along the pore surface, the inhomogeneous distribution of the surface forces associated with the Laplace pressure gives rise to an inhomogeneous stress field. These inhomogeneous stresses should produce diffusion fluxes in the interior and on the surface of the host crystal, and the fluxes should be of the same type as those observed in an inhomogeneous field of external stresses (Sec. 5).

The second effect is the change in the chemical potential μ_S of the surface atoms as a result of the distortion of the pore surface. This change in the chemical potential gives rise to additional surface diffusion fluxes. In order to determine the dependence of μ_S on the curvature of the pore surface, it is necessary to supplement the Gibbs free energy Φ with the term

$$\Phi_S = \int \gamma dS, \tag{7.4}$$

which is associated with the surface energy. When an atom reaches a given point on the surface of the pore, the change in the volume of the crystal $\delta\Omega$ is equal to the atomic volume ω and the change in the surface area of the pore δS follows from the definition of the curvature

$$2K = \frac{\delta S}{\delta \Omega} \tag{7.5}$$

and is $\delta S = -2K\omega$. Therefore, when the number of atoms in a crystal N changes by unity, the term (7.4) in the expression for Φ changes by $-2K\gamma\omega$ and the expression for $\mu_S = \delta\Phi/\delta N$ acquires an additional term $-2K\gamma\omega$ (see also [66, 67]).

The chemical potential of the surface atoms on the distorted surface of a pore is, therefore, given by[†]

$$\mu_S = \mu_{0S} - \frac{1}{3}\sigma_{ii}\omega - 2\gamma\omega K, \tag{7.6}$$

where μ_{0S} is the chemical potential of the atoms on a plane surface in the absence of distortions. The last term in Eq. (7.6) gives rise, in accordance with Eq. (5.11), to surface diffusion fluxes

$$\vec{I}_S = \frac{2D_S a\gamma}{f_S kT} \nabla_S K, \tag{7.7}$$

which are associated with the inhomogeneous distortion of the pore surface.

[†] Strictly speaking, an allowance should be made for the fact that the introduction of a new atom alters also the elastic energy of a crystal around a pore (outside this atom). However, this elastic energy is much smaller than the surface energy [the ratio of these energies is of the order of $(P_0 - 2\gamma K)/G$ or $\sim a/R$ for an empty pore] and, therefore, it can be ignored. For the sake of simplicity, we shall also ignore the dependence of the surface energy on the surface area.

We must bear in mind that, in accordance with Eq. (7.3), a change in the curvature of any part of the pore surface gives rise to stresses throughout the whole region surrounding the pore and, in particular, to stresses in all parts of the pore surface. Therefore, the effect of such a distortion does not remain local. This nonlocal aspect is allowed for by the term $-(1/3)\sigma_{ii}\omega$ in Eq. (7.6), which — in accordance with the boundary conditions of Eq. (7.3) — depends on the function $K(\vec{r}_S)$ that is specified not only at the point in question but throughout the pore surface. Therefore, the flux given by Eq. (5.13) makes a contribution to the surface diffusion fluxes resulting from the distortion of the pore surface.

In general, when a metal is subjected to an inhomogeneous electric field and a temperature gradient, it follows from Eqs. (2.1), (3.3), and (5.6) that the diffusion fluxes in the interior of a cubic crystal are given by the expression

$$\vec{I} = N_0 D \left(\frac{ez}{fkT} \dot{E} - \frac{\alpha}{T} \nabla T + \frac{1}{fkT} \nabla \delta \mu_v^\sigma \right), \qquad (7.8)$$

where $\delta \mu_v^\sigma$ is the change in the chemical potential μ_v which is associated solely with the stresses and not with the temperature. The fluxes on the pore surface, computed on the assumption that the properties of the surface are isotropic, are equal to the sum of the contributions given by (2.2), (3.6), (5.13), and (7.7):

$$\vec{I}_S = N_0 D_S a \left(\frac{ez_S}{f_S kT} \dot{E}_S - \frac{\alpha_S}{T} \nabla_S T + \frac{\omega}{3f_S kT} \nabla_S \sigma_{ii} + \frac{2\omega\gamma}{f_S kT} \nabla_S K \right). \qquad (7.9)$$

It is important to note that even in the absence of external stresses the formulas given above include the terms which are proportional to $\nabla \sigma_{ii}$ and are due to internal stresses resulting from the inhomogeneous Laplace pressure.

Distortion of a Pore in an Inhomogeneous Field of Internal Forces

If we substitute the expressions (7.8) and (7.9) for the volume and surface diffusion fluxes into Eqs. (1.1) and (1.9) and if we assume that, at the temperatures under consideration, the evaporation from the surface of a pore can be ignored $(\vec{l}\,'=0)$., we find that

$$\vec{n}\vec{v}\,'\left(\vec{r}_S\right) = D\left[\frac{ez}{fkT}\,\vec{n}\vec{E}\left(\vec{r}_S\right) - \frac{\alpha}{T}\,\vec{n}\nabla T\left(\vec{r}_S\right) + \frac{1}{fkT}\,\vec{n}\nabla\delta\mu_v^\sigma\left(\vec{r}_S\right)\right] -$$

$$- D_S a\left[\frac{ez_S}{f_S kT}\,\Delta_S\varphi_S + \frac{\alpha_S}{T}\,\Delta_S T_S - \frac{\omega}{3f_S kT}\,\Delta_S\sigma_{ii}\left(\vec{r}_S\right) - \frac{2\omega\gamma}{f_S kT}\,\Delta_S K\right]. \qquad (7.10)$$

Equation (7.10) enables us to determine the velocity of the translational motion of a pore $\vec{v}\,'$, as well as its distortion during steady-state motion, including the change in the volume of the pore. Strictly speaking, the surface Laplace operators in the above formula should be calculated on the surface of a distorted pore (this surface is generally quite complex); $\vec{n} = \vec{n}\left(\vec{r}_S\right)$ are the unit vectors normal to the pore surface.

However, in the linear theory, the distortion of a pore is proportional to the external forces. The distortion and velocity of a pore, as well as the forces acting on it, can be regarded as small quantities. Therefore, if the pore is spherical in the absence of external forces, we can ignore terms of higher orders of smallness in Eq. (7.10) and assume that the vector $\vec{n}$ is directed, as on the surface of a sphere, along the radius $(\vec{n} = \vec{r}_S/r_S)$, so that the Laplace operators can be calculated for a spherical surface. Similarly, when the boundary conditions for the stresses are given by Eq. (7.3), we may assume that the vector $\vec{n}$ is directed along the pore radius so that these conditions become

$$\sigma_{rr} = -\left(P_0 - 2\gamma K\right), \quad \sigma_{r\theta} = 0, \quad \sigma_{r\varphi} = 0 \quad (r = R). \qquad (7.11)$$

If the vectors $\vec{E}_\infty, \nabla T_\infty$, and $\nabla\sigma_{ii\infty}$ are parallel, we find that Eq. (7.10) is modified so that the values of $\vec{v}\,'(\vec{r}_S)\vec{n}$, φ_S, T_S, K, $\vec{n}(\vec{E}(\vec{r}_S) - \vec{E}_\infty)$, $\vec{n}(\nabla T(\vec{r}_S) - \nabla T_\infty)$, and $\vec{n}(\nabla\sigma_{ii} - \nabla\sigma_{ii\infty})$ on the pore surface depend only on the angle θ, measured relative to the direction of the external forces. It is convenient to expand all these quantities as series of the Legendre polynomials $P_n(\cos\theta)$. For example, an expansion of this type applied to the curvature of the pore surface gives

$$K = K\left(\theta\right) = \sum_{n=0}^{\infty} K_n P_n\left(\cos\theta\right). \qquad (7.12)$$

It is important to note that the coefficient K_1 of the first harmonic in the expansion of the curvature vanishes identically (as indicated by simple geometrical considerations):†

$$K_1 = 0. \tag{7.13}$$

The velocity $\vec{v}'(\vec{r_S})$ can be represented as the sum of a constant term $\vec{v}'$, representing the translational motion of a pore as a whole and the term $\dot{R}\vec{n}$, which is parallel to the vector $\vec{n}$ and describe a change (increase or decrease) in the radius of the pore. Therefore, the expansion of $\vec{n}\vec{v}'(\vec{r_S})$ in terms of spherical harmonics contains only the zeroth and first terms:

$$\vec{n}\vec{v}'\,(\vec{r}_S) = \dot{R}P_0(\cos\theta) + (\vec{e}\vec{v}')\,P_1(\cos\theta) \tag{7.14}$$

($P_0 = 1$; $P_1 = \cos\theta$; $\vec{v}'$ is parallel or antiparallel to the direction of the external forces $\vec{e}$).

If we expand the various quantities in Eq. (7.10) in accordance with Eq. (7.12) and if we assume that on the surface of a sphere of radius R we have

$$\Delta_S P_n(\cos\theta) = -\frac{n(n+1)}{R^2}P_n(\cos\theta), \tag{7.15}$$

we can separate Eq. (7.10) into equations corresponding to different harmonics. The equation corresponding to the zeroth har-

---

†This follows, for example, from the equation encountered in differential geometry

$$\delta K = -\frac{\delta r}{R^2} - \frac{1}{2}\Delta_S\,\delta r\,, \tag{7.12a}$$

which relates — in the first approximation — the changes in the spherical coordinate of the pore δr to the changes in the curvature δK. It follows from the above equation and from Eq. (2.19) that for any change δr of the type $\delta r = \text{const} \cdot \cos\theta$ (which simply represents a displacement of the center of gravity of the pore), the change in the curvature $K_1\cos\theta$ vanishes. It follows from Eq. (7.12a) that the coefficients δr_n and K_n in the expansions of the radius and the curvature in terms of $P_n(\cos\theta)$ are related by the expression

$$\delta r_n = \frac{2}{n^2 + n - 2}K_n R^2.$$

monic gives the rate of change of the pore radius

$$\dot{R} = D\left[\frac{ez}{fkT}\left(\vec{n}\vec{E}\left(\vec{r}_S\right)\right)_0 - \frac{\alpha}{T}\left(\vec{n}\nabla T\left(\vec{r}_S\right)\right)_0 + \frac{1}{fkT}\left(\vec{n}\nabla\delta\mu_v^\sigma\left(\vec{r}_S\right)\right)_0\right], \quad (7.16)$$

where the subscripts to the parentheses indicate the number of the coefficient in the expansion of the relevant quantity in terms of the Legendre polynomials.

The equation corresponding to the first-harmonic form of Eq. (7.10) determines the velocity of the translational motion of a pore in the system of coordinates linked to the lattice of the host crystal. When Eqs. (7.13) and (7.14) are used, this equation becomes

$$\vec{e}\vec{v}' = D\left[\frac{ez}{fkT}\left(\vec{n}\vec{E}\left(\vec{r}_S\right)\right)_1 - \frac{\alpha}{T}\left(\vec{n}\nabla T\left(\vec{r}_S\right)\right)_1 + \frac{1}{fkT}\left(\vec{n}\nabla\delta\mu_v^\sigma\left(\vec{r}_S\right)\right)_1\right] +$$

$$+ \frac{2D_S a}{R^2}\left[\frac{ez_S}{f_S kT}\varphi_{S1} + \frac{a_S}{T}T_{S1} - \frac{\omega}{3f_S kT}\sigma_{Si i1}\left(\vec{r}_S\right)\right]. \quad (7.17)$$

This expression for $\vec{v}'$ is in agreement with the expressions obtained in Secs. 2, 3, and 5 for the velocities of a pore moving under the action of various homogeneous fields of external forces.

The equations corresponding to the higher harmonics of Eq. (7.10) yield the coefficients K_n in Eq. (7.12) and they can be written in the form

$$D\left[\frac{ez}{fkT}\left(\vec{n}\vec{E}\left(\vec{r}_S\right)\right)_n - \frac{\alpha}{T}\left(\vec{n}\nabla T\left(\vec{r}_S\right)\right)_n + \right.$$

$$\left. + \frac{1}{fkT}\left(\vec{n}\nabla\delta\mu_v^\sigma\left(\vec{r}_S\right)\right)_n\right] + \frac{n(n+1)D_S a}{R^2}\left[\frac{ez_S}{f_S kT}\varphi_{Sn} + \right.$$

$$\left. + \frac{a_S}{T}T_{Sn} - \frac{\omega}{3f_S kT}\sigma_{iin}\left(\vec{r}_S\right) - \frac{2\omega\gamma}{f_S kT}K_n\right] = 0 \quad (7.18)$$

$$(n = 2, 3, ...).$$

If the above equation is used to find the coefficients K_n, it must be remembered that a dependence on K_n is exhibited not only only by the last term in the equation but also by the quantities $\sigma_{iin}(\vec{r}_S)$ and $\nabla\mu_v^\sigma(\vec{r}_S)$, which are implicit functions of K_n because of the boundary conditions given by Eq. (7.11).

If there are no external stresses, the internal stresses resulting from the Laplace pressure should decrease away from the pore. If we expand the solutions of the elasticity-theory equations in terms of P_n (cos θ), we can show [65] that the coefficients in the expansion of the diagonal part of the stress tensor, satisfying the boundary conditions (7.11), are related to the coefficients K_n in the expansion of the curvature:

$$\left.\begin{array}{l} \sigma_{iin}(\vec{r}) = 6\Gamma_n \gamma K_n \left(\dfrac{R}{r}\right)^{n+1}, \\[3mm] \Gamma_n = \dfrac{1}{3}\,\dfrac{n(2n-1)(1+v)}{n^2+n+1-2nv-v}. \end{array}\right\} \tag{7.19}$$

Since the change in the chemical potential $\delta\mu_v^\sigma$ obeys the Laplace equation and, therefore, can be represented in the form

$$\delta\mu_v^\sigma = \sum_{n=0}^{\infty} \mu_{vn}^\sigma\, \frac{R^{n+1}}{r^{n+1}}\, P_n\,(\cos\theta), \tag{7.20}$$

the coefficients of $(\vec{n}\nabla\delta\mu_v^\sigma\,(\vec{r_s}))_n$ can be related to the coefficients of μ_{vn}^σ in the expansion of $\delta\mu_v^\sigma$ on the surface of a sphere. The latter coefficients can be expressed in terms of K_n by the application of Eqs. (5.8) and (7.11) or Eqs. (5.9) and (7.11). Therefore,

$$\left(\vec{n}\nabla\mu_v^\sigma\,(\vec{r_s})\right)_n = -2\Gamma_n'\,(n+1)\,\frac{\omega\gamma}{R}\,K_n, \tag{7.21}$$

where $\Gamma_n' = 1$ in the case of perfect crystals obeying $l \gg R$ [μ_v is given by Eq. (5.8)] and $\Gamma_n' = \Gamma_n$ in the case of crystals with a high density of vacancy sources and sinks, in which $l \ll R$ [μ_v is given by Eq. (5.9)].

Substituting Eqs. (7.19) and (7.21) into Eq. (7.18), we find that the distortion in an inhomogeneous electric field and under an inhomogeneous temperature gradient is

$$K_n = \frac{R}{2\omega\gamma}\, \frac{\dfrac{ezD}{f}\left(\vec{n}\vec{E}\,(\vec{r_s})\right)_n + \dfrac{n(n+1)\,eD_S\,az_S}{f_S R^2}\,\varphi_{Sn}}{\dfrac{(n+1)D}{f}\,\Gamma_n' + n(n+1)(1+\Gamma_n)\,\dfrac{D_S a}{f_S R}} -$$

$$- \frac{\alpha Dk\left(\vec{n}\nabla T\,(\vec{r_s})\right)_n + \dfrac{n(n+1)\,\alpha_S D_S\,ak}{R^2}\,T_{Sn}}{\dfrac{(n+1)D}{f}\,\Gamma_n' + n(n+1)(1+\Gamma_n)\,\dfrac{D_S a}{f_S R}}. \tag{7.22}$$

Let us assume that, far from a pore, the electric field varies with distance and corresponds to the potential

$$\varphi_\infty = -\vec{E}_0\,\vec{r} - \frac{1}{2}\,Ar^2 P_2\,(\cos\theta). \tag{7.23}$$

Near the pore the solution of the Laplace equation, which satisfies the boundary conditions (3.8) on the pore surface and which is identical with Eq. (7.23) far from the surface, is

$$\left.\begin{array}{l} \varphi\,(\vec{r}) = -\vec{E}_0\vec{r}\left(1 + \varkappa_e\,\dfrac{R^3}{r^3}\right) - \dfrac{1}{2}\,Ar^2\left(1 + \varkappa_e'\,\dfrac{R^5}{r^5}\right)P_2\,(\cos\theta) \quad (r > R), \\[3mm] \varphi\,(\vec{r}) = -\left(1 + \varkappa_e\right)\vec{E}_0\,\vec{r} - \dfrac{1}{2}\,(1 + \varkappa_e')\,Ar^2 P_2\,(\cos\theta) \quad (r < R); \\[3mm] \varkappa_e' = \dfrac{2\,(\lambda_e - \lambda_{0e})}{3\lambda_e + 2\lambda_{0e}}. \end{array}\right\} \tag{7.24}$$

Hence, it follows that

$$\left.\begin{array}{l} \varphi_{S2} = -\dfrac{1}{2}\,(1 + \varkappa_e')\,AR^2, \\[3mm] (\vec{n}\vec{E}\,(\vec{r}_S))_2 = \dfrac{\lambda_{0e}}{\lambda_e}\,(1 + \varkappa_e')\,AR. \end{array}\right\} \tag{7.25}$$

The coefficients of higher harmonics $(n > 2)$ of the potential of the normal component of the field all vanish.

If we substitute Eq. (7.25) into Eq. (7.22), we find that the curvature of a pore subjected to an inhomogeneous electric field, whose potential is given by Eq. (7.24), is

$$\left.\begin{array}{l} K = K_0 + \delta K\,(\theta), \quad \delta K\,(\theta) = K_2 P_2\,(\cos\theta), \\[3mm] K_2 = \dfrac{(1 + \varkappa_e')\,e}{6\omega\gamma\lambda_e}\,\dfrac{\lambda_{0e}\,z f_S\,DR - 3\lambda_e\,z_S\,f D_S\,a}{f_S\,\Gamma_2'\,DR + 2\,(1 + \Gamma_2)\,f D_S\,a}\,AR^2. \end{array}\right\} \tag{7.26}$$

In particular, if the pore has a small radius so that the surface diffusion fluxes play the dominant role, we find that

$$K_2 = -\frac{(1 + \varkappa_e')\,e z_S}{4\,(1 + \Gamma_2)\,\omega\gamma}\,AR^2 \quad (D_S\,a \div DR). \tag{7.27}$$

In this case, the order of magnitude of the relative distortion of the pore is

$$\frac{|\delta r_S|}{R} \sim \frac{|\delta K|}{K_0} \sim \frac{e z_S}{\omega\gamma}\,AR^3 \tag{7.28}$$

and is proportional to the cube of the pore radius (if A is independent of R).

It follows from Eq. (7.26) that, in the case of large-radius conducting pores (for example, filled with a liquid metal), the coefficient K_2 in Eq. (7.26) is

$$K_2 = \frac{1 + \varkappa_e'}{6\Gamma_2'} \cdot \frac{\lambda_{0e}}{\lambda} \frac{ez}{\omega\gamma} AR^2 \qquad \left(\frac{\lambda_{0e}}{\lambda_e} DR \gg D_S a \right). \qquad (7.29)$$

In this case, the relative distortion is again proportional to R^3 and is of the same order of magnitude as the distortion given by Eq. (7.28).

However, in the case of large-radius nonconducting pores and if $\lambda_{0e} = 0$ (gas-filled pores), we find that

$$K_2 = -\frac{1 + \varkappa_e'}{2\Gamma_2'} \frac{f}{f_S} \frac{ez_S}{\omega\gamma} \frac{D_S a}{D} AR \qquad \left(DR \gg D_S a \gg \frac{\lambda_{0e}}{\lambda_e} DR \right). \qquad (7.30)$$

In this case, the relative distortion of the pore is proportional to R^2 and is $DR/D_S a$ times smaller than that given by Eq. (7.28).

Similarly, an inhomogeneous temperature gradient, which varies linearly with the distance far from the pore and which corresponds to the temperature distribution

$$T_\infty(\vec{r}) = T_0 + \vec{r}_\nabla T_0 - \frac{1}{2} Ar^2 P_2 (\cos\theta), \qquad (7.31)$$

produces a distortion $\delta K(\theta) = K_2 P_2(\cos\theta)$, where K_2 is given by an expression of the same type as Eq. (7.16):

$$\left. \begin{aligned} K_2 &= \frac{1 + \varkappa'}{6} \frac{k}{\omega\gamma\lambda} \frac{\lambda_0 \alpha DR - 3\lambda\alpha_S D_S a}{\Gamma_2' DR f^{-1} + 2\left(1 + \Gamma_2\right) D_S a f_S^{-1}} AR^2, \\ \varkappa' &= \frac{2(\lambda - \lambda_0)}{3\lambda + 2\lambda_0}. \end{aligned} \right\} \qquad (7.32)$$

The distortion of pores in inhomogeneous fields may be quite considerable. For example, if we use Eq. (7.28) and substitute $A \sim 10^2$ V/cm^2, $R \sim 10^{-4}$ cm, $z_S \sim 10$, $\omega \sim 10^{-23}$ cm^3, $\gamma \sim 1$ J/m^2 (10^3 ergs/cm^2), the relative distortion in an electric field is

$|\delta K|/K_0 \sim 10^{-1}$. The same order of magnitude of $|\delta K|/K_0$ is obtained in an inhomogeneous temperature gradient if $\alpha \sim 1$, $\omega \sim 10^{-23}$ cm^3, $\gamma \sim 1$ J/m^2 (10^3 ergs/cm^2), A $\sim 10^5$ deg/cm^2, R $\sim 5 \times 10^{-4}$ cm, $D_S a \gg$ DR. In this case, the stresses on the surface of the pore, calculated using Eq. (7.19) [$\sigma_{ii2}(R) \sim \gamma|\delta K| \sim 10^5$ J/m^3 (10^6 ergs/cm^3)], are quite considerable.

Strongly inhomogeneous fields are most easily established not by external sources but by neighboring inclusions or other internal inhomogeneities located near a pore which is subjected to a homogeneous external field. It follows from Eq. (2.9) or Eq. (3.9) that an inclusion subjects a neighboring pore to an electric field gradient

$$A \sim |\nabla E| \sim (R'/L)^3 |\dot{E}_\infty|/L,$$

where R' is the radius of the neighboring inclusion; L is the distance from the pore in question. Alternatively, such an inclusion gives rise to the following temperature gradient:

$$A \sim (R'/L)^3 |\nabla T_\infty|/L.$$

If L is only several times larger than R', the inhomogeneities of the electric field or the temperature gradient can be quite considerable.

Thus, closely spaced pores subjected to homogeneous external fields, which give rise to diffusional migration, should be distorted and the degree of distortion should increase with decreasing the distance between such pores. The distortion is proportional to the external fields and is superimposed on the distortion resulting from the diffusion−elastic interaction (Sec. 9), which acts even in the absence of external forces.

A pore becomes distorted when, as a result of inhomogeneous fluxes, parts of its surface shift relative to the center of gravity by a distance $|\delta r_S| \sim R|\delta K|/K_0$. The ratio of the rate of this displacement to the velocity of the center of gravity of a pore should be of the order of $\delta v'/v' \sim AR|\vec{E}_\infty|^{-1}$ or $\delta v'/v' \sim AR|\nabla T_\infty|^{-1}$ for migration in an electric field or in a temperature gradient, respectively. Therefore, the relaxation time of the distortion process is of the order of

$$\tau \sim \frac{|\delta r_S|}{|\delta v'|} \sim \frac{R}{v'} \frac{|\delta K|}{K_0} \frac{|\vec{E}_\infty|}{AR}$$

or

$$\tau \sim \frac{R}{v'} \frac{|\delta K|}{K_0} \frac{|\nabla T_\infty|}{AR}. \tag{7.33}$$

If $|\delta K|/K_0 \sim 1$, the relaxation time is usually longer than the characteristic time $\sim R/v'$ during which the pore travels a distance equal to its radius.

A pore moving in an inhomogeneous field of elastic stresses should also be distorted. Even if axisymmetric stresses in a perfect crystal vary linearly with distance (far from the pore), the angular distribution of these stresses is governed not only by the factor $P_1(\cos \theta)$ (as in the case of a temperature gradient or an electric field) but also by the term proportional to $P_3(\cos \theta)$. When expressed in spherical coordinates, these stresses can be found from the theory of elasticity:

$$\left. \begin{array}{l} \sigma_{rr}^\infty = \sigma_1 r P_1 + 6\sigma_3 r P_3, \\[2mm] \sigma_{r\theta}^\infty = -\dfrac{1}{2} \sigma_1 r P_1' + 2\sigma_3 r P_3', \\[2mm] \sigma_{\theta\theta}^\infty = 2\sigma_1 r P_1 - 9\sigma_3 r P_3 - \sigma_3 r P_3' \cot \theta, \quad \sigma_{r\varphi}^\infty = 0, \\[2mm] \sigma_{\varphi\varphi}^\infty = 2\sigma_1 r P_1 + 3\sigma_3 r P_3 + \sigma_3 r P_3' \cot \theta, \quad \sigma_{\theta\varphi}^\infty = 0, \end{array} \right\} \tag{7.34}$$

$$P_n = P_n(\cos \theta), \quad P_n' = \frac{dP_n(\cos \theta)}{d\theta},$$

where σ_1 and σ_3 are constants.

Near a pore, the stress field of Eq. (7.34) and the fields proportional to higher powers of r are distorted and supplemented by the stresses σ', which decrease rapidly with distance away from the pore. At low temperatures when vacancies cannot diffuse, the values of σ' should be selected so that the following conditions are satisfied on the surface of a spherical pore:

$$\sigma_{rr} = -P_0, \quad \sigma_{r\theta} = 0, \quad \sigma_{r\varphi} = 0 \quad (r = R). \tag{7.34a}$$

At high temperatures, a pore is distorted by diffusion and the sum $\sigma' + \sigma^\infty$ should satisfy the boundary conditions of Eq. (7.11) which make allowance for the Laplace pressure (this modifies the stress field).

In this case, Eq. (7.18) for the curvature includes the stresses σ^∞ and σ' and the curvature occurs in the boundary conditions of Eq. (7.11). Therefore, the problems of the determination of the stresses and the curvature are interrelated and it is necessary to solve a self-consistent diffusion−elastic problem. The solution of this problem can be obtained by applying the procedure used in dealing with an inhomogeneous electric field or an inhomogeneous temperature gradient field.

In particular, if the stress fields at infinity are given by Eq. (7.34) and the predominant process is the diffusion at the surface of the pore and not in its interior, the velocity of the translational motion of a pore is given by Eq. (5.24). The distortion of the pore surface is proportional to the third spherical harmonic and is given by the formula [63]:

$$K = K_0 + \delta K\,(\theta), \quad \delta K\,(\theta) = K_3 P_3\,(\cos\theta);$$
$$K_3 = \frac{35}{2}\,\frac{1+\nu}{9-\nu}\,\frac{\sigma_3 R}{\gamma}\,. \tag{7.35}$$

In this case the relative distortion is of the order of

$$\frac{|\delta r_S|}{R} = \frac{1}{5}\,\frac{|K_3|}{K_0}\,|\,P_3\,(\cos\theta)\,| \sim \frac{1}{\gamma}\,\sigma_3 R^2 \tag{7.36}$$

and is proportional to R^2 (if σ_3 is independent of R). For example, if $\sigma_3 \sim \sigma_1/L_3 \sim \sigma_1/L_3^2 \sim 10^{17}\,\mathrm{J/m^5}\,(10^{14}\,\mathrm{ergs/cm^5})$, $\gamma \sim 1\,\mathrm{J/m^2}\,(10^3$ ergs $\cdot$ cm^{-2}), we find that $|\,\delta r\,|/R \sim 1$ for R $\sim 10^{-5}$ cm, i.e., for small-radius pores. As in the case of an electric field and a temperature gradient, a considerable distortion of the pores should result from the distortion of the field of external stresses due to the presence of a neighboring inclusion.

In the case of crystalline inclusions, in which no diffusion or plastic deformation takes place, the distortions of shape are unlikely to be large. This is because a displacement of the boundary of such an inclusion by an amount δr strains it by $\delta R/R$ and gives rise to stresses $\sigma_{rr} \sim G'\delta r/R$, where G' is Young's modulus of the inclusion. These stresses are comparable with the maximum Laplace pressure $\sim \gamma / R$ (for $|\delta K| \sim K_0$) when the displacement is $\delta r \sim a$ (because $\gamma \sim aG$) and, therefore, the distortion is negligible. On the other hand, the atoms which diffuse during the initial stage of relaxation to the surface of the inclusion and form an excess layer (or a deficiency layer) of variable thickness $\sim a$ produce an inhomogeneous stress field. The diffusion fluxes associated with these internal stresses are responsible (in combination with the fluxes generated by the external forces) for the equality of the velocities of all parts of the surface of the inclusion, which, therefore, moves as one unit. The harmonics of the stress field are still given by the expressions in Eq. (7.18), where we must substitute $K_n = 0$, and by the boundary conditions on the surface of the inclusion

$$\sigma_{rr}^+ = \sigma_{rr}^-, \quad \sigma_{r\theta}^+ = \sigma_{r\theta}^- = \sigma_{\theta\varphi}^+ = \sigma_{\theta\varphi}^- = 0 \quad (r = R) \qquad (7.37)$$

(these apply in the case of easy slip) and the equations of the theory of elasticity. Here, σ^+ and σ^- refer, respectively, to the host and the inclusion. In the case of crystalline inclusions, the stresses are of the same order of magnitude as those in the case of pores.

Migration and Distortion of a Pore in a Ferromagnet or a Pyroelectric Material

When an atom in a ferromagnet is located near the top of a potential barrier and is subjected to an inhomogeneous magnetic field $\vec{H}(\vec{r})$, it experiences a force

$$\vec{f}_m = \nabla(\vec{\mu}\,\vec{H}), \qquad (7.38)$$

where μ is of the order of magnitude of the average atomic magnetic moment. The force $\vec{f}_m$ gives rise to diffusion fluxes in the interior and on the surface of a pore so that the pore migrates in an inhomogeneous external field. The normal components of these fluxes are not proportional to $\cos \theta$ and, therefore, the pore should become distorted.

These effects are strongest in ferromagnets in which the magnetic moments $\vec{\mu}$ are oriented by the exchange forces along some axis z and the average magnetic moment μ is of the same order of magnitude as the atomic moment μ_0 (in a paramagnet $\mu \sim \frac{\mu_0 H}{kT} \mu_0 \ll \mu_0$). Even in the absence of an external field, inhomogeneous magnetic fields arise near defects (for example, pores) in a ferromagnet. Although such fields do not result in the translational motion of a pore, they distort it and the distortion can be quite considerable.

The distribution of a static magnetic field can be described by a magnetostatic potential $\varphi_m(\vec{r})$ $(\vec{H} = -\nabla \varphi_m)$. We shall assume that far from a spherical pore the magnetic field varies linearly with distance and corresponds to the potential

$$\varphi_m^\infty(\vec{r}) = -\vec{H}_0 \vec{r} - \frac{1}{2} A r^2 P_2(\cos \theta) \qquad (\vec{H}_0 \| z). \qquad (7.39)$$

On the surface of a pore, the potential $\varphi_m(\vec{r})$ is continuous and its normal component satisfies the condition

$$B_n^+ = B_n^- \quad \text{or} \quad \chi \frac{\partial \varphi^+}{\partial r} - 4\pi M \cos \theta = \frac{\partial \varphi^-}{\partial r}, \qquad (7.40)$$

where $\vec{B} = \chi \vec{H} + 4\pi \vec{M}$, $\vec{M}$ is the spontaneous magnetic moment per unit volume of the host crystal $(\vec{M} \| z)$; $\chi \approx 1$ is the magnetic susceptibility of the host crystal; φ^+ and φ^- are the potentials in the host and in the pore, respectively.

The potential $\varphi_m(\vec{r})$, which satisfies the boundary conditions of Eqs. (7.39) and (7.40), is of the form

$$\varphi_m(\vec{r}) = -\left(1 + \frac{\chi - 1}{2\chi + 1} \frac{R^3}{r^3}\right) \vec{H}_0 \vec{r} - \frac{4\pi}{2\chi + 1} \frac{R^3}{r^3} \vec{M} \vec{r} -$$
$$- \frac{1}{2} A r^2 \left(1 + 2 \frac{\chi - 1}{3\chi + 2} \frac{R^5}{r^5}\right) P_2(\cos \theta) \quad (r > R). \qquad (7.41)$$

The component of the magnetic field H_z on the surface of the pore which corresponds to this potential is

$$H_z(\vec{r}_s) = H_{0z} + ARP_1(\cos \theta) - \frac{2}{1 + 2\chi} [4\pi M + (\chi - 1) H_{0z}] P_2(\cos \theta) -$$
$$- 3 \frac{\chi - 1}{3\chi + 2} ARP_3(\cos \theta) \quad (r = R + 0). \qquad (7.42)$$

It is evident from Eq. (7.42) that the potential corresponding to the force of Eq.(7.38) includes the first, second, and third spherical harmonics ($\mu \| z$). In accordance with the results obtained earlier, the first harmonic determines the velocity of the translational motion of the pore $\vec{v}$. If the dominant mechanism is the surface diffusion, we find that

$$\vec{v} = -\frac{2D_S a\vec{\mu}}{Rf_S kT} A. \tag{7.43}$$

This velocity is proportional to A, which is the gradient of the external field.

The velocity given by Eq.(7.43) is of the same order of magnitude as the velocity given by Eq. (6.14) in the case of diffusion of ferromagnetic particles (this is true also in the case when the dominant mechanism is the volume diffusion). As in the cases considered earlier, the gradient A can be very large if an inclusion is located near a pore.

The second and third harmonics of the potential, corresponding to the force of Eq. (7.38), determine the distortion of the pore. If the dominant mechanism is the surface diffusion, the distortion is given by

$$K(\theta) = K_0 + \delta K(\theta); \quad \delta K(\theta) = \frac{4\pi M^2 + (\chi - 1)H_{02} M}{(1 + \Gamma_2)(1 + 2\chi)\gamma} P_2(\cos\theta) +$$
$$+ \frac{3(\chi - 1)}{2(1 + \Gamma_3)(3\chi + 2)\gamma} MARP_3(\cos\theta). \tag{7.44}$$

It is assumed here approximately that $\mu = \omega M$. Since $\chi \approx 1$, we can ignore the terms which depend on the external field and, therefore,

$$\delta K(\theta) = \frac{4\pi M^2}{3(1 + \Gamma_2)\gamma} P_2(\cos\theta). \tag{7.45}$$

Thus, the distortion of the pore is determined by the magnetization of the ferromagnet and it should occur even in the absence of an external field (in this case, the center of gravity of the pore does not move).

If $M \sim 0.1$ T (10^3 G), $\gamma \sim 1$ J/m^2 (10^3 ergs/cm^2), we find that the relative distortion of a pore calculated from Eq. (7.45) is of the order of $|\delta K|/K_0 \sim 3 \cdot 10^3$ R, i.e., it may be considerable if $R \sim 10^{-5} - 10^{-4}$ cm.

This distortion is due to the appearance of inhomogeneous magnetic fields which result from the spontaneous magnetization of the ferromagnet and which give rise to diffusion fluxes. Under steady-state conditions, these fluxes should be compensated by the fluxes associated with the inhomogeneous distortion and with the Laplace stresses.

Formula (7.45) for the distortion of a pore in the absence of an external field can be applied also to pores in pyroelectric crystals (for example, to ferroelectrics above the Curie point). In this case, the magnetization $\vec{M}$ should be replaced by the spontaneous polarization $\vec{P}$.

Distortion of a Pore with Anisotropic Properties in a Homogeneous Field

If the anisotropy of the surface or bulk properties of a crystal is taken into account, it is found that a pore should be distorted even in a homogeneous field of external forces. We shall analyze this distortion by considering the simplest case of isotropic surface energy when a pore is spherical (in a cubic crystal) in the absence of external forces [65]. We shall also assume that the motion of a pore in the field of external forces, for example, in the field of a temperature gradient, is entirely due to the surface diffusion fluxes.

In this case, we need allow only for the anisotropy of the transport coefficient $\alpha_S D_S$ in Eq. (2.2), which gives the surface thermal diffusion flux. This anisotropy is always encountered in real crystals, even if they have the cubic symmetry. We shall simplify the problem and reduce the number of theoretical parameters by ignoring the tensor nature of the quantity $\alpha_S D_S$, i.e., we shall assume that even in the anisotropic case the thermal diffusion flux at a given point is parallel to the temperature gradient $\nabla_S T$. However, we shall assume that the coefficient of proportionality between these two quantities is different for different points on the surface. The anisotropy of the diffusion coefficient on the surface of a pore is then due to the dependence of this coefficient on the orientation of a given part of the surface relative to the crystallographic axis. We shall consider the case when the angle-dependent deviation $\alpha'_S D'_S$ from the static value $\alpha^0_S D^0_S$ is slight and, in the case of cubic crystals, is given by the formula

$$\alpha'_S D'_S = \alpha^0_S D^0_S \, \varepsilon \left(n^4_x + n^4_y + n^4_z - \frac{3}{5} \right). \tag{7.46}$$

The coordinate axes are assumed to lie along the cubic symmetry axes and ε is a small parameter.

Bearing in mind the dependence $\alpha_S D_S$ on the coordinates in Eq. (2.2) and making allowance for the diffusion fluxes (7.7) and (5.13) on the distorted surface, we shall rewrite Eq. (1.9) for the velocities of different parts of the pore surface [in the steady-state case $v(\vec{r_S}) = \vec{v}$] in the form:

$$\vec{v} \, \vec{n} + \frac{\alpha^0_S}{T} \, D^0_S a \Delta_S T - \frac{2\omega\gamma D^0_S \, a}{f_S \, kT} \, \Delta_S K - \frac{a\omega D^0_S}{3 f_S \, kT} \, \Delta_S \, \sigma_{ee} = $$

$$= -\frac{\alpha'_S}{T} \, D'_S a \Delta_S T - \frac{a}{T} \, \nabla_S \left(\alpha'_S D'_S \right) \nabla_S T. \tag{7.47}$$

The above equation is derived on the assumption that the change in the curvature δK is proportional to the small anisotropy parameter ε (this point will be discussed later), and terms of higher orders of smallness, proportional to the product $\varepsilon\delta K$, are ignored. Within the same limits of accuracy, the operators Δ_S and ∇_S can be taken over the surface of a sphere and it may be assumed that $\vec{n}$ is the normal to the surface of this sphere.

If we substitute into the right-hand part of Eq. (7.47) the expression (7.46) for $\alpha'_S D'_S$ and if we consider separate terms of the expansion in terms of the spherical harmonics $Y^m_n(\theta, \varphi)$, we obtain an expression for the curvature of the pore surface. This curvature $K(\vec{n})$ is found to depend on the orientation of the unit vector $\vec{n}$, normal to the part of the pore surface in question, relative to the cubic axes, and on the orientation — relative to the same axes — of the unit vector $\vec{e}$, which is directed along the temperature gradient (in the earlier treatment, the curvature was found to depend only on the angle θ between $\vec{n}$ and $\vec{e}$). An explicit expression for the curvature is obtained in the form [65]:

$$K(\vec{n}) = K_0 + \frac{(1+\varkappa)\,\varepsilon f_s\,ka_S^0}{4\omega\gamma}\,R\left[\frac{31-11\nu}{5\,(23+3\nu)}\,Q_5 - \frac{13-7\nu}{9\,(9-\nu)}\,Q_3\right](\vec{e}\,\nabla T_{\infty}),$$

(7.48)

where

$$Q_3 = \sum_{i=1}^{3} e_i\,n_i^3 - \frac{3}{5}\,\vec{e}\vec{n},$$

(7.49)

$$Q_5 = \vec{e}\vec{n}\sum_{i=1}^{3} n_i^4 - \frac{4}{9}\sum_{i=1}^{3} e_i\,n_i^3 - \frac{1}{3}\,\vec{e}\vec{n}.$$

According to Eq. (7.48), the relative distortion of a pore due to its anisotropy is of the order of

$$\frac{|\delta K|}{K_0} \sim \frac{a_S^0 K}{10\omega\gamma}\,|\nabla T_{\infty}|\,R^2\,|\varepsilon| \sim \frac{1}{10}\,\frac{\nu R^2}{aD^2}\,\frac{Rk}{\gamma\omega}\,|\varepsilon|.$$

(7.50)

The distortion is proportional to the external force, which is the temperature gradient (or to the velocity of the pore). In contrast to δK in an inhomogeneous field of external forces [see Eq. (7.27)], the distortion is now proportional not to the cube but to the square of the pore radius and has a more complex dependence on the direction of the vector $\vec{n}$ and on the direction of motion of the pore. For example, if we assume that $\varepsilon \sim 1$, $a_S^0 \sim 1$, $\omega \sim 10^{-23}$ cm^3, $\gamma \sim 1$ J/m^2 (10^3 ergs/cm^2), $|\nabla T| \sim 10^3$ deg/cm, we find that the distortion is of the order of $|\delta K|/K_0 \sim 10^6 R^2$ and it may be considerable if $R \sim 3 \times 10^4$ cm.

It terms with higher powers of ε (terms proportional to ε^3) are included, the anisotropy of the surface diffusion gives rise to a dependence of the velocity of a pore in a cubic crystal on its direction relative to the crystallographic axes.

Similar expressions are obtained for the curvature of a pore moving in an external electric field if $z_S'\,\vec{D}_S' = z_S^0\,D_S^\gamma\,\varepsilon\,(n_x^4 + n_y^4 + n_z^4 - 3/5)$. These expressions are obtained quite simply if Eq. (7.48) is modified by replacing the factor $(1+\varkappa)f_s ka_S^0\,\vec{e}\nabla T_{\infty}$ with $(1+\varkappa_e)\vec{e}z_S^0\,\vec{e}\vec{E}_{\infty}$ [or if Eq. (7.50) is modified by replacing $ka_S^0\,|\nabla T_{\infty}|$ with $ez_S^0\,|\vec{E}_{\infty}|$].

In real crystals, a pore is aspherical even in the absence of external forces because of the anisotropy of the surface tension $\gamma = \gamma(\vec{n})$. The external forces simply produce an additional distortion of the pore surface. In this case, the distortion is determined not only by the parameter ε, which represents the anisotropy of the transport coefficients, but also by the parameter representing the deviation of the initial shape of the pore from a sphere in the absence of external forces.

The linear theory predicts another mechanism of distortion of moving pores. This mechanism applies when a pore is in contact with some defect, for example, a dislocation line or a grain boundary, which is capable of supplying vacancies rapidly to the pore. In this case, the diffusion fluxes on the pore surface are inhomogeneous and the surface must be strongly distorted in order to equalize the velocities of different parts of the surface.

Other distortion mechanisms are active when a pore moves at high velocities so that the condition (2.13) is violated and effects which are nonlinear in respect of the velocity become active. In one-component crystals, this condition is practically always satisfied for realistic values of the external forces and, therefore, the distortion due to high velocities is of little importance. However, it may be significant in solid solutions for which the criterion of validity of the linear theory is much more stringent (Sec. 14).

In considering the distortion of pores under the action of external forces, we have limited our discussion mainly to those pores whose surfaces are initially spherical or nearly spherical. However, if a pore or an inclusion in a cubic crystal is initially strongly elongated, the diffusion fluxes which appear in the absence of external forces alter the shape of the pore or the inclusion so that it approaches its equilibrium form. This has been considered for the case of a cylindrical inclusion [68] and it has been shown that the nature of the process will depend on the ratio of the length $L_\parallel$ to this diameter $L_\perp$ of the cylindrical inclusion. If $L_\parallel/L_\perp < \Lambda^*$ (Λ^* is some limiting value of the ratio $L_\parallel/L_\perp$), a cylindrical inclusion approaches the spherical shape because of the diffusion fluxes which flow from the ends to the lateral surface of the cylinder. If $L_\parallel/L_\perp > \Lambda^*$, the same inclusion breaks up into two or more smaller parts and each of these becomes spherical. These features of the spontaneous transformation of nonequilibrium inclusions are illustrated by the sequence of photographs given in Fig. 23 (Sec. 2), which show lithium inclusions in LiF single crystals. These inclusions become spherical or split into smaller parts. It follows from the relevant experiments that $\Lambda^* \sim 3$.

Gradual changes in the initial nonequilibrium shape of an inclusion should be observed also when such an inclusion moves in a field of external forces. In this case, the approach to the steady-state spherical shape may be a more complex process because the diffusion fluxes resulting from the distortion of an inclusion are superimposed on fluxes in the field of external forces.

The "spheroidization" of moving lithium inclusions in LiF single crystals (these inclusions were elongated along the [100] axis) was observed by one of the authors of the present monograph. The sequence of shapes obtained in these experiments is shown in Fig. 23. It is evident from this figure that different parts of the inclusion surface travel at different velocities. Sometimes, the front moves in the direction opposite to the direction of motion of the center of gravity of the inclusion. This is the consequence of the superposition of two types of diffusion flux in the field of a temperature gradient: the flux from the front to the rear of the inclusion, which determines the motion of the center of gravity, and the flux from the front and rear ends to the lateral surface, which governs the spheroidization process.

The anisotropy of the surface tension should modifiy spheroidal pores in cubic crystals to the polyhedral shape. During the motion in the field of external forces the nature of the faceting of a pore may change considerably so that the dynamic shape of a pore can differ considerably from its equilibrium shape. These great changes in the faceting may be due to, for example, a strong dependence of the evaporation coefficient χ on the orientation [5]. If a pore moves sufficiently rapidly, only the faces with relatively small values of χ manage to "survive" and the pore acquires a faceting which depends on the orientation of the external forces relative to the crystallographic axes. This was observed experimentally in the motion of pores in NH_4Cl [5]. The same effect was responsible for the formation of lenticular pores in UO_2 oriented along the (111) planes: these pores give rise to a special columnar structure of grains in reactor materials (nonlinear effects associated with the temperature dependences of the rate of evaporation may also be active; these effects are considered in [69]).

The field of a temperature gradient should become distorted near the dihedral angles of an equilibrium or a dynamic polyhedron. Consequently, the diffusion fluxes (proportional to ∇T) should become strongly inhomogeneous. This should give rise to further distortion of the pore and, consequently, to additional distortion of the temperature gradient field and of the diffusion fluxes. The result is the appearance of several new effects, in particular, the dynamic breakup of an elongated pore subjected to an external field [5], which is different from the breakup of a similar pore in the absence of external fields (discussed in the preceding paragraphs).

This latter effect is illustrated by a sequence of isotherms (Fig. 26) near a moving deformable pore [5]. It is evident from this figure that the motion of a plane A, associated with the presence of the normal component of the flux $\vec{I}'$ near this plane, gives rise to a new face B. The condensation of atoms on this face produces a third face C. The neck between the faces B and C grows until the pore breaks in two. This explains the dynamic shapes of the inclusions shown in Figs. 24 and 25.

A special case of distortion of the surface of liquid inclusions occurs when the velocity of migration is governed not by the differential fluxes in the liquid but by the rate of dissolution of the host crystal at the front of the inclusion and the rate of precipitation at the rear end [178]. In this case, the chemical potential can have discontinuities which will differ with the part of the surface, as indicated by Eq. (15.21). If these discontinuities differ by $\sim\varepsilon$, the resultant distortion is

$$\frac{\delta K}{K} \sim \frac{\epsilon}{\gamma\omega K}.$$

This type of deformation was observed experimentally in studies of the migration of liquid inclusions in KCl single crystals subjected to a temperature gradient [179]. A typical sequence of shapes of one of the migrating liquid inclusions is shown in Fig. 30. In our own studies of the migration of gas-filled bubbles we subjected an NaCl single crystal to a temperature gradient ∇T

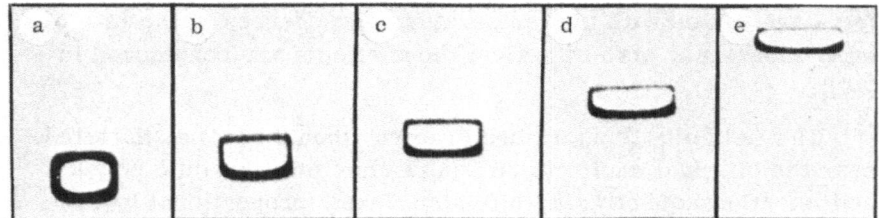

Fig. 30. Deformation of a liquid inclusion in a KCl single crystal subjected to a temperature gradient ∇T = 22 deg/cm [175]: a) 0 min; b) 20 min; c) 40 min; d) 70 min; e) 122 min. $\nabla T \parallel <100>$. Magnification 500.

inclined at an angle with respect to the normal of the (100) plane, which had the lowest surface energy. We observed dynamic shapes of the migrating bubbles (Fig. 31). These shapes were the result of competition between two processes: elongation resulting from the fact that $\varepsilon \neq 0$ and faceting because of the tendency to form planes with the lowest values of the surface energy.

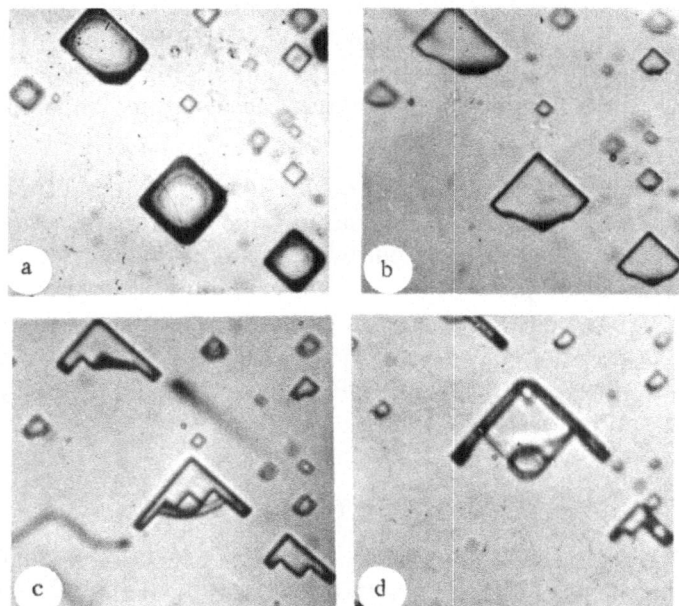

Fig. 31. Deformation of gas-filled bubbles in an NaCl single crystal subjected to a temperature gradient ∇T = 100 deg/cm (T = 720°C): a) 0 h; b) 2 h; c) 14 h; d) 25 h. $\nabla T \parallel <110>$. Magnification 330.

If the sum of the discontinuities in the chemical potentials at the front and rear ends of the inclusion $\delta\mu_i + \delta\mu_k$ is greater than $R|\nabla\mu_\infty|$ and if $\delta\mu = R|\nabla\mu_\infty| - (\delta\mu_i + \delta\mu_k)$ is negative, the migration of inclusions in an external field becomes a threshold process. In this case inclusions whose radius is below the threshold value $R^* = (\delta\mu_i + \delta\mu_k)/|\nabla\mu_\infty|$ should not migrate at all.

The existence of the threshold radius of inclusions was observed experimentally in the migration of liquid inclusions in KCl single crystals subjected to a temperature gradient [179] and the gravitational field [178]. In the latter case it was found that $R^* \approx 2 \times 10^{-3}$ cm [178].

8. DIFFUSION–ELASTIC INTERACTION BETWEEN

A PORE AND A BOUNDARY

The migration of inclusions in an inhomogeneous stress field (Sec. 5) is caused not only by external but also by internal stresses, which may appear near defects or imperfections, such as the boundary of a crystal. An important special case of the motion of inclusions in the field of internal stresses is the directional migration of a pore near the boundary of a crystal [63]. Such migration can be due to two causes.

The first cause is associated with elastic stresses, which are generated around a pore because of the Laplace pressure (or the internal pressure of the enclosed gas), acting on the surface of the pore. If the pore is near the boundary of a crystal or the boundary of a grain in an elastically anisotropic crystal, the stress field around the pore is distorted and it is not spherically symmetrical even if the host crystal is an elastically isotropic body. The image forces, which appear near the boundary of the crystal, give rise to a stress gradient on the surface of the pore. Such an inhomogeneous stress field produces surface diffusion fluxes which are responsible for the directional migration of the pore as a whole.

The second cause of the migration of a pore is associated with the difference between the chemical potentials of vacancies and the consequent difference between their concentrations on the almost planar boundary of a crystal and on the surface of the pore. The difference between the boundary values of the chemical potentials gives rise to diffusion fluxes of vacancies in the interior of the host crystal: depending on the gas pressure in the pore, these fluxes are directed either from the pore to the boundary of the crystal or in the opposite sense. The rates of vacancy loss from the half of the pore closer to the boundary of the host crystals and from the half which is further from this boundary are naturally different. Consequently, not only the volume of the pore changes but also the position of its center of gravity, i.e., the pore moves as one unit. Which of the two mechanisms of pore migration is the dominant one depends on the distance from the pore to the boundary of the host crystal and on the ratio of the diffusion coefficients D_S and D.

In each of the pore migration mechanisms, the numbers of atoms reaching different parts of the pore surface depend initially in a complex manner on the angular coordinates of the point in question and are not proportional to $\cos\theta$. Therefore, in considering the motion of a pore near the boundary of a crystal, we must also make allowance for the changes in the shape of the pore and for the changes in the stress field (Sec. 7) after the establishment of steady-state motion. It is convenient to consider the two pore migration mechanisms separately.

Diffusion-Elastic Migration of a Pore

near the Boundary of a Crystal as a

Result of Surface Diffusion

We shall consider a pore in an elastically isotropic crystal with isotropic surface properties (the effects associated with the elastic anisotropy will be discussed only qualitatively). In this case, a pore located in an infinite crystal is spherical and the stresses at a distance r from the center of the pore are given by

$$\sigma^0_{ij} = \frac{1}{2} P \frac{R^3}{r^3} \frac{r^2 \delta_{ij} - 3r_i r_j}{r^2} , \tag{8.1}$$

where the total pressure on the surface of the crystal

$$P = P_0 - \frac{2\gamma}{R} \tag{8.2}$$

is the algebraic sum of the gas pressure in the pore P_0 and the Laplace pressure $P'_L = -2\gamma/R$. If a crystal is subjected to an external pressure, the quantity P_0 should be understood to be the difference between the gas pressure within the pore and the external pressure.

In a finite crystal, the stress field is no longer given by Eq. (8.1) because this equation predicts the existence of a surface density of forces

$$f_i' = - \sigma_{ij}^0 \, n_j' \tag{8.3}$$

($\vec{n}'$ is a unit vector along the outer normal to the surface) at the boundary of the crystal as a result of internal stresses. However, in the absence of an external pressure, the forces acting on the boundary should be zero. This means that the boundary of a crystal distorts strongly the stress field and, in particular, produces inhomogeneous spherically asymmetrical stresses in the vicinity of the pore.

The stress field around a pore near a planar boundary can be expressed in a simple analytic form if the distance L from the center of the pore to the boundary is large compared with the pore radius R (Fig. 32). In this case, the stress field σ_{ij}^∞ in the region of the pore but far from its center ($R \ll r \ll L$) can be determined by the method of successive approximations. In this calculation, we must compensate the forces f_i' given by Eq. (8.3) by forces of the same magnitude but opposite sign, $-f_i'$, which act on the surface of the crystal so that — in the first approximation — the boundary conditions are satisfied on the surface. If we then determine the stresses σ_{ij}^∞ which are set up by the virtual forces $-f_i'$ far from the center of the pore in the region where $R \ll r \ll L$, we can find — in the first approximation — the change in the stresses which results from the presence of the crystal surface and which must be added to σ_{ij}^0. Near the pore, i.e., where $r \sim R$, the presence

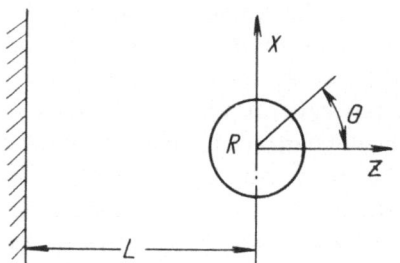

Fig. 32. Pore near the boundary of a crystal.

of an almost spherical free surface distorts considerably the field $\sigma_{ij}^0 + \sigma_{ij}^\infty$ (this field cannot satisfy the conditions on the internal surface). Consequently, the expression for the density of forces on a planar boundary, given by Eq. (8.3), must be modified somewhat. However, the necessary corrections are of high orders of smallness relative to R/L and they correspond to higher orders in the method of successive approximations, which are unimportant in the asymptotic case

$$2L \gg R. \tag{8.4}$$

If the field near a pore in the region $r \sim R$ is determined using the asymptotic value σ_{ij}^∞, the results are found to depend strongly on the selection of the boundary conditions (7.34a) or (7.11), which correspond to a perfectly spherical pore or a pore distorted because of the need to match the diffusion fluxes. In dealing with the quasisteady-state motion of a pore [the motion established after a relaxation time given by Eq. (7.33)], we must use the boundary conditions of Eq. (7.11). Then, the stress field in the region $r \sim R$ and the curvature of the pore are found to be related (Sec. 7) and a self-consistent diffusion—elastic problem must be solved in order to determine these parameters.

A calculation of this type (for details see [63]) shows that the stress field σ_{ij}^∞ generated by a boundary of a crystal (image forces) in the region $R \ll r \ll L$ is given by Eq. (7.34) to within terms which are unimportant in the situation being considered. The constants σ_1 and σ_3 in Eq. (7.34) are governed by the total pressure P on the surface of a pore in an infinite crystal, the radius of this pore R, and its distance L from the boundary of the crystal:

$$\sigma_1 = \frac{12}{5}(1 + v) P \frac{R^3}{(2L)^4}, \quad \sigma_3 = \frac{1}{10}(11 - 4v) P \frac{R^3}{(2L)^4}. \tag{8.5}$$

The inhomogeneous stresses given by Eq. (7.34) depend linearly on r and include the first and third harmonics $P_1(\cos \theta)$ and $P_3(\cos \theta)$. Under the influence of these stresses, a pore moves at right angles to the boundary of the crystal and its shape becomes distorted. If the migration mechanism is controlled by surface diffusion, we find that — as shown in Sec. 7 — the velocity of a pore is given by Eq. (5.24) and its curvature by Eq.(7.35). Substituting into these equations the values of σ_1 and σ_3 given by Eq.

(8.5), we obtain

$$\vec{v} = -\frac{1}{2}(1+\nu)\frac{D_S\,a\omega}{f_S\,kT}\,P\,\frac{R^2}{L^4}\,\vec{e}_\perp \,,$$ (8.6)

$$\frac{\delta r_S}{R} = \frac{1}{5}\frac{\delta K(\theta)}{K_0} = \frac{1}{5}\frac{K_3}{K_0}P_3(\cos\theta) =$$
$$= \frac{7}{320}\frac{(1+\nu)(11-4\nu)}{9-\nu}\frac{P}{\gamma}\frac{R^5}{L^4}P_3(\cos\theta).$$ (8.7)

Here, $\vec{e}_\perp$ is a unit vector along the inner normal to the boundary of the crystal.

If the pore in question is empty, so that $P = -2\gamma/R$, the distortion of the pore given by Eq. (8.7) is of the order of $0.1(R/L)^4$ for $L \gg R$, i.e., the distortion is slight (the distortion becomes much larger when $R \sim L$). If the pore is filled with a gas at high pressure and if $P \approx P_0 \gg 2\gamma/R$, the distortion of the pore near the boundary of the host crystal can be much greater. It follows from Eq. (8.7) that the distortion results in such a change in the pore's shape that it becomes flattened, the blunt end facing the direction of motion.

The results obtained in Sec. 7 can be used in Eqs. (8.6) and (8.7) only if the distortion of the pore is relatively weak. According to Eq. (8.7), the distortion is weak if

$$\frac{|P|R}{\gamma}\left(\frac{R}{2L}\right)^4 \ll 1.$$ (8.8)

In the case of empty pores, the above condition reduces to Eq. (8.4); however, in the case of gas-filled pores, Eq. (8.8) represents a more stringent criterion.

It follows from Eq. (8.6) that, in the asymptotic case when $R \ll 2L$, an empty pore $(P = -2\gamma/R)$ should move away from the boundary of the host crystal and a gas-filled pore, in which the pressure exceeds the Laplace value $(P > 0)$, should move toward this boundary. The velocity of the pore should be proportional to the boundary diffusion coefficient and should depend strongly on the distance from the boundary L, varying in inverse proportion to L^4. Larger pores migrate faster: the velocity of empty pores is $v \propto R$ and the velocity of pores filled with a gas at a high pressure $(P \gg 2\gamma/R)$ is $v \propto R^2$.

For example, if $D_S \sim 10^{-6}$ cm^2/sec, $T \sim 30°$K, $\omega \sim 10^{-23}$ cm^3, $P \sim 0.1$ GN/m^2 ($\sim 10^9$ dyn/cm^2), $R \sim 10^{-5}$ cm, $L \sim 10^{-4}$ cm, we find that $v \sim 10^{-9}$ cm/sec and a pore travels a distance of the order of L in a time $\sim 10^5$ sec. At relatively low temperatures, we can ignore the coalescence and migration of pores as a result of volume vacancy fluxes in a crystal (these processes are discussed in the next subsection) because the rates of these processes are proportional to the self-diffusion coefficient $D \ll D_S$. Therefore, the migration of pores toward or away from the boundary of a crystal, as described in the preceding paragraphs, may be important.

The asymptotic formulas (8.6) and (8.7) are inapplicable to pores whose distance from the boundary is $L \sim R$. However, if a pore is not too close to the boundary and its spherical shape is not greatly distorted, Eq. (8.6) still gives the correct direction and order of magnitude of the velocity. If the thickness of the layer separating the pore from the boundary is large compared with elastic displacements generated by the pressure P, the stresses in this layer acting on an empty pore in the case when $P < 0$ correspond to tension $(\sigma_{ii} > 0)$, whereas the stresses acting on the surface of a gas-filled pore in the case when $P > 0$ correspond to compression $(\sigma_{ii} < 0)$. The stresses acting on the part of the pore closer to the surface should be greater than those acting on that part which is further from the boundary of the host crystal. Therefore, in accordance with Eq. (5.13), the surface diffusion fluxes should give rise — as in the asymptotic case of large values of L — to the migration of empty pores away from the boundary, whereas the pores filled with a gas at high pressure should move toward the boundary.

The situation becomes much more complex when the layer between the pore and the boundary of the host crystal becomes so thin that it is comparable with the elastic displacements resulting from the pressure P. Very thin layers become stretched irrespective of the sign of P. In this case, the motion of a pore may be affected considerably not only by the diffusion fluxes on the pore's surface but also by those on the outer boundary of the crystal. Therefore, the shape of the pore may become strongly aspherical.

The behavior of pores located next to the boundary of the host crystal has not yet been investigated. It is possible that a pore very close to the boundary slows down very rapidly and becomes practically immobile before compression changes to tension in that part of the surface of a gas-filled pore which is closer to the boundary. Then, the stresses in the part of the pore closer to the boundary become nearly equal to those in the part further from the boundary, and the surface diffusion fluxes practically cease to flow in the pore. The diffusion fluxes on the boundary of the crystal near the pore may result in the expulsion of the pore (and the surrounding shell) from the crystal. Pores of this kind may gradually disappear by emerging on the surface of the crystal (this may happen as a result of the Brownian motion considered in Sec. 6), or they may collapse as a result of plastic deformation of the thin layer separating the pore from the boundary of the crystal or as a result of coalescence due to volume diffusion of vacancies from the pore to the boundary.

Gas-filled pores (P > 0), which are attracted to the boundary of the host crystal as a result of the migration discussed above, may reach the boundary layer of thickness $L \sim R$ and then emerge from the crystal or collect in the boundary layer ($L \sim R$). Empty pores (P < 0) are repelled by the boundary and migrate into the interior of the crystal. In both cases, the net result is a strong reduction in the concentration of pores near the boundary of the host crystal.

Since $\vec{v} = \frac{dL}{dt}\,\vec{e}_{\perp}$, the distance of a pore from the boundary $L(t)$ satisfies the following equation when $L \gg R$ [this follows from Eq. (8.6)]:

$$\frac{dL}{dt} = -\frac{1}{2}\,(1 + v)\,\frac{D_S\,a\omega}{f_S\,kT}\,P\,\frac{R^2}{L^4}.$$ (8.9)

Hence, when P > 0, the time τ_L taken by a pore to travel from a distance L_0 to the boundary is

$$\tau_L = \frac{2}{5}\,\frac{f_S\,kT}{(1 + v)\,D_S\,a\omega PR^2}\,L_0^5.$$ (8.10)

In this case, the departure from Eq. (8.9) in the region $L \sim R$ is unimportant if $L_0 \gg R$. Thus, we find that $\tau_L \propto L_0^5$. Similarly, the time τ'_L in which an empty pore travels a distance $L_1 - L_0$ into the interior of the crystal, i.e., from the point L_0 to the point L_1, is

$$\tau'_L = \frac{2}{5} \frac{f_S \, kT}{(1 + \nu) \, D_S \, a\omega PR^2} \left(L_1^5 - L_0^5 \right) \tag{8.11}$$

and is of the same order of magnitude as τ_L given by Eq. (8.10) provided $L_1 - L_0 \sim \frac{1}{2} L_0$.

The migration of pores toward or away from the boundary of their host crystal should produce a zone in which the concentration of pores is much lower than elsewhere. According to Eqs. (8.9)-(8.11), the order of magnitude of the thickness Λ of this zone is

$$\Lambda \sim \left(\frac{D_S \, a\omega PR^2}{kT} \right)^{1/5} t^{1/5} \tag{8.12}$$

and it increases with time in accordance with the law $\Lambda \propto t^{1/5}$. If the pores migrate into the interior of the crystal, they may accumulate in a region located at a distance $L \sim \Lambda$ from the boundary.

An allowance for the elastic anisotropy does not alter the order of magnitude of the inhomogeneous stresses acting on a pore located near the boundary of a crystal. Consequently, this allowance does not alter the order of magnitude of the velocity of migration (although it may greatly change its actual value). On the other hand, the elastic anisotropy may have a strong influence on the direction of motion of a pore. If the boundary of an anisotropic crystal does not coincide with a high-symmetry plane, a pore will not move at right angles to this boundary but at some angle with respect to the normal.

Moreover, the elastic anisotropy gives rise to the important phenomenon of attraction or repulsion of pores not only near the crystal—vacuum boundary but also near the boundary between two grains in a polycrystalline sample. In elastically isotropic crystals in the absence of easy slip on the boundary, the stress field around a pore is not distorted by the boundary and there is no interaction between the pore and the boundary if $L > R$. This interaction does appear in strongly anisotropic crystals with large disorientation angles and the energy of this interaction is of the same order of magnitude as that in the case of the crystal—vacuum boundary. Therefore, the velocity of a pore near a grain boundary is of the same order of magnitude as the velocity given by Eq. (8.6). In the case of weakly anisotropic crystals, the expression for the velocity v should be supplemented by an additional factor ζ, which is the elastic anisotropy parameter. If the angle of disorientation of a grain boundary is small, the velocity v is proportional to this angle if the boundary of one of the grains does not coincide with a high-symmetry plane, and proportional to the square of this angle if the boundaries of both grains coincide with such a plane.

If the surface tension γ is isotropic, the diffusion-elastic interaction between a pore and a boundary vanishes if the pressure of the gas in the pore P_0 is compensated by the Laplace pressure $2\gamma/R$ so that $P = 0$. However, if an allowance is made for the orientational dependence of γ (this gives rise to aspherical distortion of the pore far from the boundary), only the average pressure will be compensated when $P_0 = 2\overline{\gamma}/R$. Under these conditions, the interaction of a pore with a boundary will be observed. This interaction increases as the orientational dependence of γ increases, and the velocity decreases with distance L more rapidly than in direct proportion to L^{-4}.

Volume-Diffusion-Induced Changes

in the Volume and Shape of a Pore

and the Migration of Its Center of

Gravity near a Boundary

Volume fluxes of vacancies directed toward or away from a pore in an infinite crystal result in the symmetrical growth or contraction of the pore. If this pore is close to a boundary, a departure from the spherical symmetry leads to certain special features in the growth or contraction of the pore, which alter the shape and cause migration of the pore's center of gravity. We shall analyze these features by considering the behavior of a single pore near the boundary of the host crystal in the case when volume diffusion is important. We shall restrict our analysis to the case when the vacancy concentration has its equilibrium value c_V^0 far from the pore in the interior of the crystal, as well as on its planar boundary. We shall also assume that there are only a few defects

in the host crystal and that the distances between the vacancy
sources and sinks are large compared with the pore radius and
with the distance between the pore and the boundary. Once again,
we shall consider the asymptotic case $2L \gg R$.

In a perfect crystal, the chemical potential of vacancies μ_c
near the surface of a pore is given by Eq. (5.8), in which the normal
stresses are related, in accordance with Eq. (7.11), to the curvature of the pore's surface, i.e.,

$$\mu_v = \mu_{v0} - \omega(P_0 - 2\gamma K). \tag{8.13}$$

On the other planar boundary of the crystal and far from the pore
in the interior of the crystal, we have

$$\mu_v = \mu_{v0}. \tag{8.14}$$

This difference between the chemical potentials near to and
far from the pore gives rise to two types of diffusion flux of vacancies. The first type is the flux between the pore and the boundary or between the pore and the interior of the crystal due to the
difference between the chemical potentials (8.13) and (8.14), which
is governed by the average curvature of the pore $1/R$, and the gas
pressure in the pore P_0. The second flux is the diffusion of vacancies between different parts of the pore which may have different curvatures. However, we can show that, in the asymptotic
case, this flux can be ignored in the calculation of the rate of
change of the average radius R and of the velocity of the center of
gravity. This is achieved by dropping the small terms $\sim R/L$ from
the final expression for the velocity of the center of gravity of the
pore and by dropping the terms $\sim(R/L)^3$ from the expression for
$\dot{R}$. Then, in the first approximation, we can replace the curvature
K in the boundary condition (8.13) by the average constant value
$K \approx K_0 = 1/R$. The diffusion fluxes between different parts of the
surface of a given pore are important only in the determination of
the distortion of the pore.

The solution of the Laplace equation (5.7) for the chemical
potential subject to the boundary conditions (8.13) and (8.14) can
be obtained in bispherical coordinates, which are suited to this
problem. We can use this solution to find the velocity of different
parts of the pore's surface and, consequently, the changes in the
volume of the pore, the velocity of its center of gravity, and its
curvature (for details, see [63]).

The rate of change of the volume or radius of a pore in an infinite crystal with an equilibrium concentration of vacancies at infinity are given by the following expressions [compare Eqs. (4.5) and (4.2)]:

$$\frac{dR}{dt} = \frac{DP\omega}{RkT}, \qquad \frac{d\Omega}{dt} = 4\pi R^2 \frac{dR}{dt} = \frac{4\pi DP\omega R}{kT}. \qquad (8.15)$$

These rates are governed by the volume self-diffusion coefficient of atoms D and by the difference P between the gas pressure in the pore and the Laplace pressure. In the case of pores which are located at a distance L ≫ R from the boundary of the crystal, the expression for the rate of change of the volume of the pore

$$\frac{d\Omega}{dt} = \frac{4\pi DP\omega R}{kT} \left(1 + \frac{R}{2L} \right) \qquad (8.16)$$

differs little from the expression in Eq. (8.15), i.e., the presence of the boundary simply gives rise to a small correction, which is of the order of R/L.

In addition to the constant part of the normal component of the volume flux of atoms $\vec{I}_n$, which is responsible for the change in the average radius of the pore, the pores located at a finite distance L from the boundary of a crystal are affected also by the term in $\vec{I}_n$ which is proportional to cos θ (θ is the angle between $\vec{n}$ and $\vec{e}_\perp$) and which gives rise to the motion of a pore as a whole, i.e., to the motion of its center of gravity. The velocity $\vec{v}$ of this motion is

$$\vec{v} = -\frac{3}{4} \frac{DP\omega}{RkT} \left(\frac{R}{L} \right)^2 \vec{e}_\perp . \qquad (8.17)$$

In contrast to the velocity (8.6) governed by the surface diffusion fluxes, the velocity (8.17) is proportional to the volume self-diffusion coefficient of atoms.

In the case of small values of R/L (which is the case considered here), the velocity (8.17) is ~ (L/R)² times smaller than the rate of change of the average radius dR/dt = Ṙ. Therefore, a pore characterized by P < 0 disappears in a time

$$\tau_r \approx \frac{kT}{2D|P|\omega} R_0^2 \qquad (8.18)$$

during which it travels (away from the boundary) a distance of the order of $R_0(R_0/L)^2$, which is much smaller than the initial radius of the pore R_0. The radius of a pore characterized by $P > 0$ increases and, at the same time, the pore's center of gravity travels toward the boundary. In a time

$$\tau'_L \sim \frac{kT}{DP\omega} L_0^2 , \tag{8.19}$$

during which the radius of the pore increases to $\sim L_0$ and the front of the pore reaches the boundary of the crystal, the center of gravity of this pore travels a distance comparable with L_0.

If the concentration of vacancies in the interior of the host crystal differs from the equilibrium concentration c_v^0, a vacancy concentration gradient is established and — in addition to the motion due to the interaction between the pore and the boundary whose velocity is given by Eq. (8.17) — the pore should also move in the field of this gradient. This motion was discussed in Sec. 4.

The distortion of the pore is due to terms in the normal component of the flux of atoms $\vec{In.}$, which are proportional to higher powers of $\cos \theta$. The distortion depends on the ratio of the surface to the volume diffusion coefficient, which determines the nature of the relaxation of the distortion as a result of surface or volume diffusion fluxes between different parts of the pore's surface which have different curvatures. If the surface diffusion coefficient is sufficiently large so that $D_S a \gg RD$ [but $D_S a \ll RD(L/R)^2$] and the relaxation of the distortion is due to surface fluxes, the change in the curvature is proportional to the second harmonic $P_2(\cos \theta)$ and is given by the expression

$$\delta K(\theta) = K_2 P_2(\cos\theta), \quad K_2 = \frac{5}{8}\frac{DRP}{D_S a\gamma}\left(\frac{R}{2L}\right)^3 ,$$

$$\frac{\delta r_S}{R} = \frac{1}{2} K_2 RP_2(\cos\theta), \quad RD\left(\frac{L}{R}\right)^2 \gtrsim D_S a \gg RD. \tag{8.20}$$

The distortion given by Eq. (8.20) increases rapidly as the distance between the pore and the boundary grows and the value of the ratio D_S/D rises.

In pores of sufficiently large radius, for which $D_S a \ll RD$, the distortion relaxes as a result of volume diffusion fluxes. In this case, the order of the change in the curvature is

$$|\delta K(\theta)| \sim \frac{P}{\gamma}\left(\frac{R}{2L}\right)^3, \quad \frac{|\delta r_S|}{R} \sim \frac{|P|R}{\gamma}\left(\frac{R}{2L}\right)^3, \quad D_S a \ll RD. \qquad (8.21)$$

Since $|P|R/\gamma = 2$ for the "empty" pores, the distortion given by Eqs. (8.20) and (8.21) can be considerable for $R \ll 2L$ only if the pore is filled with a gas at a high pressure.

A comparison of Eqs. (8.6) and (8.17) demonstrates that the ratio of the velocities of the pores migrating under the influence of surface and volume diffusion is a quantity of the order of

$$\frac{D_S a}{DR} \frac{R^2}{L^2}. \qquad (8.22)$$

At moderate temperatures, when the ratio D_S/D is very large, and for moderately large values of L, the ratio given by Eq. (8.22) exceeds unity, i.e., the migration of a pore is governed primarily by surface diffusion.

The migration of a pore as a result of diffusion—elastic interaction with the boundary of a crystal becomes important if, during the time τ_L needed to travel a distance $\sim L$, the pore does not disappear (for $P < 0$) or its radius does not increase to $\sim L$. If the migration is primarily due to surface diffusion fluxes, it follows from Eqs. (8.11) and (8.18) that when $P < 0$ the condition for the existence of the diffusion—elastic interaction reduces to

$$\frac{\tau_L}{\tau_r} \sim \frac{DR}{D_S a}\left(\frac{L}{R}\right)^5 < 1. \qquad (8.23)$$

If $P > 0$, we find from Eqs. (8.10) and (8.19) that this condition becomes

$$\frac{\tau_L}{\tau_L'} \sim \frac{DR}{D_S a}\left(\frac{L}{R}\right)^3 < 1. \qquad (8.24)$$

The conditions of Eqs. (8.23) and (8.24) are more stringent than the condition obtaining for a large value of the ratio of the velocities, as given by Eq. (8.22). However, at moderate temperatures, when the ratio D/D_S is sufficiently small, these conditions should be satisfied. For example, if $D/D_S \sim 10^{-8}$ and $R/a \sim 10^3$, we find that the condition (8.23) is satisfied in a layer of thickness $L < 10R$, and when $P < 0$ the pores can migrate from

this layer to a depth ~L in the interior of the crystal before they
disappear. The condition (8.24) for gas-filled pores is satisfied
if L < 50R.

The filling and migration of large-radius pores near the boundary of a crystal with a high dislocation density may occur not only
as a result of diffusion processes but also due to plastic deformation. This mechanism is discussed in [70] but, unfortunately, incorrect expressions are used in that paper for the displacement
fields.

9. DIFFUSION- ELASTIC INTERACTION

BETWEEN PORES

The diffusion of pores should occur not only near the boundary
of a crystal but also in the field of other pores [63]. In an infinite
elastically isotropic crystal, the stress fields (8.1) around the
pores do not give rise to dilatation because $\sigma_{ii} = 0$. If a given pore
experiences only these stresses as a result of the presence of a
second pore, we find that the stresses σ_{ii} giving rise to surface
diffusion fluxes in the first pore must vanish. Finite values of σ_{ii}
arise not because of the influence of the field of the second pore but
because of the image forces exerted by the first pore on the second. However, the image-force stresses σ_{ii} decrease rapidly with
the distance between pores L (the decrease is proportional to $1/L^6$)
and the diffusion—elastic interaction is quite weak if L ≫ R.

A much stronger interaction between pores at distances L ≫ R
should occur in elastically anisotropic crystals because, in this
case, $\sigma_{ii} \neq 0$ [71] for the elastic fields around even a single spherical defect, and there is no need to make allowance for the secondary effects associated with the image forces. In real crystals,
the elastic anisotropy is usually strong and, therefore, it is this
anisotropy that is responsible for the diffusion—elastic interaction between the pores separated by large distances L. Simple
analytic expressions for the velocities of interacting pores can be
obtained in restricting the analysis to the case of weak elastic anisotropy in a cubic crystal for which the anisotropy parameter

$$\zeta = \frac{c_{11} - c_{12} - 2c_{44}}{c_{11}}$$

$$\tag{9.1}$$

(c_{ij} are the elastic moduli) is small. As in Sec. 8, an analytic approach is possible only in the asymptotic case when the distance between the centers of the pores L is large compared with their radii R_1 and R_2.

We shall select a system of coordinates coinciding with the cubic axes of the crystal in question. As in the isotropic case (Secs. 5 and 7), the velocity of the translation of a pore as a whole is governed by the coefficients σ_j in the expansion of the trace of the stress tensor σ_{ii} (on the surface of the pore), in which only three first-order harmonics are used:

$$\sigma_{ii} = \sum_{j=1}^{3} \sigma_j r \frac{x_j}{r} + \cdots \quad (r = R). \tag{9.2}$$

This expression does not include the terms with higher harmonics. These terms determine only the change in the curvature of the pore, which will not be considered here.

If the motion of the pore is due to surface diffusion fluxes and the coefficient D_S is isotropic, the expansion of Eq. (9.2) combined with Eqs. (5.24) and (5.17) yields the velocity of a pore of radius R_2:

$$v_j = -\frac{2}{3} \frac{D_S a\omega}{f_S kTR_2} \sigma_j. \tag{9.3}$$

Thus, the calculation of the velocity reduces to the determination of σ_{ii} on the surface of a pore in an elastically anisotropic crystal.

If the anisotropy is infinitesimal ($\zeta \ll 1$), the dilatation u_{ii} around a spherical pore of radius R_1, whose surface is subjected to a pressure P_1, is given by the formula (for details see [71])

$$u_{ii} = -\frac{15}{8} \frac{P_1 R_1^3}{G} \zeta \frac{x_1^4 + x_2^4 + x_3^4 - \frac{3}{5} r^4}{r^7}. \tag{9.4}$$

If the two pores under consideration are separated by a large distance L, the field u_{ii} exerted by the first pore in the region $R_2 \ll r_2 \ll L$ near the second pore (r_2 is the distance from the center of that pore) varies slowly with distance and can be regarded as the asymptotic external field acting at large distances from the center of the second pore. If the origin of the coordinates is placed at the center of the second pore, this asymptotic field can be written in the form

$$u_{ii}^{\infty} = u_{ii}^{c} + \frac{15}{8} \frac{P_1 R_1^3}{GL^4} \zeta \sum_{j=1}^{3} \left[\frac{9}{5} e_i + 4e_i^3 - 7e_i \left(e_1^4 + e_2^4 + e_3^4 \right) \right] x_i , \quad (9.5)$$

where u_{ii}^{c} is a constant and $\vec{e}$ is a unit vector directed from the second pore to the first.

In the determination of the actual strains and stresses near the second pore, we must make allowance for the fact that the stress field near this pore is distorted. The distortion is due to the fact that the boundary conditions (7.3) must be satisfied on the surface of the second pore and these conditions include the curvature given by Eqs. (7.10) and (7.13). However, a calculation of this distortion of the field in a weakly anisotropic crystal [72] has shown that the first harmonics of σ_{ii} on the surface of the second pore are exactly the same as the harmonics corresponding to the asymptotic expression (9.5) for the dilatation. Therefore, if ζ is small, the velocity of a pore can be determined directly from Eqs. (9.3) and (9.5) if we apply the standard relationship between σ_{ii} and u_{ii} in an elastically isotropic crystal.

The formulas for the components v_j of the velocity of a pore of radius R_2 in the field of a pore of radius R_1 (the velocity is measured relative to the host crystal) can be represented in the form

$$v_j = \frac{1}{6} v_0 \left[9 + 20 e_j^2 - 35 \left(e_1^4 + e_2^4 + e_3^4 \right) \right] e_j , \quad (9.6)$$

where

$$v_0 = -\frac{D_S a\omega}{f_S kT} \frac{P_1 R_1^3}{R_2 L^4} \frac{1+\nu}{1-2\nu} \zeta. \quad (9.7)$$

In general, the velocity of a pore is directed at an angle with respect to the vector $\vec{e}$. In this case, the projection of $\vec{v}$ onto $\vec{e}$ is positive or negative, depending on the orientation of $\vec{e}$ relative to the crystallographic axes and on the signs of P_1 and ζ. If the projection is positive, the pores are attracted toward each other but if it is negative they are repelled.

If the vector $\vec{e}$ is directed along one of the high-symmetry axes in a cubic crystal, the velocity $\vec{v}$ is parallel to $\vec{e}$. It follows from Eq. (9.6) that

$$\vec{v} = -v_0\vec{e} \quad \text{for } \vec{e} \parallel [100], \tag{9.8}$$

$$\vec{v} = -\frac{1}{4}v_0\vec{e} \quad \text{for } \vec{e} \parallel [110], \tag{9.9}$$

$$\vec{v} = -\frac{2}{3}v_0\vec{e} \quad \text{for } \vec{e} \parallel [111]. \tag{9.10}$$

It is evident from Eqs. (9.6) and (9.8)-(9.10) that along certain definite orientations, pores separated by large distances L should approach each other. Depending on the signs of P_1, P_2, and ζ, the pores will approach each other when the line joining them is oriented along the [110], [111], or [100] axes.

However, when these pores have approached each other to within a distance $L \approx R_1 + R_2$, the asymptotic method used in the derivation of Eqs. (9.6), (9.8)-(9.10) is no longer applicable. We must now make allowance for the image forces, which are just as important in elastically anisotropic crystals as in elastically isotropic ones. This allowance will alter the results considerably.

For example, if the radius of one of the pores R_2 is considerably less than the radius of the other pore, we find that, for short distances between the pores, the physical situation is the same as that in the interaction between a pore and a planar boundary, as discussed in Sec. 8. The direction of the motion of the more rapidly moving pore of radius R_2 is now determined not by the signs of P_1 and ζ (and the orientation of $\vec{e}$), as is the case for large distances L, but by the sign of the difference $P_2 - P_1$, which replaces the pressure P_0 in the formulas (Sec. 8). If both pores are empty or both are filled with a gas at the same pressure, we find that $P_2 - P_1 < 0$ and in the region where $R_2 \ll L' \ll R_1$ the second pore is attracted by the first, and this is true up to distances $L' \sim R_2$ at which the nature of the migration may change (Sec. 8). However, if $P_2 > P_1$, we find that the smaller and more mobile pore is repelled by the first pore beginning from distances $L' \sim L \sim R_1$, which are of the order of the radius of the larger pore.

It follows from Eqs. (9.6) and (9.7) that the velocity of interacting pores is proportional to the pressure P_1 or P_2, and that it vanishes if the gas pressure in each pore is compensated by the Laplace pressure, i.e., if $P_1 = P_2 = 0$. However, this result is valid only if we ignore the dependence of the surface tension on the orientation. An allowance for this dependence shows that complete compensation of the Laplace pressure by the pressure of the gas pore does not occur, and even when $2\bar{\gamma}/R = P_0$ some elastic stresses appear around a pore. As in the case of a pore located close to the planar boundary of a crystal (Sec. 8), these stresses give rise to a diffusion−elastic interaction between pores, i.e., they give rise to a relative migration of these pores (the velocity of such migration decreases very rapidly with increasing L).

10. COALESCENCE IN AN ENSEMBLE

OF MOVING INCLUSIONS

It follows from the results given in Secs. 2-5 that the velocity of the migration of inclusions in the fields of external forces usually depends on the radii of the inclusions. For example, when the migration is primarily due to surface diffusion fluxes, we find that

$$\vec{v} = \frac{C}{R} \, \vec{e}. \tag{10.1}$$

In an ensemble of particles or pores, the radius always varies over a certain range and, therefore, the velocity of different inclusions will be different. The faster inclusions will collide with the slower ones. The probability $\mathfrak{N}$ that colliding inclusions will coalesce is quite high (particularly in the case of gaseous or liquid inclusions). Thus, collisions can increase the average radius of inclusions in an ensemble and the rate of this coalescence will depend strongly on the external forces responsible for migration. In contrast to the conventional type of coalescence (the diffusion of vacancies or impurity atoms across a crystal from small to large inclusions), collision coalescence is governed not only by the value of the diffusion coefficient in the bulk of the host crystal but also by the diffusion coefficient on the surface and in the interior of the inclusion. Therefore, collision coalescence will predominate in a certain range of temperatures, inclusion radii, times, and external forces.

Even in the absence of external forces, inclusions can migrate in a crystal as a result of the Brownian motion (Sec. 6) or the diffusion−elastic interaction (Sec. 9). The velocity of small-radius inclusions can be quite high. Consequently, as a result of their fairly random motion, inclusions may collide and coalesce. In a certain range of temperatures and inclusion radii, the coalescence mechanism associated with the Brownian motion may predominate.

These mechanisms of the coalescence of inclusions in solids are similar to the well-known mechanisms of the coagulation of droplets in colloidal emulsions in liquids or gases [74-78]. The difference between solids, liquids, and gases lies only in the nature of the migration of inclusions: in solids, inclusions migrate primarily as a result of diffusion on the surface or in the interior of the inclusion. An allowance for the coalescence in the diffusional migration of pores in crystals helps one to understand the behavior of pore ensembles in solids which are formed during sintering, and the swelling of materials subjected to irradiation (Sec. 18). Coalescence of this type is considered in [10, 56, 79-81].

The coalescence mechanism in any specific case may be governed by the forced motion of inclusions in a field of external forces or by the Brownian motion. It is convenient to consider these two mechanisms separately.

Coalescence of Inclusions in a Field of External Forces

An ensemble of spherical inclusions in a solid can be described by specifying the distribution $f(R, t)$ of their radii. By definition, $f(R, t)dR$ is equal to the number (per unit volume) of inclusions whose radii are in the range from R to R + dR. The time dependence of the distribution function can be found by solving the transport equation.

In deriving the transport equation, we shall assume that the interaction between inclusions and the correlation between their positions can both be ignored (this is approximately valid for sufficiently large external forces, low inclusion concentrations, and weak interactions between them). We shall also assume that the coalescence of inclusions is limited by their velocity and not by the rate at which adjacent inclusions merge (we shall assume that the probability $\mathfrak{N}$ is close to unity). Then, the number of collisions $\mathfrak{N}(R_1, R_2)n_1 n_2$ between n_1 particles of radius R_1, moving at a velocity $\vec{v}_1$, and n_2 particles of radius R_2, moving at a velocity $\vec{v}_2$ (per unit volume and unit time), is given by the total volume of cylinders of area $\pi(R_1 + R_2)^2$ and length $|\vec{v}_1 - \vec{v}_2|$, multiplied by $n_1 n_2$:

$$\mathfrak{N}(R_1, R_2)\, n_1\, n_2 = \pi(R_1 + R_2)^2 \, |\vec{v}_1 - \vec{v}_2| \, n_1\, n_2. \tag{10.2}$$

A collision between two inclusions of radii R_1 and R_2 produces a larger inclusion of radius R. In the case of solid and liquid inclusions and in the case of empty pores, the volume of the new inclusion is equal to the sum of the volumes of the two original particles or pores, i.e.,

$$R^3 = R_1^3 + R_2^3. \tag{10.3}$$

In this case, the transport equation for the distribution function $f(R, t)$ is of the form

$$\frac{\partial f(R, t)}{\partial t} = \frac{1}{2} \int_0^\infty \int_0^\infty \mathfrak{N}(R_1, R_2) f(R_1, t) f(R_2, t) \delta [R -$$

$$- (R_1^3 + R_2^3)^{1/3}] \, dR_1 \, dR_2 - f(R, t) \int_0^\infty \mathfrak{N}(R, R_1) \, f(R_1, t) dR_1. \tag{10.4}$$

Here, the first term on the right-hand side represents the increase in the number of inclusions of a given radius R as a result of the coalescence of inclusions of smaller radii. The integrand in this term contains the product of the distribution functions $f(R_1, t)$ and $f(R_2, t)$ [in accordance with Eq. (10.2)] and a δ function, which represents the law of conservation of the volume of the inclusions given by Eq. (10.3). The second term on the right-hand side of Eq. (10.4) represents the reduction in the number of inclusions of radius R as a result of their coalescence with inclusions of all other radii.

The transport equation becomes somewhat different in the case of pores filled with inert gases, formed as a result of nuclear fission in irradiated materials. If the concentration and mobility of vacancies in such a material are sufficiently high, it is found (Sec. 2) that the radius of a gas-filled pore rapidly reaches its steady-state value as a result of vacancy diffusion. This steady-state value of R can be found by equating the pressures in Eq. (2.34) and neglecting the external pressure P^0 (in the case of small-radius pores): this value is $R = 2\gamma/P_0$. Since $P_0 = n_i kT \left(\frac{4\pi}{3} R^3\right)^{-1}$, where n_i is the number of inert-gas atoms in a given pore, the expression for the steady-state value of the radius can be written in the form

$$R^2 = \frac{3kT}{8\pi\gamma}\, n_i. \tag{10.5}$$

When pores of this kind coalesce, the total number of inert-gas atoms is conserved but the volume of a new pore increases relatively rapidly (as a result of the arrival of vacancies) to a value corresponding to the total number of atoms n_i. Therefore, in accordance with Eq. (10.5), we find that not the cube but the square of the radius is conserved when gas-filled pores coalesce, i.e.,

$$R^2 = R_1^2 + R_2^2. \tag{10.6}$$

If we use Eq. (10.6), we find that the transport equation for a system of gas-filled pores in a fissionable material becomes

$$\frac{\partial f(R,\,t)}{\partial t} = \frac{1}{2} \int_0^\infty \int_0^\infty \mathfrak{R}\,(R_1,\,R_2)\, f\,(R_1,\,t)\, f\,(R_2,\,t)\, \delta\,[R -$$

$$- \left(R_1^2 + R_2^2\right)^{1/2}]\, dR_1\, dR_2 - f\,(R,\,t) \int_0^\infty \mathfrak{R}(R,\,R_1)\, f\,(R_1,\,t) dR_1, \tag{10.7}$$

which differs from Eq. (10.4) by the argument of the δ function. We note that, in the case of large-radius gas-filled pores or in the case of high external pressures, when $P^0 \gg 2\gamma/R$, it follows from Eq. (2.34) that R^3 is conserved in the coalescence of pores, so that Eq. (10.4) applies again.

The solution of the nonlinear integral differential equations
(10.4) or (10.7) is usually difficult. However, the characteristic
time τ_0, during which the inclusions double in size, can be found
without solving these equations. For example, let us assume that
the motion of inclusions is solely due to surface diffusion fluxes
so that, in accordance with Eqs. (10.1) and (10.2), we obtain

$$\mathfrak{N}(R_1, R_2) = \pi (R_1 + R_2)^2 C \left| \frac{1}{R_1} - \frac{1}{R_2} \right|. \qquad (10.8)$$

Let us also assume that the width δR of the initial distribution of
the inclusion radii about the most probable value R_m is quite wide
$(\delta R \sim R_m)$. Then, bearing in mind that $\int f(R, t)dR \sim p/R_m^3$, where
p is the initial volume concentration of the inclusions, we find that
Eqs. (10.4) and (10.7) yield the following estimate of the characteristic time τ_0 [10]:

$$\tau_0 \sim \frac{R_m^2}{pC} \sim \frac{R_m}{pv}. \qquad (10.9)$$

For example, if $R_m \sim 10^3$ Å, $p \sim 10^{-2}$ and $v \sim 10$ Å/sec, we
find that τ_0 is of the order of several hours, i.e., the coalescence
mechanism considered here is fairly rapid. The initial rate of
coalescence depends strongly on the nature of the original distribution of the inclusion radii. In the case of very narrow distributions, the rate of coalescence can be quite low (in the limiting
case of inclusions of the same radii, all the velocities are the same
and the inclusions do not collide). Obviously, this case of very
narrow initial distributions is hardly ever encountered in practice.

In particular, it follows from Eq. (10.8) that inclusions corresponding to the "wings" of the distribution, whose radii are considerably smaller or larger than the most probable value R_m,
collide more frequently than other inclusions and gradually disappear. We may expect that, after a long time, $t \gg \tau_0$, the distribution becomes simpler, approaching some standard asymptotic
form which is independent of the initial distribution. This hypothesis is supported by the results of a numerical solution of the
transport equation (10.7) obtained on a computer [81] and by those
of an analytic treatment [56] based on the use of the similarity
principle (this principle has been applied to the coagulation of colloidal systems [77, 78]).

The similarity principle can also be applied to Eq. (10.7) if the order of homogeneity h of the function

$$\mathfrak{N}(bR_1, bR_2) = b^h \, \mathfrak{N}(R_1, R_2) \qquad (10.10)$$

is less than 2. Then, Eq. (10.7) has the solution

$$f(R, t) = t^{-\frac{3}{2-h}} F(\eta), \quad \eta = R t^{-\frac{1}{2-h}}, \qquad (10.11)$$

in which the function $t^{3/(2-h)} f(R, t)$ depends on the variable η (and on t and R separately) and the power exponents are selected so that the sum of the values of R^2 for all the pores is independent of time, in accordance with Eq. (10.6) [the function $f(\eta)$ is explicitly independent of time]. This standard distribution $f(R, t)$ is approached by pores filled with inert gases after a long time $t \gg \tau_0$. When this limit is reached, all the moments of the distribution function depend on time in accordance with the simple power law:

$$f_n = \int_0^\infty f(R, t) R^n dR = A_n t^{-\frac{2-n}{2-h}}, \quad A_n = \int_0^\infty \eta^n F(\eta) d\eta, \qquad (10.12)$$

and it follows from Eq. (10.6) that the second moment is invariant.

In the case we are considering, the rate of collisions is given by Eq. (10.8) and we find that h = 1. Therefore, the zeroth moment of the distribution $f_0 = N_i$, equal to the number of inclusions per unit volume, decreases proportionally to t^{-2}, and the average radius of the inclusions $\bar{R} = f_1/f_0$ and their total relative volume $V = (4\pi/3)f_3$ increases proportionally to t. The numerical coefficients A_n in Eq. (10.10) can be determined provided we know the explicit form of the function $F(\eta)$. An approximate derivation of this function [56] yields results which are in good agreement (to within a few percent) with the results of numerical calculations on a computer [81]:

$$N_i(t) = 1.3 \frac{\gamma}{N_g kTC^2 t^2} = 1.3 N_i^0 \left(\frac{\tau}{t}\right)^2, \qquad (10.13)$$

$$\bar{R}(t) = \frac{0.25 N_g kTCt}{\gamma} = 0.72 R_0 \frac{t}{\tau}, \qquad (10.14)$$

$$V(t) = 0.25 \left(\frac{N_g kT}{\gamma}\right)^2 Ct = 6.06 N_i^0 R_0^3 \frac{t}{\tau}. \qquad (10.15)$$

Here, N_g is the total number of inert-gas atoms in all the pores contained in unit volume; $N_i^0 = N_i(0)$ is the initial number of inclusions; R_0^2 is the mean-square value of the initial radius of the inclusions; and the characteristic time τ ($\tau \sim \tau_0$) is

$$\tau = \frac{1}{C} \left(\frac{\gamma}{N_g N_i^0 kT} \right)^{1/2} = \left(\frac{2\pi}{3} \right)^{1/2} \frac{R_0^2}{Cp} = 1.45 \frac{R_0^2}{Cp}. \qquad (10.16)$$

In the special case when the function $f(R, 0)$ at t = 0 is described by a normal logarithmic distribution, we can also obtain an approximate solution which applies at any time. In this case, the number of inclusions is given by the following approximate formula

$$N_i(t) = N_i^0 \left(0.87 \frac{t}{\tau} + 1 \right)^{-2}. \qquad (10.17)$$

It is evident from this expression that the asymptotic solution is reached at a time τ.

In the case of empty pores or liquid inclusions, it follows from Eq. (10.3) that the third (not the second) moment of the distribution function should be invariant and that the transport equation of Eq. (10.4) is applicable. This invariance condition is satisfied if Eq. (10.11) is replaced by a solution which satisfies the similarity principle:

$$f(R, t) = t^{-\frac{4}{3-h}} F(\eta), \quad \eta = Rt^{-\frac{1}{3-h}}. \qquad (10.18)$$

The moments of the distribution function given by the above solution are

$$f_n = A_n' t^{-\frac{3-n}{3-h}}, \quad A_n' = \int_0^\infty \eta^n F(\eta) \, d\eta. \qquad (10.19)$$

After a sufficiently long time t $\gg \tau$ [here, τ is given by the second expression in Eq. (10.16)], the formulas for $f(R, t)$ and f_n reduce to the asymptotic expressions of Eqs. (10.18) and (10.19). It follows from Eq. (10.8) that, in this case, we should substitute h = 1 in Eqs. (10.18) and (10.19). The number of inclusions per unit volume now decreases proportionally to $1/t^{3/2}$ and it is of the order of

$$N_i\,(t)\,\propto N_i^0\,\left(\frac{\tau}{t}\right)^{3/2}. \tag{10.20}$$

The average radius of an inclusion $\overline{R} = f_1/f_0$ increases in proportion to $t^{1/2}$:

$$\overline{R}\,(t)\,\propto R_0\,(t/\tau)^{1/2}. \tag{10.21}$$

The total volume of inclusions $V(t)$ remains constant with time. However, the average volume of a single inclusion $V(t)/N_i(t)$ increases proportionally to $(t/\tau)^{3/2}$, i.e., it increases more slowly than in the case of gas-filled pores. Formulas (10.20) and (10.21) describe also the behavior of an ensemble of large-radius gas-filled inclusions (or of inclusions with high internal pressures) for which $P^0 \gg 2\gamma/R$ and which are described by the transport equations (10.4).

Coalescence Due to the Brownian Motion of Inclusions

We shall consider only one special case of the coalescence of pores in the absence of external forces, namely, the case when the Laplace pressure is compensated by the pressure of a gas inside a pore. Then, even if the volume and surface properties of the host crystal are strongly anisotropic, the diffusion-elastic interaction between pores (Sec. 9) is absent and the pores are subjected only to the Brownian motion.

As in the presence of external forces, the distribution of the radii of the pores is governed by the transport equation (10.7) or by Eq. (10.4) in the case of large-radius pores. However, the expression for the number of collisions $\mathfrak{N}(R_1, R_2)n_1n_2$ between n_1 pores of radius R_1 and n_2 pores of radius R_2 is different. In the Brownian motion mechanism, the number of collisions is given by [75]

$$\mathfrak{N}\,(R_1, R_2) = 4\pi\,[D_i\,(R_1) + D_i\,(R_2)]\,(R_1 + R_2). \tag{10.22}$$

Here, $D_i(R)$ is the diffusion coefficient of the inclusions or pores given by Eqs. (6.18)-(6.21). Equation (10.22) is derived on the assumption that the pores do not interact even if they are in contact with each other, and we assume that $R^2 \ll D_i t$.

The characteristic time necessary for doubling the average radius of the inclusions can be estimated as easily as in the case of coalescence in a field of external forces. For example, if the motion of inclusions is due to surface diffusion fluxes and D_i is given by Eq. (6.9), it follows from Eqs. (10.7), (10.22), and (6.19) that the order of magnitude of this time is

$$\tau' = \frac{f_S}{10\Gamma} \frac{1}{D_S a\omega \left(N_i^0\right)^{5/2}} \left(\frac{3N_i\, kT}{8\pi\gamma}\right)^{3/2} = \frac{4\pi f_S}{30\Gamma} \frac{R_0^6}{D_S\, a\omega p}. \qquad (10.23)$$

The time τ' depends strongly on the surface diffusion coefficient D_S, the initial radius of the particles (pores) R_0, and their volume concentration p. For example, if $D_S \sim 10^{-6}$ cm^2/sec, p $\sim 10^{-2}$, $R_0 \sim 10^{-6}$ cm, the time constant is of the order of $\tau' \sim 10^2$ sec, whereas if $R_0 \sim 10^{-5}$ cm and the other parameters are the same, we find that $\tau' \sim 10^8$ sec and the coalescence mechanism discussed here is important only over periods of the order of a few months or even years.

After a time considerably longer than τ', the solution of the transport equation (10.7) becomes simpler; it is given by the asymptotic formula (10.11) and the moments of the distribution are given by the formulas in Eq. (10.12). The difference between the coalescence due to the Brownian motion and that due to a field of external forces appears only in the order h of the homogeneity of the function $\mathfrak{N}(R_1, R_2)$. It follows from Eq. (10.22) and (6.19) that in the Brownian motion mechanism the relevant order is h $= -3$. It follows from Eq. (10.12) that after a long time we find that $N_i(t) \propto t^{-2/5}$, $R \propto t^{1/5}$, and $V \propto t^{1/5}$. Using the approximate values of the coefficients A_n obtained in [56], we can express the above quantities in the form

$$N_i(t) = 0.74 N_i^0 \left(\frac{\tau'}{t}\right)^{2/5}; \qquad (10.24)$$

$$\overline{R}(t) = 1.13 R_0 \left(\frac{t}{\tau'}\right)^{1/5}; \qquad (10.25)$$

$$V(t) = 5.2 N_i^0\, R_0^3 \left(\frac{t}{\tau'}\right)^{1/5}. \qquad (10.26)$$

If the initial distribution of the pore radii can be represented by a δ function (in this case, all the pores have the same radius

R_0 at t = 0), we can obtain an approximate interpolation expression for the function $N_i(t)$ which applies at all times [56]:

$$N_i(t) = N_i^0 \left[2.15 \frac{t}{\tau'} + 2 \exp\left(-0.57 \frac{t}{\tau'}\right) - 1 \right]^{-2/5}. \qquad (10.27)$$

After a long time t ≫ τ', Eqs. (10.27) and (10.24) become identical.

In the case of pores filled with gases at high pressures ($P^0 \gg 2\gamma/R$) or in the case of liquid inclusions, when the transport equation of Eq. (10.4) is valid (the interaction between the inclusions is ignored), we can use Eqs. (10.18) and (10.19) which define $f(R,t)$ and the moments, provided we substitute h = −3 (this is permissible only after a long time has elapsed from the beginning of the coalescence process). In this case, the number of inclusions and their average radius are given by the expressions

$$N_i(t) \propto N_i^0 \left(\frac{\tau'}{t}\right)^{1/2}, \quad \overline{R}(t) \propto R_0 \left(\frac{t}{\tau'}\right)^{1/6}, \qquad (10.28)$$

which differ slightly from Eqs. (10.24) and (10.25).

These results show that the coalescence mechanism governed by the Brownian motion of inclusions may be important in the case of small-radius inclusions, particularly during the initial stages of the formation and growth of an ensemble of pores in an irradiated material. For example, numerical calculations are reported in [81] for an initial state in which each pore contains just one inert-gas atom [i.e., $N_i^0 = N_g$ and $R_0 \sim 1$ Å, in accordance with Eq. (10.5)] and for $D_S = 10^{-5}$ cm^2/sec, $\omega = 1.2 \times 10^{-23}$ cm^3, $\gamma = 1.7$ J/m^2 (1700 ergs/cm^2). These parameters correspond to copper at T = 1000°K and $N_g = 10^{20}$ cm^{-3}. It is reported in [81] that, in this case, $\tau' = 4 \times 10^{-12}$ sec, i.e., the characteristic time is negligible. After a long time has elapsed (t = 3 × 10^6 sec, i.e., of the order of one month) from the beginning of irradiation, the average radius of a pore, estimated in [56, 81], should be $\overline{R}(t)$ = 1300 Å and the total relative volume of the pores should be V(t) = 0.06. If the initial radius of the pores is large, the time τ' needed to establish the asymptotic distribution is much longer. For example, if initially each pore contains $n_g = 10^5$ inert-gas atoms ($R_0 \sim 10^2$ Å), we find that $\tau' = 1200$ sec.

The Brownian motion cannot be the dominant mechanism of the coalescence of pores if R_0 is large or if $t \gg \tau$. It is evident from Eqs. (10.25) and (10.26) that the average radius and the volume of pores increase very slowly (proportional to $t^{1/5}$) when $t \gg \tau$. The coalescence of pores under the action of external forces is much more important if the initial radius is large or when the later stages of the coalescence process are considered. It follows from Eq. (10.16) that, in this case, the characteristic time depends much less strongly on R_0 than the time given by Eq. (10.23) and the average radius and the total volume of the pores, given by Eqs. (10.14) and (10.15), increase much more rapidly (proportional to t) than in the case of the Brownian motion.

The external forces responsible for the migration and coalescence of pores can be due to a temperature gradient, external stresses, forces exerted on pores by grain boundaries moving during recrystallization (Sec. 11), or internal stresses around dislocations (Sec. 17). In particular, coalescence of this type should be very rapid near slip planes or other regions, in which local stresses are high.

The coalescence mechanism of pores resulting from their migration in the field of an external force or from their Brownian motion completes with the conventional coalescence mechanism due to the volume diffusion of vacancies from the small to the large pores. The corresponding time for doubling the average volume of empty pores is of the order of [44, 45]

$$\tau'' \sim \frac{kTR_0^3}{D\gamma\omega} \tag{10.29}$$

and is inversely proportional to the volume diffusion coefficient D. The ratio of τ'' to the characteristic time for the coalescence of pores in a field of external forces, given by Eq. (10.16), is of the order of

$$\frac{\tau''}{\tau} \sim \frac{kTR_0\,Cp}{D\gamma\omega}. \tag{10.30}$$

Since, at low temperatures and for large forces, we find that $C \sim D_S$ (and D_S/D is large), the ratio given by Eq. (10.30) may exceed unity. The principal coalescence mechanism will then be that due to their motion in an external force field.

The Brownian motion of inclusions and their migration in a field of external forces may exert a considerable influence not only on the coalescence of pores but also on the kinetics of phase transitions in the initial stages of aging (this applies if, for example, the boundary diffusion coefficient of the second-phase particles is large).

Experimental Investigations of the

Coalescence of Moving Inclusions

We have shown theoretically that the coalescence of inclusions traveling in an ordered manner in an external force field occurs when, for some reason, the inclusions in an ensemble travel at different velocities. Under real conditions, the velocities may be different due to the different dimensions and shapes of the inclusions moving in an external field, or to the different forces acting on the inclusions. In particular, a considerable inhomogeneity of the force field may give rise to an inhomogeneity in the distribution of stresses or in the concentration of a solution.

In real crystals, the kinetics of the coalescence of inclusions moving in a field of external forces may be governed, to a considerable degree, by the interaction between fields of stresses around adjacent inclusions. In particular, this interaction may stop almost completely the coalescence of moving inclusions.

The coalescence of moving inclusions was demonstrated clearly in an electron-microscopic study of the motion of helium-filled bubbles in copper [16]. The experiments were carried out at 800°C. At this temperature, the mobility of the vacancies in the copper was high and, therefore, the stresses which could appear around gas-filled bubbles were easily relieved.

Spherical gas-filled bubbles of $R \sim 5 \cdot 10^{-6}$ cm radius migrated in the field of a temperature gradient as a result of mass transfer along the bubble's surface, i.e., the velocity was $v \propto R^{-1}$. Consequently, the small bubbles caught up with the large ones and merged with them. The coalescence of these bubbles is illustrated in the succession of frames shown in Fig. 33.

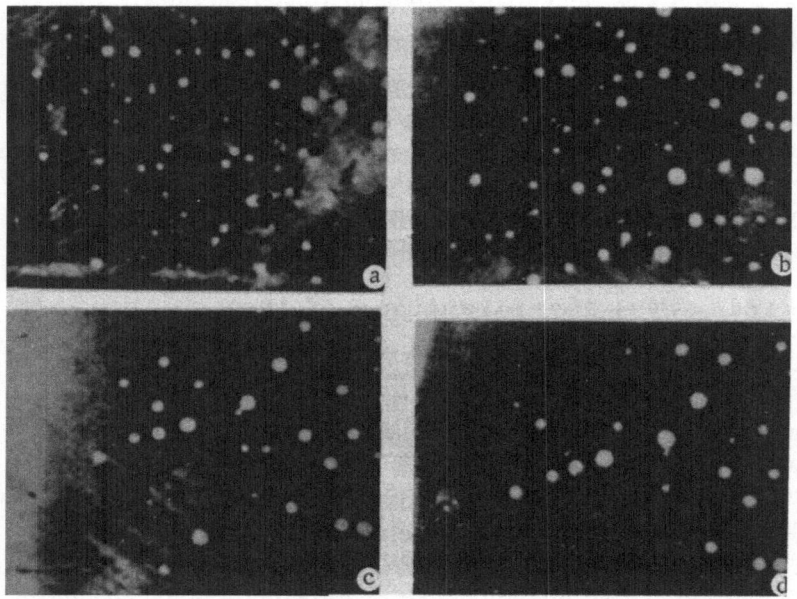

Fig. 33. Sequence of electron photomicrographs (a-d) illustrating the coalescence of bubbles moving in the field of a temperature gradient [16].

The condition (10.6) was always obeyed whenever the bubbles merged as a result of direct collisions. This indicated that the colliding bubbles were in equilibrium, i.e., that no elastic stresses were generated by them in the elastically isotropic crystal.[†]

An important feature of these experiments was the almost complete insolubility of helium in copper. Therefore, the observed coalescence could only be due to direct collisions between bubbles. The diffusional coalescence of the gas-filled bubbles did not occur because the gas was insoluble in the host crystal.

The migration and coalescence of gas-filled bubbles may be seriously restricted by the presence of insoluble impurities in the

---

† Some departure from the condition (10.6) could be due to the fact that the gas contained in bubbles of radius $R \sim 5 \times 10^{-6}$ cm was held under a considerable pressure $P \sim 300$ GN/m^2 (300 atm). Consequently, the behavior of this gas could depart from the laws governing ideal gases, which were assumed in the derivation of Eq. (10.6). However, this did not mean that the bubbles were not in equilibrium.

host crystal, which act as traps for such bubbles. The localiza-
tion of a bubble at an impurity inclusion may be due to a reduction in the surface energy of the bubble—inclusion—host system.
The slowing down of the coalescence by impurities may have an
important practical effect. When a gas-filled bubble coalesces
under conditions such that the bubble surface remains constant,
the volume occupied by the gas increases, i.e., the sample swells.
This is why small amounts of insoluble impurities in uranium may
give rise to swelling of the fuel elements (Sec. 18).

The coalescence of second-phase inclusions and of bubbles,
resulting from their direct collisions, has been observed experimentally by many workers who studied the motion of bubbles under the influence of a concentration gradient in a diffusion zone
[82], and an electric field gradient [83], and other forces. This
effect will be considered later.

The coalescence of colliding inclusions or gas-filled cavities
is a nonthreshold process because it is accompanied by a mono-
tonic decrease in the free energy of the system.

When two material inclusions coalesce, their total volume is
conserved provided the condition (10.3) is satisfied and, consequently, $R_1^2 + R_2^2 > R^2$. The free energy of the system decreases
by an amount $\delta F \sim \gamma '[R^2 - (R_1^2 + R_2^2)]$, where $\gamma '$ is the energy of
the inclusion—host interface. The coalescence of equilibrium gas-
filled bubbles is subject to the condition (10.6), which expresses
the invariance of the interfacial energy. In this case, the increase
in the total volume $(R_1^3 + R_2^3) < R^3$ leads to a reduction in the free
energy as a result of the expansion of the gas occluded in the bubbles. However, under real conditions, the coalescence of adjacent
inclusions or gas-filled bubbles may have a threshold for the following reasons. If the matter in the host crystal or in the inclusion
contains surface-active impurities, the adsorption of these impurities on the inclusion—host interface reduces the energy of this
interface. Immediately after coalescence, when the boundary separating the inclusions breaks down, a surface not covered by the
impurities in question should become exposed. This circumstance
may give rise to some increase in the free energy of the system
and, consequently, the coalescence can occur only when an energy
threshold is overcome.

The results of the following model experiments demonstrate the validity of the predicted features of the coalescence of adjacent inclusions or bubbles. Hypodermia needles of different diameters are used to introduce low-pressure air into transparent technical-grade glycerin. The gas-filled bubbles formed in this way float up to the surface at a velocity which depends on the bubble radius, $v \propto R^2$, in accordance with the Stokes law. During this motion, collisions take place between bubbles of different radii. However, such collisions do not usually give rise to coalescence. The bubbles become deformed but coalesce only in rare cases. A similar effect should occur also in a solid matrix. This was confirmed in an electron-microscopic study [16, 17] of the coalescence of gas-filled bubbles in copper and uranium oxide: this study revealed the existence of pairs of adjacent bubbles which did not merge for a long time.

Let us now consider the final stage in a coalescence event, which is the merging of two inclusions into one of larger volume. At this stage, the colliding inclusions intermingle and the transfer of mass can only result from various diffusion mechanisms.

The problem of the kinetics of the merging of solid inclusions in solids has not yet been analyzed theoretically. The complexity of the problem is due to the need to allow not only for the kinetics of the reduction in the surface energy but also for changes in the elastic energy of a system comprising two inclusions and the host crystal. It is simpler to solve the problem of the coalescence of gas-filled inclusions by assuming that this process is analogous to the sintering of particles, particularly spherical particles. A gas-filled bubble can be regarded as a "negative" particle. A partial solution of this problem is given in [84, 85].

The relative importance of the mechanisms which determine the kinetics of coalescence of adjacent gas-filled inclusions depends on the linear dimensions of these inclusions.

Under isothermal conditions, the relative importance of the various mechanisms may depend on time because of changes in the characteristic distances of the mass-transfer process. In particular, it is possible to follow the evolution of the relative importance of the surface and volume self-diffusion by comparing the corresponding fluxes.

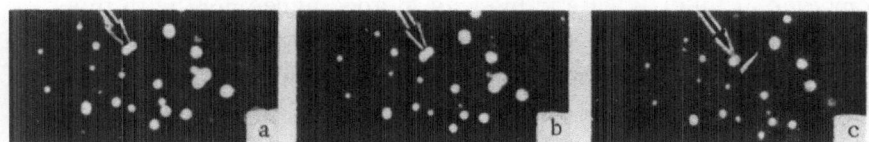

Fig. 34. Successive stages (a, b, c) of the coalescence of helium-filled bubbles in a copper plate [85].

The dimensionless ratio of the fluxes in surface and volume self-diffusion in a thin layer between two closely spaced inclusions can be written in the form

$$\frac{I_S}{I} \sim \frac{D_S}{D} \frac{2\pi x a}{2\pi x \rho} \sim \frac{D_S}{D} \frac{Ra}{x^2},$$

where x is the radius of the contact layer; ρ is the radius of curvature of the contact "neck" between two spherical inclusions whose radii are R.†

Under isothermal conditions the role played by surface diffusion in reaching a given degree of merging of the gas-filled bubbles, characterized by the ratio x/R = C, increases with decreasing linear size of the bubbles in proportion to $1/C^2R$. Surface diffusion predominates in the initial stage of the process irrespective of the radii of the bubbles and it remains important in the range of values

$$\frac{x}{R} < \left(\frac{D_S}{D}\frac{a}{R}\right)^{1/2}. \qquad (10.30a)$$

It follows from Eq. (10.30a) that when R ≪ $(D_S/D)a$ the diffusion mechanism should govern the mass transfer right up to the moment of complete merging of the bubbles.

Information on changes in the shape of coalescing gas-filled bubbles can be used to find the surface self-diffusion coefficient if this coefficient governs the changes in shape. Nichols [85] found the surface self-diffusion coefficient from the data on the merging of gas bubbles. He calculated D_S using electron photomicrographs [16] of the successive stages of the merging of two helium bubbles in copper (Fig. 34). He found that $D_S = 1.4 \times 10^{-5}$

---

† The quantities x, ρ, and R are related by $\rho \propto x^2/R$ if x ≪ R.

cm^2/sec at 900°C. This value was in good agreement with the published values [20]. This agreement suggested that the coalescence of colliding helium bubbles in copper [16] was due to surface self-diffusion.

Coalescence in a Two-Dimensional Ensemble of Moving Inclusions

In the preceding subsection we considered the coalescence of inclusions distributed in three dimensions. However, in some cases, inclusions form only on certain surfaces, for example, on the external surface of a crystal or on grain boundaries, or they accumulate on these surfaces during some heat treatments. We then find that the inclusions are bound strongly to such surfaces (Sec. 11) and, in some cases, they can move only along these surfaces (this is true of the migration under the action of external forces or as a result of the Brownian motion). In these cases, it is necessary to study the coalescence in a two-dimensional ensemble of moving inclusions, which has been discussed by M. N. Botvinko and M. A. Krivoglaz [180].

The kinetics of the coalescence in a two-dimensional ensemble of inclusions subjected to a field of external forces is described by the same equations (10.4) and (10.7) which have been derived for the three-dimensional ensembles. The two-dimensional nature of the motion affects only the expression for the number of collisions $\mathfrak{N}(R_1, R_2)$. In contrast to the three-dimensional case [see Eq. (10.2)], this expression is not governed by the volume of a cylinder of base area $\pi(R_1 + R_2)^2$ and height $|\vec{v}_1 - \vec{v}_2|$, but by the area of a rectangle of base $2(R_1 + R_2)$ and of the same height, i.e.,

$$\mathfrak{N}(R_1, R_2) = 2(R_1 + R_2)|\vec{v}_1 - \vec{v}_2|. \tag{10.31}$$

Here, R is the radius of the circle formed by the intersection of the inclusion (which is not necessarily spherical) with the surface on which the motion takes place.

Following the three-dimensional case and using Eqs. (10.4), (10.7), and (10.31), we can easily estimate the characteristic time τ_{0S} for the doubling of the average radius of the inclusions. We shall assume that the velocity of the diffusional migration depends strongly on the radius, i.e., $|dv/dR| \sim v/R$ (this applies, for example, to the motion of inclusions associated with surface diffusion fluxes), and that the initial distribution of the radii of the inclusions is fairly wide. Then, bearing in mind that $\int f(R, t)dR \sim S_0 R_m^{-2}$, where S_0 is the initial relative area occupied by the inclusions, we find that

$$\tau_{0S} \sim \frac{R_m}{S_0 v_m}. \qquad (10.32)$$

Here, v_m is the velocity of those inclusions which have the most probable value of the radius R_m. For example, if $R_m \sim 3 \times 10^3$ Å, $S_0 \sim 3 \times 10^{-2}$, $v_m \sim 10$ Å/sec, we find that $\tau_{0S} \sim 10^4$ sec.

Obviously, the time τ_{0S} should be compared with the charteristic time τ_S'' of the competing coalescence mechanism, which is due to the surface diffusion of impurity atoms or vacancies between inclusions of different radii. It follows from the results reported in [87] that τ_S'' is of the order of

$$\tau_S'' \sim \frac{kT}{D_S' \gamma} \left(\frac{R_m}{a} \right)^4, \qquad (10.33)$$

where D_S' is the surface diffusion coefficient of impurity atoms far from an inclusion.

For example, if $R_m \sim 3 \times 10^3$ Å, $T \sim 1000°K$, $\gamma \sim 0.1$ mJ/cm² ($\sim 10^3$ ergs/cm²), and $D_S' < 10^{-8}$ cm²/sec, we find that $\tau_S'' > 10^4$ sec, i.e., it exceeds the value of τ_{0S} estimated above. Therefore, the coalescence of inclusions as a result of their motion is the dominant effect. This mechanism is favored by relatively large radii of inclusions and very strong external forces. It is also favored when the surface diffusion coefficient of the host atoms D_S at the inclusion–host interface is much greater than the surface diffusion coefficient D_S' (if the motion of inclusions is due to surface diffusion fluxes). Other favorable conditions include easy dissolution or evaporation of the host material and its easy transfer across the inclusion; in this case, the velocity of an inclusion should be limited by the conditions at the inclusion–host interface (the velocity of the inclusion is then $v \propto R$).

If surface diffusion fluxes play the dominant role in the motion of inclusions and if, in accordance with (10.1), the velocity of the inclusions is inversely proportional to their radii, we find that after a long time $t \gg \tau_{0S}$ the solution of the equations for the distribution function $f(R, t)$ tends to an asymptotic limit, which is obtained by applying the similarity principle. Applying the same reasoning as in the preceding subsections but using Eq. (10.31) for $\Re(R_1, R_2)$ and assuming that $h = 0$, we find that the volume of coalescing solid, liquid, or empty inclusions is conserved. The number of inclusions per unit area $N_{iS}(t)$, their average radius $\bar{R}_S(t)$, and the area $S(t)$ occupied by the inclusions are then given by the asymptotic formulas

$$\left.\begin{array}{c} N_{iS}(t) = \dfrac{1.51}{Ct} = 1.51 N_{iso}\dfrac{\tau_S}{t}, \quad \bar{R}_S(t) = 0.81 R_{S0}\left(\dfrac{t}{\tau_S}\right)^{1/3}, \\[2mm] S(t) = S_0\left(\dfrac{\tau_S}{t}\right)^{1/3}, \quad \tau_S = \dfrac{1}{N_{iso}C}. \end{array}\right\}(10.34)$$

Here, the characteristic time τ_S is of the same order of magtinude as τ_{0S}. In the formulas of Eq. (10.34), the quantities N_{iS0}, S_0, and $R_{S0} = (S_0/\pi N_{iS0})^{1/2}$ represent, respectively, the number of inclusions per unit area, the relative area occupied by the inclusions, and the initial mean-square radius.

Thus after a long time, the number of inclusions per unit area decreases in inverse proportion to the time, the average radius increases as $t^{1/3}$, and the relative area occupied by the inclusions decreases as $t^{-1/3}$ because the radius increases but the total area remains constant.

The following formulas apply to inclusions which are filled with an inert gas at the Laplace pressure (in this case, the square of the radius and not the volume remains invariant after coalescence):

$$\left.\begin{array}{c} N_{iS}(t) = \dfrac{0.95}{Ct} = 0.95 N_{iso}\dfrac{\tau_S}{t}, \\[2mm] \bar{R}_S(t) = 0.94 R_{S0}\left(\dfrac{t}{\tau_S}\right)^{1/2}, \\[2mm] S = \text{const}, \quad \tau_S = \dfrac{1}{N_{iS0}C}. \end{array}\right\}(10.35)$$

These formulas are valid for $t \gg \tau_S$ and show that the average radius increases proportionally to $t^{1/2}$.

Coalescence due to the Brownian motion may be important in the absence of external forces. An expression for the number of collisions $\mathfrak{N}(R_1, R_2)$ applicable to the two-dimensional Brownian motion is more complex because — in contrast to the three-dimensional case — the diffusion flux of the probability density in the direction of an inclusion does not tend to a constant limit for $D_i t \gg R^2$ but depends logarithmically on time. Nevertheless, if the condition $|\ln S_0| \gg 1$ is satisfied with logarithmic accuracy, $\mathfrak{N}(R_1, R_2)$ is given by the expression

$$\mathfrak{N}(R_1, R_2) = 4\pi \left[D_i(R_1) + D_i(R_2)\right] \ln^{-1}\left[\frac{D_i(R_1) + D_i(R_2)}{(R_1 + R_2)^2}\tau_{0S}\right]. \quad (10.36)$$

Equations (10.4), (10.7), (10.36), and (6.19) yield an estimate of the characteristic time τ_{0S} for the Brownian mechanism of coalescence when the motion of inclusions is due to surface diffusion fluxes:

$$\tau_{0S} \sim \frac{R_m^2}{S_0 D_S} \frac{R_m^4}{a^4}. \quad (10.37)$$

The ratio of this characteristic time to the time constant of the conventional coalescence of Eq. (10.33), $\dfrac{\tau_{0S}}{\tau_S^{''}} \sim \dfrac{R_m^2 \gamma}{S_0 kT} \dfrac{D_S'}{D_S}$, may be less than unity only if the ratio of the surface diffusion coefficients D_S' and D_S is very small. In this case, the coalescence mechanism associated with the Brownian motion may predominate if R_m is sufficiently small (i.e., when τ_{0S} is not too large). For example, if $R_m \sim 3 \times 10^{-6}$ cm, $S_0 \sim 10^{-2}$, $D \sim 10^{-5}$ cm^2/sec, it follows from Eq. (10.37) that $\tau_{0S} \sim 10^4$ sec. Asymptotic expressions for $N_{iS}(t)$, $S(t)$, $R_S(t)$, which apply after a long time has elapsed from the beginning of the coalescence process, are power functions of t [180].

11. INTERACTION BETWEEN INCLUSIONS

AND GRAIN BOUNDARIES

The directional motion in the field of external forces may drive some inclusions onto grain boundaries. In the absence of external forces, second-phase particles or pores can also migrate to grain boundaries as a result of the diffusion—elastic interaction between pores and grains (discussed in Sec. 8), or as a result of the Brownian motion (discussed in Sec. 6). Moreover, in some cases, the formation of a heterogeneous system (for example, the solidification of melts containing insoluble inclusions or the aging of solid solutions) may be accompanied by a preferential precipitation of inclusions on the grain boundaries.

When an inclusion comes into contact with a grain boundary, it interacts strongly with this boundary because of the grain-boundary surface tension forces (this interaction is different from the elastic interaction between a boundary and a distant inclusion, which has been discussed in Sec. 8). Therefore, an inclusion may become pinned to a grain boundary and a considerable force may be needed to detach it. Thus, a gradual accumulation of inclusions is likely at the grain boundaries.

The appearance of inclusions on grain boundaries may alter considerably the properties of these boundaries; for example, their effective viscosity may change or the boundaries may become brittle. Moreover, the migration of particles or pores from the interiors of grains to their surfaces will alter the properties of a crystal as a whole. The appearance of obstacles, in the form of inclusions attached to the grain boundaries, may hinder considerably the motion of these boundaries, i.e., the rate of recrystallization. If the inclusions are highly mobile, they will move together with the boundaries. This simultaneous motion of the inclusions and grain boundaries is sometimes responsible for the predominant influence of inclusions on the rate of recrystallization. On the other hand, the drag of inclusions by a grain boundary may have a considerable effect on the behavior of an inclusion ensemble because it may give rise to zones which are free of inclusions or it may enhance the rate of collisions and coalescence of the inclusions near a moving boundary, etc.

Interaction of Inclusions with Grain Boundaries and the Effective Viscosity of Grain Boundaries in Heterogeneous Systems

We shall analyze the interaction between a spherical inclusion and a grain boundary by considering the simplest case in which we can ignore the dependence of the surface energy γ' of the inclusion—host interface on the orientation relative to the crystallographic axes. We shall also assume that the contribution of the energy of the elastic stresses around an inclusion is unimportant (this is true, for example, in the case of gas-filled pores for which the Laplace pressure is balanced by the pressure of the enclosed gas so that the total pressure P is zero).

The magnitude of the force of interaction of an inclusion with a grain boundary depends on the mobility of this boundary and on the duration of the process being considered. If the boundary is sufficiently mobile, it should become distorted near the inclusion in such a way that it becomes perpendicular to the surface of the inclusion.

The force of interaction can be estimated approximately by assuming that a unit length of the contact between the grain boundary and the inclusion is subject to the grain-boundary surface tension force γ', which is directed normal to the surface of the inclusion. The projection of this force onto the normal to the grain boundary $\vec{e_z}$ (the unit vector $\vec{e_z}$ is directed from the center of the spherical inclusion to its surface) is $\gamma' \cos \theta$. The sum of the forces applied to all parts of the line of contact of length $2\pi R \sin \theta$ is [88]

$$\vec{F} = \pi R \gamma' \sin 2\theta \vec{e_z}. \qquad (11.1)$$

This force has its maximum value $F_m = \pi R \gamma'$ for $\theta = 45°$.

The force $\vec{F}$ can be estimated more rigorously by calculating the change in the grain-boundary surface energy U resulting from the contact of an inclusion with this boundary. Then, the force can be found as the derivative of U. The change in the surface energy is due to a reduction in the area of the surface separating the grains by an amount equal to the circular area occupied by the inclusion, and to an increase in the area of the same surface as a result of its distortion. A calculation carried out on the assumption that the cross section of the distorted surface is hyperbolic [89] yielded

$$F_m = 3.96 R\gamma', \qquad (11.2)$$

which is somewhat larger than the value given by Eq. (11.1).

If, in a given range of temperatures, a grain boundary is practically immobile, an estimate of the force of interaction must be modified by an allowance for the fact that the appearance of a spherical particle whose center is located at a distance $U = \pi \gamma '(R^2 - L^2)$ ($U = 0$ for $L \geq R$). Consequently, an inclusion near a grain boundary is acted on by a force $\vec{F} = -\dfrac{\partial U}{\partial L}\vec{e}_z$ which is equal to

$$\vec{F} = 2\pi\gamma'L\vec{e}_z = 2\pi\gamma' R\cos\theta\vec{e}_z \quad (L \ < \ R),$$
$$\vec{F} = 0 \quad (L > R).$$
(11.3)

Equations (11.1)-(11.3) have been derived on the assumption that the thickness of the grain boundary is considerably less than the dimensions of the inclusion. It is evident from these formulas that the maximum force F_m is of the same order of magnitude for both flexible and rigid grain boundaries.

When an allowance is made for the elastic energy in the host crystal, the force $\vec{F}$ does not vanish when $L > R$. For example, in the case of empty pores ($P = -2\gamma/R$), in an elastically anisotropic crystal, the elastic interaction of a pore with a grain boundary between strongly disoriented grains (considered in Sec. 8) gives rise, for $L \sim R$, to a force $|\vec{F}| \sim \gamma R$, which is comparable with the force given by Eq. (11.2). However, this force decreases rapidly with increasing L in direct proportion to $(R/2L)^4$.

A considerable external force is needed to overcome the force given by Eq. (11.2) or (11.3) and to detach an inclusion from an immobile grain boundary. Thus, if the external force is due to a field of inhomogeneous stresses, the force F_n can be overcome by stress gradients of the order of $|\nabla\sigma_{ii}| \sim F_m/4R^3 \sim \gamma'/R^2$. If $\gamma' \sim 1$ J/m^2 (10^3 ergs/cm^2) and $R \sim 10^{-5}$ cm, we find that $|\nabla\sigma_{ii}| \sim 10^{14}$ J/m^4 (10^{13} ergs/cm^4). Therefore, for grains of dimensions $L_g \sim 10^{-4}$ cm in the case when σ_{ii} are of the order of $|\nabla\sigma_{ii}|L_g$, the stresses $\sigma_{ii} \sim 10^8$ J/m^3 (10^9 ergs/cm^3) are relatively high and may exceed the flow stress.

If the forces are due to a temperature gradient, a force F_m can be overcome if the gradient $|\nabla T|$ is of the order of $|\nabla T| \sim \dfrac{\omega}{10\alpha k}\dfrac{F_m}{R^3} \sim \dfrac{\omega\gamma'}{\alpha k R^2}$. For example, if $\gamma' \sim 1$ J/m^2 (10^3 ergs/cm^2),

$\omega \sim 10^{-23}$ cm^3, $\alpha \sim 10$, R $\sim 10^{-5}$ cm, the value of $|\nabla T| \sim 10^5$ deg/cm is greater than the usual temperature gradient (however, if R $\sim 10^{-4}$ cm, $|\nabla T| \sim 10^3$ deg/cm, which is a more reasonable value). Therefore, a considerable fraction of particles or pores may not be detached from the grain boundaries by the diffusional migration of inclusions in a field of external forces and the structure of a disperse system may change considerably.

Inclusions pinned to a grain boundary may hinder the slip of grains along the boundaries as well as normal translation. We shall start by considering the influence of inclusions on the slip of grains, which is governed by the effective viscosity of the grain boundaries [10].

The difference between the stresses acting on the front and rear boundaries of a grain, $\Delta\sigma$, gives rise to a tangential shear force $\sim L_g^2 \Delta\sigma$, which acts on the lateral boundaries (we shall assume that the dimensions of the grains are of the same order of magnitude along all directions) and to tangential stresses $\sigma_i \sim \Delta\sigma$. If the crystal is free of inclusions, the action of these stresses at high temperatures causes a grain to slip at a velocity v_g relative to the next grain. The velocity v_g is proportional to σ_t, where

$$\sigma_t = \frac{\eta_S}{a} v_g . \tag{11.4}$$

Here, η_S is the viscosity of the grain boundary which is of the following order for a strongly disoriented grain [48]:

$$\eta_S \sim kT/Da. \tag{11.5}$$

In heterogeneous systems, the inclusions pinned to grain boundaries hinder the slip of the grains and may increase considerably the effective viscosity η_S. If the viscosity of the boundary, given by Eq. (11.5), can be ignored, the tangential shear forces $\sim L_g^2\Delta\sigma$ acting on the lateral boundaries are distributed among $\sim L_g^2/\rho_0^2$ inclusions pinned to these boundaries (ρ_0 is the average distance between the inclusions). Each inclusion is subject to a force $F' \sim \rho_0^2\Delta\sigma$. If this force exceeds the force F_m representing the interaction between an inclusion and the grain boundary, the solid inclusions become detached from the boundaries and easy slip may take place.

In the case of liquid or gas-filled inclusions, we find that $F' > F'_m$, where $F'_m \sim U'/R \sim F_m$ $(U' \sim \pi R^2 \gamma ')$, that a gliding boundary may intersect the inclusions, and that easy slip can still take place. However, in some cases, the condition $F' > F_m$ is satisfied only if the stresses $\Delta\sigma$ are sufficiently high. For example, if $\gamma' \sim 1$ J/m^2 (10^3 ergs/cm^2), $R \sim 10^{-5}$ cm, $\rho_0 \sim 10^{-4}$ cm, the necessary stresses are $\Delta\sigma \sim 10^8$ J/m^3 (10^9 ergs/cm^3).

If the stresses are not very high and $F' < F'_m$, inclusions will neither be expelled nor intersected, and at low temperatures it may become impossible for a grain boundary to slip. At high temperatures, such slip may reduce to diffusion of the host matter around the inclusions. Bearing in mind that $|\nabla\sigma| \sim F'/R^3 \sim \Delta\sigma\rho_0^2/R^3$ near an inclusion and using Eqs. (5.22) and (5.24), we can readily show that the velocity of a grain boundary due to such diffusion around inclusions and the corresponding effective viscosity of a grain boundary $\eta_S \sim a\Delta\sigma/v_g$ are of the order of

$$v_g \sim D \frac{\omega\Delta\sigma}{kT} \frac{\rho_0^2}{R^3} , \qquad \eta_S \sim \frac{kT}{Da} \frac{R^3}{a\rho_0^2} \qquad (11.6)$$

if the volume diffusion in the host crystal is the predominant mechanism and

$$v_g' \sim D_S \frac{\omega\Delta\sigma}{kT} \cdot \frac{\rho_0^2 a'}{R^4} , \qquad \eta_S \sim \frac{kT}{D_S a} \cdot \frac{R^4}{a^2 \rho_0^2} \qquad (11.7)$$

if the grain-boundary diffusion is the principal mechanism. The effective viscosities corresponding to these two processes may exceed considerably the viscosity of a grain boundary in an inclusion-free crystal, given by Eq. (11.5), provided the dimensions of the inclusions and their density on the grain boundary are sufficiently high [in the case represented by Eq. (11.6), this is true if $R/a \gg (\rho_0/R)^2$].

Inclusions reduce not only the effective viscosity of the grain boundaries but also the grain-boundary peak of internal friction. The influence of inclusions on the effective viscosity of grain boundaries was also considered in [181-183] and the results were in order-of-magnitude agreement with the estimates given in [10].

Drag of Inclusions by a Moving Grain

Boundary

The recrystallization of polycrystalline aggregates is a result of the normal translation of grain boundaries, so that some grains grow at the expense of others. The velocity of a grain boundary under these conditions is proportional to the force ΔG which acts on a unit area of the boundary. In the recrystallization of plastically deformed crystals, the force ΔG is the difference between the energies (per unit volume) of a plastically deformed grain and a neighboring unstressed grain.

In undeformed polycrystalline aggregates, the average dimensions of the grains increase due to a reduction of the grain-boundary energy as a result of the growth of larger grains at the expense of smaller ones. Since the relative change in the area of a grain boundary resulting from its displacement is inversely proportional to the radius of the grain R_g, the force acting per unit area ΔG can be described by the approximate formula [89,90]:

$$\Delta G = \zeta' \gamma' \left(\frac{1}{R_{ga}} - \frac{1}{R_g} \right), \tag{11.8}$$

where ζ' is a dimensionless constant of the order of unity and R_{ga} is a quantity of the order of the average grain size (it differs from this grain size by a numerical factor).

In a more rigorous treatment, Eq. (11.8) should include the radii of neighboring grains and not the average radius. However, this does not affect order-of-magnitude estimates.

The presence of inclusions on grain boundaries may affect considerably the growth of the grains. If, in a given range of temperatures, the velocity of diffusional migration of inclusions is low, such inclusions act as obstacles to the motion of the grain boundaries. These boundaries remain immobile as long as the force f which tends to move the boundary is less than the maximum restoring force $n_i F'_m$ exerted by n_i inclusions which are encountered in a unit area of the boundary. For example, if the grains in an undeformed polycrystalline sample grow as a result of the force given by Eq. (11.8), sufficiently small grains can overcome the force $n_i F'_m = 2N_i RF'_m = \dfrac{3}{2\pi} pF'_m/R^2$ exerted by the inclusions distributed randomly in the crystal in such a way that their volume concentration is p [if the number of inclusions per unit volume is $N_i = p\left(\dfrac{4}{3}\pi R^3\right)^{-1}$, there are, on the average, $2RN_i$ inclusions per unit area].

However, beginning from a certain average grain radius R_g^*, the force $\Delta G = \zeta''\gamma\,'/R_g^*$ ($R_{ga}^* \sim R_g^*$ and $\zeta'' \sim \zeta'$) becomes equal to the force $n_i F'_m$ and then the grains with the usual distribution of dimensions (close to R_g^*) cease to grow. The average grain size R_g^* is then found to be related, in a simple manner, to the volume concentration of inclusions and their radius R [90, 89]:

$$R_g^* = \zeta_1 \frac{R}{p}, \tag{11.9}$$

which is derived on the assumption that $\zeta_1 \sim 1$ and that Eq. (11.2) applies.

Sometimes the distribution of the grain dimensions is not concentrated within a narrow range near R_g^* and the crystal contains grains whose size is much larger than the average value. In this case, the growth of almost all the grains is blocked by the precipitates on their boundaries and the forces ΔG exceeds the restoring force $n_i F$ for a few of the largest grains which grow at the expense of their neighbors, whereas the other grains remain unaffected (this is known as the anomalous growth and is discussed in detail in [89, 90]).

If, in a given range of temperatures, the velocity of the diffusional migration of inclusions is considerable, the mechanism of the effect of inclusions on the motion of the grain boundaries is different. This mechanism is related to the diffusion drag of inclusions by grain boundaries. At sufficiently high values of ΔG, a grain boundary becomes detached from the inclusions. Even if ΔG is small, the boundary continues to move but the velocity is now limited by the velocity of the inclusions [55, 91].

We shall consider first the case when the velocity of the inclusions is much less than that of the grain-boundary migration in an inclusion-free crystal. The velocity of the grain boundaries is limited by the presence of inclusions and, in the first approximation, we can regard the boundary as immobile. The force ΔG

which causes the motion of the boundary in the absence of inclusions is balanced out almost exactly by the force $n_i F$ which
results from the presence of n_i inclusions on a unit surface area
of the boundary (this is true in the case when $F = \Delta G/n_i < F_m$).

Each inclusion is thus acted on by a force $F = \Delta G/n_i$. The
forces acting on the inclusions and on different parts of the grain
boundary give rise to an inhomogeneous distribution of stresses,
which is responsible for the diffusional migration of inclusions.
These stresses have a complex angular distribution. The second
and higher harmonics of the stresses distort a moving pore or increase the density of the host crystal near a solid inclusion,
whereas the first harmonic causes the migration of the inclusion
as a whole. To within a numerical factor ξ', of the order of unity,
the velocity of such migration under the action of the force F can
be found from the formulas given in Sec. 6, which are derived
for the simpler case of external forces which are applied only to
an inclusion and which are characterized by a simple volume distribution.

If the motion of an inclusion is mainly due to diffusion fluxes
in the bulk of the host crystal, it follows from Eq. (6.7) that the
velocity of this inclusion — equal to the velocity of the associated
grain boundary v_g — is given by the formula

$$v_g = \frac{3\Gamma'_1 \xi'}{2\pi f} \frac{D\omega}{kT} \frac{\Delta G}{n_i R^3} \sim \frac{D\omega}{kT} \frac{\Delta G}{n_i R^3}. \tag{11.10}$$

If the inclusion migrates mainly as a result of diffusion fluxes in
its interior, it follows from Eq. (6.12) that

$$v_g = \frac{3\xi'}{4\pi} \frac{D_A c_A^0 \omega}{f_A kT} \left(1 - \frac{f_A \omega_A}{\omega'} + \frac{f_A \Gamma}{3} \frac{\omega}{\omega'} \right) \frac{\Delta G}{n_i R^3} \sim \frac{D_A c_A^0 \omega^2}{10 kT\omega'} \frac{\Delta G}{n_i R^3}. \tag{11.11}$$

Finally, if the radius of the inclusion is sufficiently small and its
migration is primarily due to surface or boundary diffusion, it follows from Eq. (6.8) that

$$v_g = \frac{\Gamma \xi'}{2\pi f_S} \frac{D_S a\omega}{kT} \frac{\Delta G}{n_i R^4} \sim \frac{1}{10} \frac{D_S a\omega}{kT} \frac{\Delta G}{n_i R^4}. \tag{11.12}$$

According to Eq. (11.10)-(11.12), the influence of temperature on the velocity of a grain boundary is determined primarily

by the exponential temperature dependences of the factors D, D_A, c_A^0, and D_S. If, in a given temperature range, several migration mechanisms are of comparable importance, the temperature dependence of the velocity of the grain boundary becomes much more complex.

If the migration of an inclusion is due to diffusion fluxes in its interior or in the interior of the host crystal, the velocity v_g is proportional to $(n_i R^3)^{-1}$ and is determined solely by the total volume of inclusions on the grain boundary. When these inclusions coalesce, the value of v_g does not change. However, if the migration of an inclusion is due to surface diffusion fluxes, we find that the velocity $v_g \propto (n_i R^4)^{-1}$ is reduced by the coalescence of inclusions.

Inclusions are usually present not only on the boundary but also in the interior of a grain and a moving boundary captures new inclusions, leaving behind an inclusion-free zone. Therefore, the surface density of inclusions n_i increases with time, in accordance with the law $n_i = n_i^0 + N_i s$, where n_i^0 is the initial density of inclusions on the boundary and s is the path traveled by the boundary. Substituting this expression for n_i into the formulas (11.10)-(11.12) for the velocity $v_g = ds/dt$, we find that this velocity gradually decreases. If the initial density n_i^0 can be ignored compared with $N_i s$, we find that the velocity and the path vary in accordance with the following simple laws [55, 92]:

$$v_g = \sqrt{\frac{B \Delta G}{2 N_i t}}, \quad s = \sqrt{\frac{2B}{N_i} \Delta G t}, \tag{11.13}$$

where B is the coefficient of $\Delta G / n_i$ in Eqs. (11.10)-(11.12) for v_g and it is assumed that ΔG is constant in the range of time intervals in question.

The capture of inclusions by moving grain boundaries and the formation of inclusion-free zones in the interior of grains may result in the embrittlement of the boundaries and a reduction in the bulk value of the flow stress. The collapse of small grains as a result of the motion of their boundaries may produce clusters of inclusions [93].

If ΔG is sufficiently large so that the values of v_g given by Eqs. (11.10)-(11.12) become comparable with v_g^0, the velocity of the inclusions becomes of the same order as the velocity of a grain boundary v_g^0 in the absence of inclusions. In this case, the force acting per unit area of the grain boundary is not ΔG but $\Delta G - n_i F$ and the force F acting on each inclusion is related to the velocity of an inclusion by (see also [91])

$$v_g = v_g^0 \frac{\Delta G - n_i F}{\Delta G} = v = BF. \qquad (11.14)$$

Hence, it follows that

$$F = \frac{v_g^0 \Delta G/n_i}{v_g^0 + B \dfrac{\Delta G}{n_i}}, \quad v = v_g^0 \frac{B\Delta G/n_i}{v_g^0 + B\Delta G/n_i}. \qquad (11.15)$$

If $\Delta G > F_m (n_i + B\Delta G/v_g^0)$, a grain boundary breaks away from the inclusions. Since inclusions usually have a range of sizes, the boundary first becomes detached from the larger inclusions and then, when the velocity has increased sufficiently, from smaller inclusions. It should be noted that at high velocities of the grain boundary $v_g \sim v_g^0$ the presence of inclusions may affect the coefficient in the dependence of v_g^0 on ΔG [94].

Experimental Investigations of the Motion

of Inclusions Pinned to Grain Boundaries

Interactions between macroscopic inclusions and grain boundaries are revealed by many indirect consequences such as the inhibition of secondary recrystallization, changes in the creep kinetics, influence on the grain-boundary peak of internal friction, etc. Some of these will be considered in Chap. III. In the present section, we shall discuss the results of direct observations of the motion of a grain boundary which carries macroscopic inclusions or encounters them during its motion.

We have already said that inclusions located on grain boundaries can impede the grain-boundary flow (the sliding or shearing of grain boundaries along one another) and grain-boundary migration (motion at right angles to the boundary, encountered, for example, in secondary recrystallization).

In grain-boundary flow, the presence of inclusions alters the effective viscosity of the boundaries. If the stresses do not exceed the linear creep threshold [95], the crystal deforms as a Newtonian body ($\varepsilon \propto \sigma$). In this case, the kinetics of deformation is controlled by continuous diffusion fluxes within the structure elements separated by boundaries, which act as vacancy sources and sinks. The necessary condition for the Nabarro−Herring−Lifshits mechanism [46-48] is the feasibility of the viscous sliding of structure elements along the boundaries separating them [48]. In the absence of inclusions, the viscosity of a boundary in the Nabarro−Herring−Lifshits mechanism is given by Eq. (11.5).

The mechanism of diffusional creep can be observed in its pure form if the viscosity of a boundary is considerably less than the diffusional viscosity of a structure element, which is related to the dimensions of this element by $\eta \propto L^2$. It follows from Eq. (11.6) that macroscopic inclusions located on a boundary can increase its viscosity quite considerably. Consequently, they may impede or practically prevent continuous deformation of structure elements in a real crystal.

This effect was observed clearly by Ziling and Grankin [96], who studied experimentally the creep of the Cu + 1% Al alloy. Disperse particles of γ-Al_2O_3 formed in this alloy as a result of internal oxidation. These particles were located mainly on grain boundaries. The creep tests were carried out at 1000°C, applying stresses of σ = 50-30 N/m^2 (5-3 kgf/cm^2), which were known to be below the linear creep threshold [95]. When pure copper was tested under similar conditions [97], the diffusional creep was observed clearly. However, in samples containing γ-Al_2O_3 inclusions, the diffusional creep was suppressed by the presence of inclusions which increased the viscosity of the grain boundaries. Obviously, the importance of diffusional creep should increase as the amount of the disperse phase decreases.

Disperse inclusions located on the boundaries between structure elements make it difficult to study diffusional creep because their presence alters the creation and absorption of vacancies by grain boundaries to processes of the threshold type [144-145].

The drag of macroscopic inclusions by a moving grain boundary was observed clearly by Koch and Aust [98], who studied the motion of carbide particles $(CrFe)_{23}C_6$ in austenitic stainless steel. In these experiments, prestrained samples were annealed at 900°C for 2 h. Such annealing gave rise to recrystallization grain growth accompanied by the motion of boundaries on which carbide in-

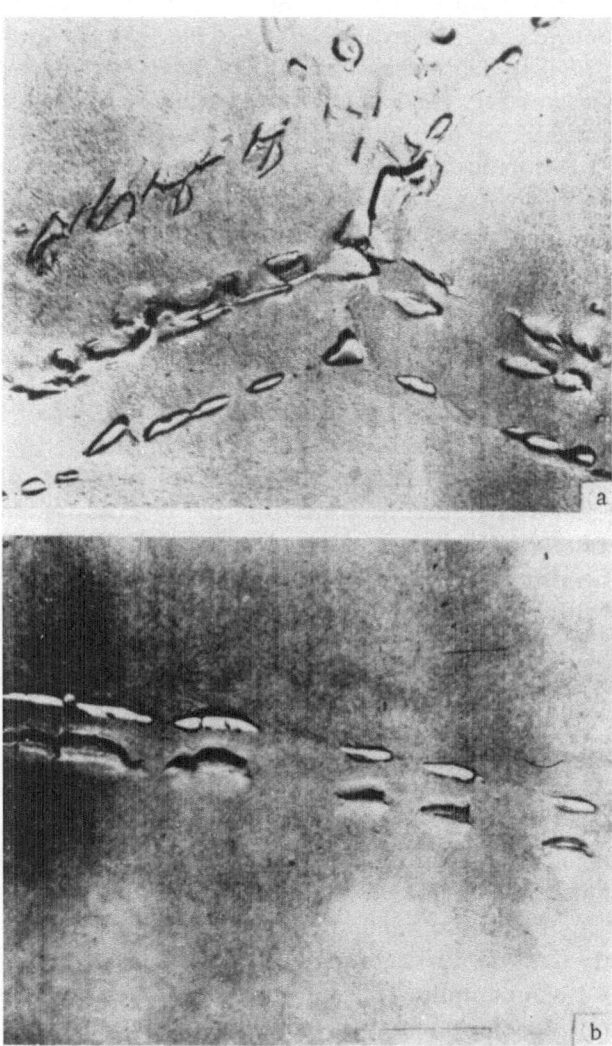

Fig. 35. Electron micrographs demonstrating the motion of a grain boundary and the associated carbide inclusions in austenitic steel [98]: a) motion of a triple contact; b) motion of a single boundary.

clusions were located (the linear dimensions of these inclusions were of the order of 5×10^{-5} cm). The structure of the boundaries with associated inclusions was investigated by the electron-microscopic replica and shadow technique. The electron micrographs (Fig. 35) indicated that a moving boundary dragged the inclusions located on it. The impurities left behind at the points occupied orginally by carbide inclusions were visible in the replicas of etched surfaces. This made it possible to determine the initial and final positions of a chain of inclusions located on a grain boundary. Unfortunately, Koch and Aust [98] did not report the data which would be useful in estimating the force responsible for the motion of the boundary and, therefore, we found it impossible to deduce the mechanism of the transport of matter responsible for the motion of carbide inclusions.

A detailed study of the drag exerted on second-phase particles by moving grain boundaries was carried out by Ashby, Centamore, and Palmer [92, 93] (see also [94]). Disperse particles of SiO_2, GeO_2, B_2O_3, and Al_2O_3 were produced by the internal oxidation of copper containing silicon, beryllium, boron, and aluminum impurities. The linear dimensions of these particles were $\sim(1-4) \times 10^{-5}$ cm and the relative volume occupied by them was $\sim(1-8) \times 10^{-3}$. At the temperatures employed in [92, 93], these particles were in different structure states: SiO_2 was amorphous, GeO_2 and B_2O_3 were liquid, and Al_2O_3 was crystalline.

The local deformation of a sample necessary to produce a region in which the motion of grain boundaries occurred during secondary recrystallization was produced by an indenter. The application of a concentrated load is a very convenient technique in investigating moving boundaries carrying foreign inclusions. Away from the point of application of such a load, the local strain decreases monotonically and the average grain size increases (Fig. 36). Consequently, a given sample will contain grain boundaries which need different forces to be set in motion and the magnitude of these forces decreases away from the indenter. Consequently, one can study the interaction between inclusions and grain boundaries moving at different velocities. The displacements of oxide particles observed experimentally in [92, 93] were small compared with the size of the deformed region (the ratio was $\sim 10^{-2}$). Therefore, it was reasonable to assume that a grain boundary exerted a practically constant drag force on the inclusions associated with it.

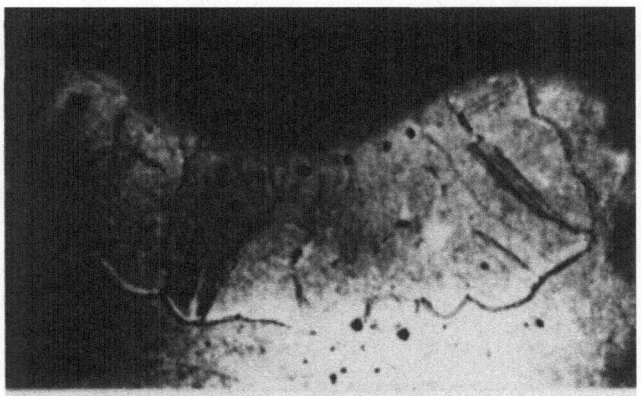

Fig. 36. Structure and schematic representation of a part of a copper sample deformed by an indenter (the structure is that obtained after annealing at 1000°C for 2 h) [92].

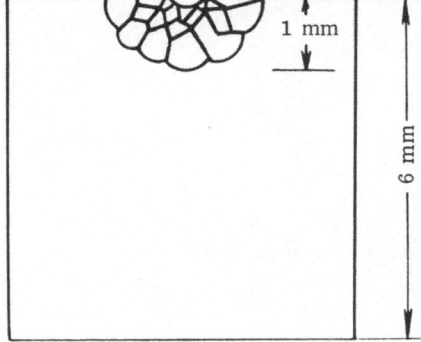

The micrographs shown in Figs. 37 and 38 indicate that high-temperature annealing produced inclusion-free zones near the grain boundaries. These zones could result from the dissolution of the inclusions located at some distance from a boundary, followed by the diffusion of the inclusion atoms and their condensation on the boundary.

This process was possible but it did not predominate in the experiments described in [92, 93] because inclusion-free zones were found to be located on one side of a grain boundary, whereas the dissolution−condensation process should have produced zones located symmetrically on both sides of the boundary. Experiments on copper containing SiO_2 particles established that an increase in the number of inclusions located on a boundary, n_B, was only slightly less than the number of inclusions in a depleted zone n_Z (Table 2). This was due to the drag exerted on the inclusions

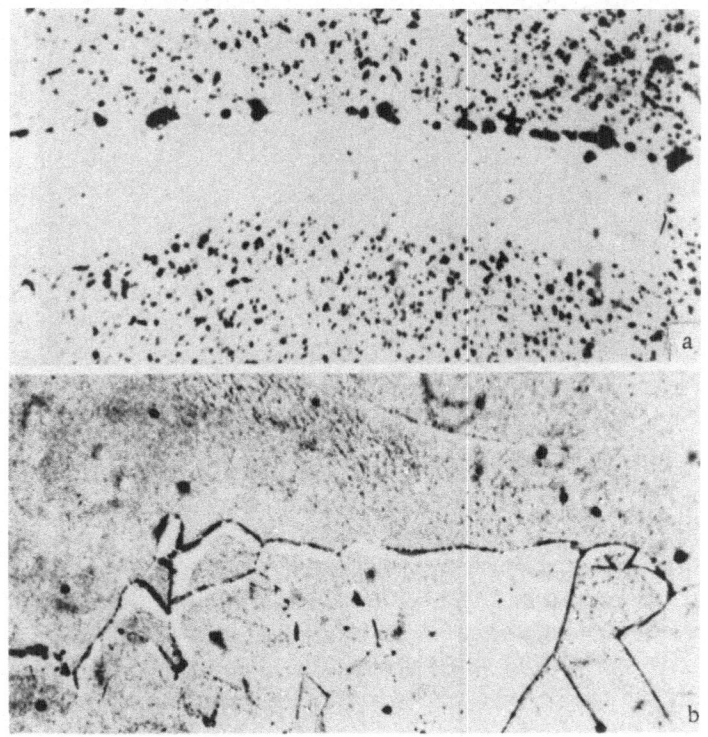

Fig. 37. Inclusion-free zone behind a moving boundary [93]: a) ×1000; b) ×120.

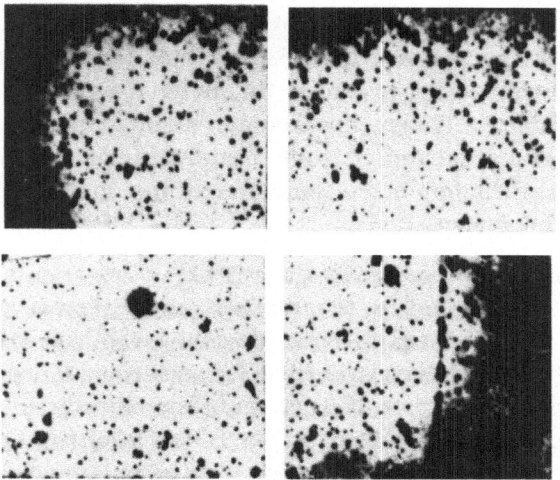

Fig. 38. Micrographs showing motion of a grain boundary in copper containing SiO_2 inclusions. T = 950°C [92].

TABLE 2. Drag of SiO_2 Inclusions by Moving Grain
Boundaries in Copper

Duration of annealing at 950°C, min	Average number of SiO_2 particles per 10 μ of boundary length	Average number of SiO_2 particles dragged away from boundary zone 10 μ long
110	22	27 ± 6
170	33	37 ± 7
260	47	46 ± 8
440	52	68 ± 13

by a moving boundary. The small difference between n_B and n_Z could be a consequence of the coalescence of inclusions located on a boundary. The possibility of this process is suggested by the results shown in Fig. 37, which demonstrates that some of the inclusions on a boundary were larger than in the interior of a grain.

Ashby and Centamore [92] found that the width S of a zone depleted of inclusions and located near a boundary varied parabolical-

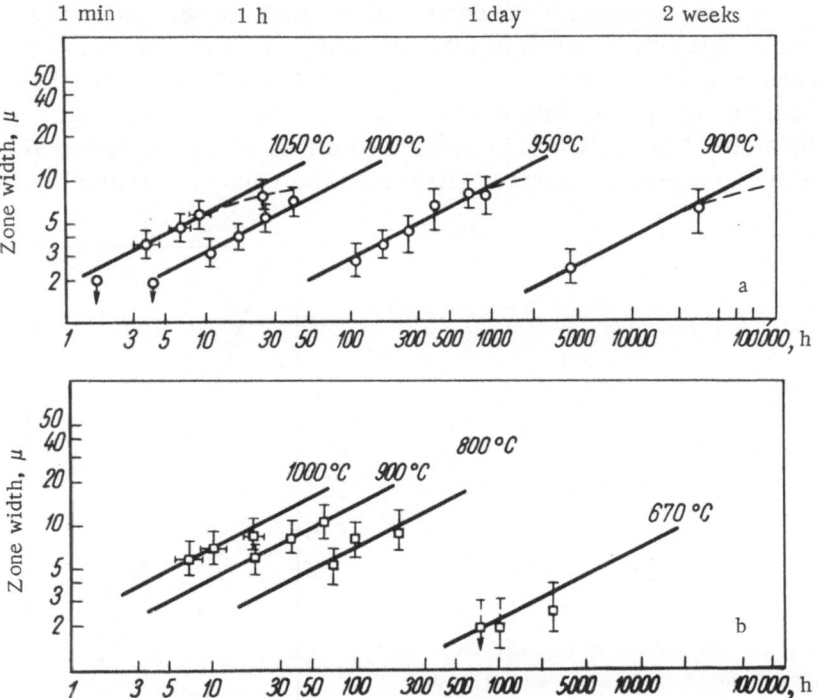

Fig. 39. Dependence of the width of an inclusion-free zone behind a moving grain boundary on the duration of isothermal annealing at various temperatures: a) Cu—SiO_2 system; b) Cu—GeO_2 system [92].

ly with time, in accordance with Eq. (11.13) (Fig. 39). The dependence of the displacement of a boundary on the radius of inclusions could be approximated by a power function (Fig. 40).

An important observation was made in [92, 93]: an inclusion-free zone was found to form at some distance from the point of application of an indenter. A boundary moving in the most strongly deformed region did not drag the inclusions; it became detached from them and left them behind. This feature of the interaction of a moving grain boundary with inclusions was considered in the preceding subsection.

The mechanism controlling the motion of inclusions can be determined by investigating the temperature dependence of their mobility. It is evident from Eq. (11.14) that the mobility of the inclusions is determined by a constant B, which — according to Eq. (11.13) — can be found from the time dependence of the displacement of a boundary S at various temperatures.

The temperature dependences of the mobility of oxide inclusions in copper, obtained in [92, 93], are plotted in Fig. 41. The shaded regions in this figure correspond to the dependences B(T) calculated on the assumption that the transport of mass in the migration of an inclusion is controlled either by the diffusion along the inclusion—host interface (B_S) or by the volume diffusion in the

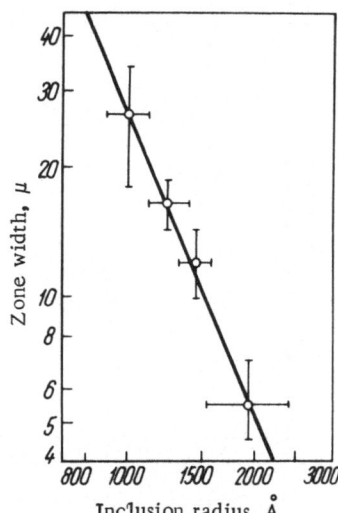

Fig. 40. Dependence of the width of an inclusion-free zone on the radius of inclusions (Cu—SiO$_2$ system) [92].

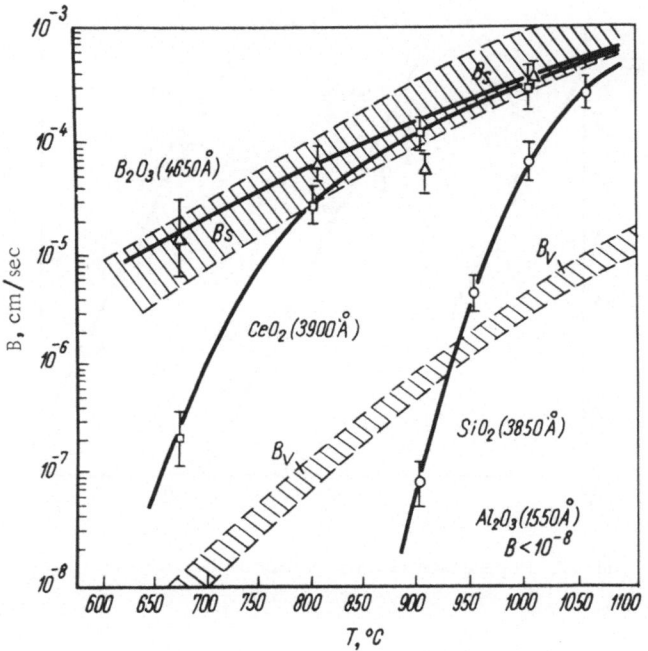

Fig. 41. Calculated and experimental temperature dependences
of the inclusion mobility B [92].

host crystal (B_V). An allowance for the small scatter of the inclusion radii is made in these calculations. We can see that neither
of these two diffusion mechanisms predominates in the motion of
B_2O_3, GeO_2, or SiO_2 inclusions.

It follows from Fig. 42 that the mobility of these particles and
the viscosity of the corresponding liquid or amorphous oxides, determined in independent experiments [99, 100], are related by B ∝
$1/\eta$. This gives us grounds for assuming that the motion of oxide
inclusions is due to the diffusion—viscous transport of matter within the inclusions.

Al_2O_3 particles were found to be practically immobile. This
could be due to the complete coherence of the inclusion—host interfaces, which would inhibit the formation and disappearance of
vacancies at this interface (vacancies are essential for the motion
of an inclusion, as discussed in Sec. 1). These interfaces were
coherent if the Al_2O_3 particles were formed by internal oxidation.

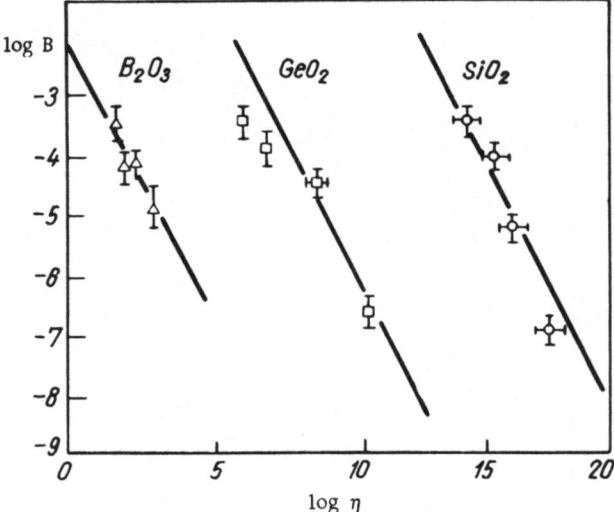

Fig. 42. Dependence of the mobility of inclusions on their viscosity [92].

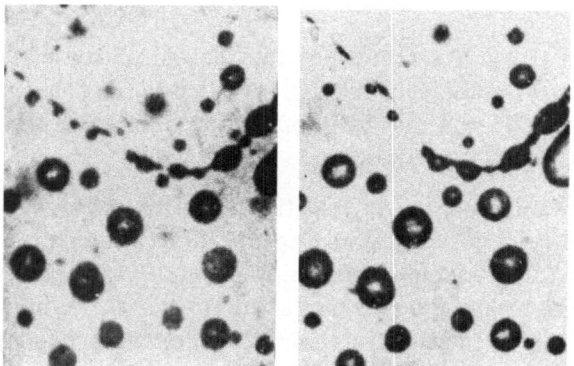

Fig. 43. Drag of grain boundaries by gas-filled bubbles in polycrystalline camphor subjected to a temperature gradient $|\nabla T| = 6.1$ deg/cm. a, b) Different stages separated by $\Delta t = 19$ h. The right-hand part of the boundary is dragged by a larger number of bubbles and it moves faster than the left-hand part. ×540.

When the Al_2O_3 inclusions were introduced into nickel or silver by the mechanical mixing of the appropriate powders, it was found that the interfaces were incoherent and the inclusions were easily dragged by moving the grain boundaries.

Other workers [184] carried out experiments similar to those described above. They studied the drag exerted by moving boundaries on 2×10^{-5} cm helium-filled bubbles in polycrystalline copper. An analysis of the time and temperature dependences of the velocity of these bubbles led to the conclusion that the kinetics of the process was to a great extent governed by the nucleation of steps on the free surfaces of the bubbles and that the motion of these steps was essential for mass transport. A similar hypothesis was put forward in [177] in connection with the Brownian migration of helium filled bubbles in copper and gold.

The difference between the free energies of atoms in neighboring grains, which is needed to ensure that the motion of a grain boundary is accompanied by the drag of foreign inclusions, may be generated by means other than mechanical strain. For example, it may be due to the neutron irradiation of crystals. If α uranium is irradiated at 300°C with a neutron flux of $\sim 10^{13}$ cm$^{-2} \cdot$ sec^{-1} for ~ 1 h, the crystal accumulates an energy of ~ 3 TJ/cm^3 ($\sim 3 \times 10^{19}$ ergs/cm^3). At the boundary between the irradiated and unirradiated regions, the motion of the grain boundaries during recrystallization is subject to considerable stresses, amounting to ~ 300 MN/m^2 ($\sim 3 \times 10^9$ dyn/cm^2). Such boundaries can drag gas bubbles of $\sim 3 \times 10^{-6}$ cm radius [91].

Simultaneous motion of inclusions and grain boundaries can also occur when inclusions subjected to an external force field drag a grain boundary with them. When a force F_i is applied to inclusions, a grain boundary with n_i inclusions per unit area experiences a force $F = n_i F_i$. If the boundary is planar, it experiences the same force as a grain boundary bent to a radius $R \sim \gamma'/n_i F_i$ during spontaneous secondary recrystallization.

The drag of grain boundaries by migrating inclusions was observed experimentally in polycrystalline camphor [175], containing gas-filled bubbles subjected to a temperature gradient. It is clear from Fig. 43 that the part of the boundary with a higher density of bubbles is migrating faster.

Two-Component Solids

12. MIGRATION OF INCLUSIONS IN THE FIELD OF EXTERNAL FORCES IN SOLID SOLUTIONS

The migration of inclusions in two-component crystals (solid solutions, ionic crystals, and other compounds) is, in many respects, different from their migration in crystals which contain atoms of one kind only. In the case of two-component crystals, an inclusion may migrate under the influence of diffusion fluxes of atoms of both types, and the mobilities of these atoms in an external force field may be different. Therefore, gradients of concentrations of atoms or defects (or an additional inhomogeneous force field) are needed to ensure continuous diffusion fluxes of different kinds of atoms so that the rate of arrival or departure of atoms of different kinds on the surface of a moving dislocation are equal.

The kinetic interaction between atoms of different kinds resulting from the appearance of concentration gradients may have a considerable influence on the motion of inclusions in external force fields in solid solutions. In particular, the velocity of inclusions may be changed considerably by the addition of a small number of impurity atoms to a solid solution. Moreover, apart from the motion of inclusions under the action of external forces, considered in the preceding chapter, inclusions can migrate in solid solutions in the field of a concentration gradient. Moreover, nonlinear effects are much more important in solid solutions than in one-component crystals. These points will be discussed in Secs. 12-14, where we shall consider metallic and covalent solid solutions. We shall discuss next the migration of inclusions in ionic crystals (Sec. 15), in which some special features associated with the charge of ions will be considered.

Diffusion Fluxes in Solid Solutions

In the case of the vacancy mechanism, fluxes of A and B atoms in a binary solid solution A−B result from the elementary processes of the exchange of atoms with neighboring vacancies. The magnitudes of these fluxes depend on external forces and on the concentration gradients of atoms or vacancies. Explicit expressions for diffusion fluxes depend on the ratio between the distance l separating vacancy sources and sinks and the characteristic lengths in which the concentration and external forces change significantly. In the problems that will be considered here, the characteristic length is usually the inclusion radius R.

If a crystal or grain has very few defects and no vacancy sources and sinks in its interior, or if the distances l are large compared with R, the concentration of vacancies at a given point is not governed by the concentration of atoms, the temperature, and the stresses at that point. Therefore, a vacancy concentration gradient should be regarded as an independent thermodynamic quantity (a generalized external thermodynamic force) which causes atoms to diffuse. In this case, the diffusion fluxes $\vec{I_1}$ and $\vec{I_2}$ of the A and B atoms in the absence of external forces (the fluxes are measured relative to the lattice) are linear functions of the concentration gradients $c_1(\vec{r})$ and $c_2(\vec{r})$ of these atoms and the vacancy concentration gradient $c_v(\vec{r})$:

$$\vec{I_i} = -N_0 D_i \nabla c_i + N_0 F_i \nabla c_v \quad (i = 1, 2). \tag{12.1}$$

Here, D_1 and D_2 are the volume diffusion coefficients of the A and B atoms in the absence of a vacancy concentration gradient. These coefficients are of the same order of magnitude but differ somewhat from the ordinary chemical diffusion coefficients $\bar{D}_i$, which are found by making allowance for the equilibrium vacancy concentration gradient in an inhomogeneous solid solution [see Eq. (12.5)]. The second term in Eq. (12.1) arises because a directional flux of vacancies, proportional to the gradient of their concentration, produces fluxes of the A and B atoms, which are directed opposite to the vacancy flux. These atomic fluxes are of the same order as the vacancy flux $\vec{I}_v$ (in a system of coordinates relative to the lattice, we have $\vec{I}_1 + \vec{I}_2 + \vec{I}_v = 0$), i.e., the forces F_i are of the order of magnitude of the vacancy diffusion coefficient D_V and are considerably greater than the diffusion coefficients D_i:

$$F_i \sim D_v, \quad D_i \sim c_v^0 D_v. \tag{12.2}$$

Similarly, vacancy fluxes in perfect crystals are governed by the vacancy concentration gradient and by the gradient of the concentration of atoms of a given type, for example, the A atoms. It follows from $\vec{I}_1 + \vec{I}_2 + \vec{I}_v = 0$, $c_1 + c_2 + c_v = 1$, and Eq. (12.1) that

$$\left.\begin{array}{l} \vec{I}_v = - N_0 D_v \nabla c_v - N_0 (D_2 - D_1) \nabla c_1, \\[2mm] D_v = F_1 + F_2 + D_2. \end{array}\right\} \tag{12.3}$$

If the density of vacancy sources and sinks in a crystal is high and the distances between them are short compared with the characteristic length ($l \ll R$), the physically infinitesimal volume used in discussing diffusion processes can be made large compared with l. The average concentration of vacancies in such a volume assumes (practically instantaneously if the vacancy sources and sinks are of sufficient strength) an equilibrium value corresponding to the concentration, temperature, and stresses in this volume. Then, ∇c_v is not an independent thermodynamic quantity but is defined uniquely by the atomic concentration, temperature, and stress gradients and is a linear function of these gradients. Consequently, the terms containing the vacancy concentration gradient can be eliminated from the formulas for diffusion fluxes by expressing ∇c_v in terms of ∇c_i, ∇T, and $\nabla \sigma_{ii}$ and by suitably renormalizing the coefficients standing in front of these gradients.

For example, if only a gradient of the concentration of atoms is present, we may assume c_v to be a function solely of the concentration c_1 of the A atoms and, in this case, we have

$$\nabla c_v = \frac{\partial c_v^0}{\partial c_1} \nabla c_1 = \frac{\partial c_v^0}{\partial c_2} \nabla c_2 \qquad (c_1 + c_2 \approx 1).$$

Therefore, the diffusion fluxes of Eq. (12.1) become proportional to the gradients of the concentrations of atoms and can be written in the form

$$\vec{I}_i = -N_0 \bar{D}_i \nabla c_i, \tag{12.4}$$

where

$$\bar{D}_i = D_i - F_i \frac{\partial c_v^0}{\partial c_i}. \tag{12.5}$$

If we consider surface diffusion fluxes in solutions, we must make allowance for the fact that, as a result of the Gibbs enrichment, the concentrations c_{Si} in a surface layer whose thickness is of the order of the atomic spacing a (this applies to metallic and covalent solid solutions) differs from concentration c_i in the bulk of a crystal. This difference is expressed by factors which depend on the temperature and concentration:

$$c_{Si} = \chi_i' c_i, \qquad \chi_i' = \chi_i' (c_i, T). \tag{12.6}$$

The difference arises because some bonds of the surface atoms are broken and, therefore, the energy considerations favor the emergence of specific atoms on the surface. In studies of surface diffusion, we must consider only the concentrations c_{Si}.

On the surface of a crystal, the density of defect sources and sinks is usually sufficiently high to ensure that $l \ll R$ and, therefore, an equilibrium concentration of defects corresponding to local conditions is established on a given part of the surface. It follows that the surface diffusion fluxes associated with the gradient of the concentration of atoms are described by formulas resembling Eq. (12.4):

$$\vec{I}_{Si} = -N_0 a D_{Si} \nabla_S c_{Si}. \tag{12.7}$$

Here, D_{Si} is the coefficient of surface chemical diffusion (resulting from a gradient in the concentration of a chemical element and not just a radioactive isotope) of atoms i in a monatomic layer of thickness a (the flux $\vec{I}_{Si}$, applies to unit length).

The expressions for the diffusion fluxes of atoms in a solid solution subjected to an external force field are obtained by generalizing the corresponding expressions for the fluxes of atoms in one-component crystals. For example, the formula for the vol-

ume fluxes of A or B atoms in a metallic solid solution subjected
to an electric field is a generalization of Eq. (3.3):

$$\vec{I}_i = N_0 \frac{c_i D_i^\bullet}{f_i kT} ez_i \vec{E}. \tag{12.8}$$

Here, $i = 1, 2$; D_1^* and D_2^* are the self-diffusion coefficients of the
A and B atoms, i.e., the diffusion coefficients of labeled atoms in
the presence of a gradient of their concentration in the case when
the gradient of the total concentration of all isotopes of a given
chemical element is zero (D_i^* differs from the chemical diffusion
coefficients D_i and $\overline{D}_i$, which apply in the presence of gradients
of the concentrations of chemical elements, as explained in Chap.
IV of [101]); z_i is the average effective charge of ions i, which is
related to the force acting directly on an ion in an electric field
and to the electron drag force; f_i is the correlation factor of these
ions in a solid solution (Sec. 3).

We must bear in mind that Eq. (12.8) is semiphenomenological
and that the parameters D_i^*, z_i, and f_i, which occur in this formula, are found by fairly complex averaging processes. Therefore,
they do not have the same simple physical meaning as the corresponding quantities in the case of one-component crystals. For
example, the probability of the appearance of a vacancy alongside
a given ion i and the probability of the diffusion jump of an ion
into this vacancy, which govern the self-diffusion coefficient D_i^*,
depend on the nature of the ions (A or B) in neighboring lattice
sites. The activation energies of elementary diffusion processes
depend on the actual configurations of neighboring ions. The self-
diffusion coefficient is found by averaging over all such configurations and, in the case of two-component crystals, such a coefficient does not have a single activation energy (which is the case
in one-component crystals). Therefore, in the case of solid solutions, the dependence of D_i^* on $1/T$ is generally not exponential
but more complex. However, an analysis shows that, when the
temperature range is not too wide, the dependence of D_i^* on $1/T$
can be approximated quite accurately by an exponential function
even though the activation energy Q_i and the pre-exponential factor D_{0i}^* in the approximate formula

$$D_i^\bullet = D_{0i}^\bullet \exp\left(-\frac{Q_i}{kT}\right) \tag{12.9}$$

are effective quantities and do not have simple physical meaning (for details see Chap. IV of [101]).

The occupancy of the outer electron shells of the i-th ion and the electron scattering cross section of this ion (in the case when the ion is near the top of a potential barrier) should also depend on the configurations of the neighboring ions. Therefore, the forces acting on the i-th ion in an electric field depend on its configuration, and the effective charge z_i (defined as the ratio of the force to $e\vec{E}$) should be found by averaging over the configurations of neighboring ions in the solid solution. It is important to note that the probabilities of different configurations of the A and B ions around a given ion depend not only on the composition of the solid solution but also on the temperature. Therefore, the values of z_i should be functions of the concentration and temperature.

The dependences on the concentration and temperature of the correlation factor f_i in some solid solutions may be even more important. This is due to the fact that the binding energies of vacancies in solid solutions may depend strongly on the nature of the atom to which a vacancy is bound. In vacancy–impurity complexes containing one or more impurities, the correlation between successive vacancy–atom interchange events may be much stronger than in one-component crystals, in which all the atoms surrounding a vacancy are identical. At high temperatures, these complexes dissociate and the situation approaches that obtained in perfect crystals.

It follows that even when the factor f for a one-component crystal is close to unity and is governed solely by the crystal structure, the factors f_i in a solid solution may be much smaller than unity and may depend strongly on the temperature and composition of the solution [102]. It should be stressed that the temperature and concentration dependences of the effective charge may be found to be much stronger than the real dependences if experimental data on the electrotransport in alloys are analyzed using a formula of the Eq. (12.8) type, which ignores the correlation factor f_i or assumes that this factor is constant.

An expression for the volume diffusion fluxes in the field of a temperature gradient is similar to Eq. (2.1):

$$\vec{I}_i = -N_0 \frac{c_i \alpha_i}{T} D_i^* \nabla T. \tag{12.10}$$

As in the case of fluxes in an electric field, the coefficients α_i are obtained by averaging over the various configurations and they may depend strongly on the composition of the solid solution and on the temperature (α_i include the reciprocal correlation factors $1/f_i$ and the terms which represent the drag of diffusing ions by the fluxes of phonons and electrons in a temperature gradient).

Obviously, the fluxes of atoms in an inhomogeneous temperature field are related not only to the temperature gradient but also to the vacancy concentration gradient. However, if the distances l between the vacancy sources and sinks are short compared with the characteristic lengths L at which the temperature and composition vary significantly far from an inclusion and the surface of an inclusion has a sufficient number of vacancy sources and sinks, it is found that in the steady state, when $\Delta T = 0$ and $\Delta c_v = 0$, we obtain the same expression as in the case of one-component crystals (Sec. 2), $\nabla c_v = \dfrac{\partial c_v^0}{\partial T}\nabla T$, and the term proportional to ∇c_v can be included in Eq. (12.10). Thus, the coefficients α_i in Eq. (12.10) are renormalized.

Similarly, the expressions for the surface diffusion fluxes $\vec{I}_{Si}$ of the A and B atoms of a solid solution A−B subjected to a temperature gradient, an electric field, or an inhomogeneous stress field, are obtained by the generalization of Eqs. (2.2), (3.6), and (5.13) to the case of two-component solids:

$$\vec{I}_{Si} = -N_0 c_{Si} \frac{\alpha_{Si}}{T} D_{Si}^* a \nabla_S T, \qquad (12.11)$$

$$\vec{I}_{Si} = N_0 c_{Si} \frac{D_{Si}^* a}{f_{Si} kT} e z_{Si} \vec{E}_s, \qquad (12.12)$$

$$\vec{I}_{Si} = N_0 c_{Si} \omega_{Si} \frac{D_{Si}^* a}{3 f_{Si} kT} \nabla_S \sigma_{ii}. \qquad (12.13)$$

Here, D_{Si}^* is the surface self-diffusion coefficient, i.e., the diffusion coefficient of labeled atoms i (in general, this coefficient differs from the coefficient of surface chemical diffusion D_{Si}); z_{Si} and f_{Si} are, respectively, the effective charge and the correlation factor; ω_{Si} are quantities of the order of the atomic volume.

As in the formulas for volume fluxes, these parameters are of a semiphenomenological nature: they are obtained by averaging over all the configurations and, in general, depend strongly on the concentration and temperature.

In all the expressions given so far, the diffusion fluxes of atoms in solid solutions have been considered in a system of co-ordinates linked to the crystal lattice. By definition, the flux of sites in this system is zero, i.e., the total flux of the A and B atoms and vacancies vanishes:

$$\vec{I_1} + \vec{I_2} + \vec{I_v} = 0. \tag{12.14}$$

On the other hand, the matter, i.e., the A and B atoms, considered in the lattice coordinate system, diffuses in an opposite direction to the flux of vacancies. The average velocity of the atoms $\vec{v_a}(\vec{r})$ is equal to the product of the total flux of atoms $\vec{I_1} + \vec{I_2}$ and the atomic volume

$$\vec{v_a}(\vec{r}) = \omega(\vec{I_1} + \vec{I_2}) = -\omega \vec{I_v}. \tag{12.15}$$

Since the boundaries of a crystal move together with the matter, Eq. (12.15) gives the velocity of the boundaries in a system of co-ordinates linked to the lattice. Then, the fluxes of vacancies should correspond to a given point on a boundary.

The formation or disappearance of vacancies in the bulk of a crystal or on its surfaces, which occurs continuously during dif-fusion, should cause the existing atomic planes to disappear and new ones to appear. This should give rise to continuous diffu-sional deformation, i.e., to the modification of the crystal lattice and its motion.

Since the lattice itself becomes modified and migrates, it is sometimes more convenient to consider diffusion fluxes not in a system of coordinates linked to the lattice but in a system linked to the matter in a crystal at a given point or at the boundaries. Since the lattice moves relative to the matter at a velocity $-\vec{v_a}$, we can transform our equations to fluxes $\vec{I_i^m}$ in the system of co-ordinates linked to the matter by subtracting the products $\vec{v_a}(\vec{r}) N_i$, from the fluxes $\vec{I_i}$ in the lattice coordinate system; here, $N_i = N_0 c_i$ is the number of atoms of type i (i = A, B) per unit volume. For example, if the density of vacancy sources and sinks is high and if $l \ll R$, we find that, in the system of coordinates linked locally to the matter at a given point, the flux of the A or B atoms follows from Eqs. (12.4) and (12.15):

$$\vec{I}_i^m = -N_0\bar{D}_i\nabla c_i - \vec{v}_a(\vec{r})N_i = -N_0\bar{D}_i\nabla c_i +$$

$$+ N_0 c_i(\bar{D}_1\nabla c_1 + \bar{D}_2\nabla c_2) = -N_0\tilde{D}\nabla c_i. \qquad (12.16)$$

In this formula, $\tilde{D}$ is the mutual diffusion coefficient

$$\tilde{D} = c_1\bar{D}_2 + c_2\bar{D}_1, \qquad (12.17)$$

where an allowance is made for the fact that $c_1 + c_2 = 1$, $\nabla c_1 = -\nabla c_2$ (if we ignore small terms in c_v and ∇c_v). In contrast to the atomic fluxes, the vacancy flux changes only by a small amount, $-\vec{v}_a(\vec{r})N_v$, in the transformation from one system of coordinates to the other (here, $N_v = N_0 c_v$ is the number of vacancies per unit volume). This amount is proportional to the square of the concentration of vacancies c_v. In most cases, this difference can be ignored.

In some situations, the most natural system of coordinates is that linked to an external boundary of a crystal, from which the positions of various points inside the crystal can be measured conveniently. Continuous modification of the lattice is represented, in this system of coordinates, by motion at a velocity $\vec{w}(\vec{r})$.

If the diffusion fluxes are inhomogeneous in space, the lattice is not only displaced but also deformed and the relative change in any element of volume $\delta V(\vec{r})$ per unit time is

$$\frac{1}{\delta V}\frac{d\delta V}{dt} = \operatorname{div}\vec{w}(\vec{r}). \qquad (12.18)$$

Since a crystal remains continuous during diffusion (for the time being we shall consider only the case when no pores are formed) and the density of the lattice sites does not vary, the change in volume given by Eq. (12.18) can only result from the fact that an excess number of vacancies diffuses into a region of volume δV and disappears at vacancy sinks within this region. Simultaneously, the same number of atoms leaves this region and its volume decreases. Conversely, if sources within a region of volume δV generate more vacancies than the number absorbed by sinks, we find that the diffusion of vacancies out of this region is accompanied by the arrival of new atoms which increase the number of sites and, therefore, the volume of the region increases. Consequently, the relative change in volume per unit time

$$\frac{1}{\delta V}\frac{d\delta V}{dt} = -\omega\frac{\partial N_v}{\partial t} = \omega\text{div}\,\vec{I}_v \tag{12.19}$$

should be equated to (12.18). This yields a differential equation for the velocity of the lattice $\vec{w}\,(\vec{r})$

$$\text{div}\,\vec{w}\,(\vec{r}) = \omega\,\text{div}\,\vec{I}_v. \tag{12.20}$$

In the simplest (one-dimensional) case when the change in the concentration and the moving forces are directed along the x axis, all the quantities depend only on one coordinate x. Equation (12.20) can then be integrated quite easily and this yields the following relationship between the velocity of the lattice and the flux of vacancies at a given point x:

$$w_x(x) = \omega\int_{x_0}^{x}\frac{dI_{vx}}{dx}\,dx = \omega\left[I_{vx}(x) - I_{vx}(x_0)\right]. \tag{12.21}$$

Here, $I_{vx}(x)$ and $I_{vx}(x_0)$ are the vacancy fluxes at the point x and on the boundary (x_0). By definition, the boundary is immobile in the selected system of coordinates.

The normal component of the velocity of any element of the surface of an inclusion migrating as a result of diffusion fluxes in the interior of the host crystal is equal to the normal component of the velocity of the host matter near this surface element. According to Eq. (1.1), the velocity of the host matter in the system of coordinates linked to the crystal lattice is given by Eq. (12.15). However, when the density of vacancy sources and sinks is high, it is more convenient to determine the velocity of inclusions not relative to the lattice but relative to the host matter, i.e., in the final analysis, relative to the external boundaries of a sample.

This can be done by considering the velocity of an inclusion $\vec{v}\,(\vec{r_s})$ relative to those parts of the sample which are at a distance r from the inclusion when this distance is large compared with the inclusion radius R but small compared with the characteristic length L at which significant changes in the concentration and moving forces occur far from the inclusion. Since r ≪ L, the velocity of the crystal lattice does not change significantly at distances $\sim r$ and $\vec{w}\,(\vec{r})$ can be regarded as constant (although the velocity of the matter changes greatly in the region $r \sim R$ because of the changes in the moving forces). In order to go over from the velocity $\vec{v}'$ in the system of coordinates linked to the lattice to the velocity $\vec{v}$ in the system linked to the host matter far from an inclusion, it is sufficient to follow Eq. (1.6) and subtract from $\vec{v}'$ the velocity of Eq. (12.15), which describes the motion of matter in this range of values of $\vec{r}$ relative to the lattice:

$$\vec{v} = \vec{v}' + \omega \vec{I}_{v\infty},	(12.22)$$

where $\vec{I}_{v\infty}$ represents the flux in the region in question, which is far from an inclusion characterized by $R \ll r \ll L$.

If, far from an inclusion, the distributions of the concentration, temperature, and external forces are homogeneous, it follows from Eq. (12.21) that the velocity of the matter relative to an external boundary at a point x in a region $R \ll r \ll L$ is

$$\vec{v}_a(x) + \vec{w}(x) = -\omega \vec{I}_v(x_0)$$

and is governed solely by the flux of vacancies near the boundary. Consequently, the velocity of an inclusion relative to the boundary $\vec{v}_i$ is

$$\vec{v}_i = \vec{v} + \vec{v}_a(x) + \vec{w}(x) = \vec{v}' + \omega\left[\vec{I}_{v\infty}(x) - \vec{I}_v(x_0)\right].	(12.23)$$

The velocity of inclusions in solid solutions is related to the total diffusion fluxes of the A and B atoms by the same expressions as in the case of one-component crystals (Sec. 1). However, in the case of solid solutions, there should be additional detailed relationships which follow from the conditions of mass balance, which must be established separately for the A and B atoms. Obviously, the total number of atoms i (i = A, B) which are carried away per unit time from some element of the surface of an inclusion of area dS by all the diffusion fluxes $\vec{I}_i$, $\vec{I}_{Si}$, $\vec{I}_i'$ of i atoms (these are the diffusion fluxes on the surface and in the interior of the host and in the interior of the inclusion) should be equal to $N_i \vec{v}' \vec{n} dS = N_0 c_i \vec{v}' \vec{n} dS$, which is the number of atoms i contained in an element of volume $\vec{v}' \vec{n} dS$, which is swept through (per unit time) by an inclusion moving at a velocity $\vec{v}'$. Hence, it follows that in a system of coordinates linked to the lattice the velocity of a surface element $\vec{v}'\,(\vec{r}_S)$ is related to the diffusion fluxes by the expression

$$\vec{v}'\,(\vec{r}_S)\,\vec{n}\,(\vec{r}_S)\,c_i = \omega\,[\vec{I}_i\,(\vec{r}_S) - \vec{I}_i(\vec{r}_S)]\,\vec{n}\,(\vec{r}_S) + \omega\,\mathrm{div}_S\vec{I}_{Si} \quad (i=1,2). \quad (12.24)$$

The above expression represents two relationships which can be satisfied simultaneously in a binary solution only if the concentrations are redistributed around a moving dislocation.

In a system of coordinates linked to matter far from an inclusion (in the region where $R \ll r \ll L$), we find that Eqs. (12.24) and (12.22) yield a generalization of Eq. (1.12) in the case of two-component solutions:

$$\vec{v}\,(\vec{r}_S)\,\vec{n}\,(\vec{r}_S)\,c_i = \omega\,[\vec{I}_i\,(\vec{r}_S) - \vec{I}_i'\,(\vec{r}_S) + c_i\vec{I}_{v\infty}]\,\vec{n}\,(\vec{r}_S) + \omega\,\mathrm{div}_S\vec{I}_{Si} \quad (i=1,2).$$

The velocities of inclusions in the fields of different forces can be found from Eqs. (12.24), (12.25), and from the expressions for diffusion fluxes given earlier in the present section. Usually, some of the terms on the right-hand sides of these equations can be ignored. We shall consider first the case when the volume diffusion fluxes in the host sample can be ignored (the importance of these fluxes will be considered in detail in Sec. 13).

Migration of Inclusions in Solid Solutions due to Surface Diffusion in the Host Crystal

To illustrate some of the features of the migration of inclusions in solid solutions, we shall consider the displacement of an inclusion in a temperature gradient when only the surface diffusion fluxes are important (in a given range of temperatures and inclusion radii) [12].

These fluxes are given by Eq. (12.11). Since different types of atom differ in their surface mobilities, the temperature gradient will result in the faster motion of the atoms of one kind on the surface of an inclusion, whereas the atoms of the other kind will lag behind. Consequently, an inhomogeneous distribution of the surface concentration will arise at the inclusion—host interface and this will give rise to the diffusion fluxes of Eq. (12.7), which must be added to the thermal diffusion fluxes of Eq. (12.11) and which will compensate the difference between the mobilities of the A and B atoms. The concentration inhomogeneity will grow until the resultant fluxes of the A and B atoms become equal so that the velocities of inclusions as a result of the influence of these atoms become equal, in accordance with Eq. (12.25). This balancing process is practically complete in an interval which is of the order of the relaxation time of Eq. (1.10). Then, the surface diffusion in the host assumes its steady-state properties and the velocity of the inclusion reaches its constant limiting value.

In the case we are considering, the volume diffusion fluxes in the host substance are unimportant and, therefore, the host matter can be regarded as immobile relative to the lattice. It follows that Eqs. (12.24) and (12.25) for the velocities of inclusions become identical $(\vec{v}' = \vec{v})$. If we substitute $\vec{I}_i = \vec{I}' = \vec{I}_{v\infty} = 0$ in Eq. (12.25) and determine the surface diffusion fluxes $\vec{I}_{si}$ from Eqs. (12.7) and (12.11), we obtain a system of equations which can be used to determine the velocity of an inclusion and the surface concentration inhomogeneity:

$$
\left.
\begin{aligned}
\vec{v}\left(\vec{r}_S\right) \vec{n}\left(\vec{r}_S\right) c_1 + aD_{S1}\Delta_S c_{S1} &= -c_{S1}\frac{\alpha_{S1}}{T}D_{S1}^{\bullet}a\Delta_S T, \\[2mm]
\vec{v}\left(\vec{r}_S\right)\vec{n}\left(\vec{r}_S\right) c_2 + aD_{S2}\Delta_S c_{S2} &= -c_{S2}\frac{\alpha_{S2}}{T}aD_{S2}^{\bullet}\Delta_S T, \\[2mm]
\left(c_{S1} + c_{S2} = 1\right).
\end{aligned}
\right\}
\qquad (12.26)
$$

If we solve the system and assume that $\Delta c_{S2} = -\Delta c_{S1}$, we find that

$$
\vec{v}\left(\vec{r}_S\right)\vec{n}\left(\vec{r}_S\right) = -\frac{a}{T}\frac{c_{S1}\alpha_{S1}D_{S1}^{\bullet}D_{S2} + c_{S2}\alpha_{S2}D_{S2}^{\bullet}D_{S1}}{c_1 D_{S2} + c_2 D_{S1}}\Delta_S T, \qquad (12.27)
$$

$$
\Delta_S c_{S1} = -\frac{1}{T}\frac{\alpha_{S1}D_{S1}^{\bullet}c_2 c_{S1} - \alpha_{S2}D_{S2}^{\bullet}c_1 c_{S2}}{c_1 D_{S2} + c_2 D_{S1}}\Delta_S T. \qquad (12.28)
$$

It is evident from Eqs. (12.27), (2.10), and (2.19) that the scalar product $\vec{v}\left(r_S\right)\vec{n}\left(r_S\right)$ is proportional to $\cos\theta$. Under steady-state conditions, when the velocities of different parts of the surface $\vec{v}\left(r_S\right) = \vec{v}$ are equal, this relationship applies only in the case of

spherical inclusions. Thus, when the inclusion—host interface has
isotropic properties, an inclusion in a solid solution remains
spherical (as in a one-component crystal) during its migration.
The velocity $\vec{v}$ of the translational motion of an inclusion in a
solution follows from Eqs. (12.27), (2.10), and (2.19):

$$\vec{v} = 2(1 + \varkappa)\, \frac{a}{R}\, \frac{1}{T}\, \frac{c_{S1}\alpha_{S1}\overset{\bullet}{D}_{S1}D_{S2} + c_{S2}\alpha_{S2}\overset{\bullet}{D}_{S2}D_{S1}}{c_1 D_{S2} + c_2 D_{S1}}\, \nabla T_\infty. \qquad (12.29)$$

It is evident from Eqs. (12.28) and (2.10) that the change in the
concentration δc_{S1} of the A atoms at the inclusion—host interface,
associated with the migration of an inclusion in the field of a temperature gradient, is also proportional to cos θ and to the difference of temperatures between the two ends of the inclusion:

$$\delta c_{S1} = -(1 + \varkappa)\, \frac{1}{T}\, \frac{\alpha_{S1}\overset{\bullet}{D}_{S1}\, c_2\, c_{S1} - \alpha_{S2}\overset{\bullet}{D}_{S2}\, c_1\, c_{S2}}{c_1 D_{S2} + c_2 D_{S1}}\, R\, |\nabla T_\infty|\cos\theta. \qquad (12.30)$$

The quantity δc_{S1} is determined by the difference between the
mobilities of the A and B atoms at the inclusion—host interface
when the whole system is subjected to a temperature gradient
(these mobilities must be weighted by the factors $c_{S1}c_2$ and $c_{S2}c_1$).
Since $|\alpha_{Si}| \sim 1\text{-}10$, and the change in temperature $R\,|\nabla T_\infty|$ at
distances of the order of the inclusion radius is usually several
orders of magnitude less than the temperature T, it follows that
the change in the concentration $|\delta c_{S1}|$ is usually much less than
unity and, therefore, we can use the linear theory.

Equation (12.29) for the velocity of an inclusion in a solid solution can be represented in a simpler but less accurate form if
we use the well-known relationship which occurs in the phenomenological theory of diffusion [103] and which relates the chemical diffusion and self-diffusion coefficients. In the case of surface diffusion, this relationship is of the form

$$D_{Si} = \overset{\bullet}{D}_{Si}\, \frac{c_{Si}}{kT}\, \frac{\partial \mu_i}{\partial c_{Si}}, \qquad (12.31)$$

where μ_1 and μ_2 are the chemical potentials of the A and B atoms
in a solution. Although Eq. (12.31) is derived ignoring the actual
diffusion mechanism and is not exact, it still provides a satisfac-

tory description of the relationship between D_i and D_i^* in many
solutions (for details see reviews [104, 105, 101]). If we substitute
Eq. (12.31) into Eq. (12.29) and use the Gibbs—Duhem relationship (where the "phase" is the surface layer in a solution):

$$c_{S1} \frac{\partial \mu_1}{\partial c_{S1}} = c_{S2} \frac{\partial \mu_2}{\partial c_{S2}} , \qquad (12.32)$$

we obtain

$$\vec{v} = 2(1 + \varkappa) \frac{a}{R} \frac{c_{S1}\alpha_{S1} + c_{S2}\alpha_{S2}}{T} D_{S\,eff}\nabla T_\infty, \qquad (12.33)$$

where

$$D_{S\,eff} = \frac{D_{S1}^* D_{S2}^*}{c_1 D_{S2}^* + c_2 D_{S1}^*} . \qquad (12.34)$$

If $c_1 = 1$, it follows that $c_{S1} = 1$, $c_{S2} = 0$; $D_{S\,eff} = D_{S1}^*$, and Eqs.
(12.29), (12.33) reduce to Eq. (12.21), which applies to one-com-
ponent crystals. In the general case of solutions of arbitrary composition, these formulas can be used to determine the concentration dependence of the mobility of an inclusion if we know the concentration dependences of the thermal diffusion ratios and of the
diffusion coefficients of the components.

It follows from Eqs. (12.29) and (12.23) that the addition of a
small number of low-mobility B atoms to a crystal consisting of
A atoms can reduce strongly the velocity of an inclusion compared
with its velocity in a one-component (A) crystal. This applies
even if the self-diffusion coefficient D_{S1}^* is not affected by such
doping. The considerable reduction in $D_{S\,eff}$ and, consequently,
in the inclusion velocity occurs if D_{S2} is so much smaller than D_{S1}
that even when c_2 is small the quantity $c_2 [(D_{S1}/D_{S2}) - 1]$ becomes
comparable with unity (or even much larger than unity). In the
range of low concentrations, where $1 \gg c_2 \gg D_{S2}/D_{S1}$, the velocity of an inclusion is inversely proportional to c_2. On the other
hand, it follows from Eqs. (12.29) and (12.33) that a considerable
increase in the velocity of an inclusion can result from the addition of a small number of impurities only if the mobility of the
impurities is sufficiently high.

The possibility of a large reduction in the velocity of an inclusion as a result of doping is due to the fact that the surface

diffusion fluxes responsible for the migration of an inclusion in a solid solution should be continuous and the velocity is limited by the slower atoms. A considerable diffusion flux of the slower atoms is possible only in the presence of a temperature gradient and a steep concentration gradient. However, the latter gradient reduces strongly the diffusion flux of the faster atoms because — in the case of these atoms — the effect of such a gradient is opposite to the effect of a temperature gradient.

Similar procedures can be adopted in calculating the velocity of inclusions migrating under the influence of surface diffusion fluxes in solid solutions subjected to an electric field or an inhomogeneous stress field. A comparison of Eqs. (12.11) and (12.12) or Eqs. (12.11) and (12.13) shows that in the intermediate formulas, as well as in the final expression (12.29) for the velocity of an inclusion, we must replace $\alpha_{Si} \nabla_S T = (1 + \chi)\alpha_{Si} \Delta T_\infty$ with $-\dfrac{3ez_{Si}}{2kf_{Si}} \vec{E}_S = -\dfrac{3}{2} \dfrac{ez_{Si}}{kf_{Si}} \vec{E}_\infty$ (for a nonconducting inclusion in an electric field) or by $-\dfrac{\omega_{Si}}{3kf_{Si}} \nabla_S \sigma_{ii} = -\dfrac{5}{3} \dfrac{\omega_{Si}}{kf_{Si}} \sigma_1 \vec{e}_z$ [σ_1 represents, in accordance with Eq. (5.14), the inhomogeneity of the stress field]. Therefore, the velocity of a nonconducting inclusion migrating under the influence of surface fluxes in a solid solution subjected to an electric field is

$$\vec{v} = -3 \frac{a}{R} \frac{1}{kT} \frac{c_{S1}z_{S1}f_{S1}^{-1}D_{S1}^{\bullet}D_{S2} + c_{S2}z_{S2}f_{S2}^{-1}D_{S2}^{\bullet}D_{S1}}{c_1 D_{S2} + c_2 D_{S1}} e\vec{E}_\infty, \qquad (12.35)$$

and the velocity in a stress field is

$$\vec{v} = -\frac{10}{3} \frac{a}{R} \frac{1}{kT} \frac{c_{S1}\omega_{S1}f_{S1}^{-1}D_{S1}^{\bullet}D_{S2} + c_{S2}\omega_{S2}f_{S2}^{-1}D_{S2}^{\bullet}D_{S1}}{c_1 D_{S2} + c_2 D_{S1}} \sigma_1\vec{e}_z. \qquad (12.36)$$

If we use the approximate relationship (12.31), we find that the above formulas can be rewritten in the form

$$\vec{v} = -3 \frac{1}{kT} \cdot \frac{a}{R} \frac{c_{S1}z_{S1}f_{S2} + c_{S2}z_{S2}f_{S1}}{f_{S1}f_{S2}} D_{s\,eff}e\vec{E}_\infty, \qquad (12.37)$$

$$\vec{v} = -\frac{10}{3} \frac{a}{R} \frac{1}{kT} \frac{c_{S1}\omega_{S1}f_{S2} + c_{S2}\omega_{S2}f_{S1}}{f_{S1}f_{S2}} D_{s\,eff}\sigma_1\vec{e}_z. \qquad (12.38)$$

If $c_1 = 1$, we find that Eqs. (12.35) and (12.37) or Eqs. (12.36) and (12.38) reduce, respectively, to Eqs. (3.19) or (5.24), which apply to one-component crystals. As in the case of migration in a temperature gradient, the addition of a small amount of low-mobility impurities can reduce strongly the velocity of an inclusion moving in an electric field or in a stress gradient.

We must bear in mind that the formulas for the velocity of inclusions in solid solutions are derived on the assumption that the enriched Gibbs layer on the surface is continuously replenished. This happens if the characteristic time for the disturbance of an enriched layer as a result of the motion of inclusions, given by $\sim a/v$, is considerably greater than the characteristic time of diffusional replenishment of the same layer, $\sim a^2/D_{\text{eff}}$, where D_{eff} is the effective volume diffusion coefficient in the solution. For example, if an inclusion migrates in a temperature gradient, it follows from Eq. (12.33) that this requirement can be reduced to

$$\frac{|\nabla T_\infty| R}{T} \ll \frac{D_{\text{eff}}}{\alpha_S D_{s\,\text{eff}}} \frac{R^2}{a^2} . \tag{12.39}$$

Since $R|\nabla T_\infty| \ll T$ and $R \gg a$, the above condition is usually satisfied (with the exception of very low temperatures). However, if the opposite inequality is satisfied, we must replace c_{Si} with c_i in Eq. (12.33) and we have to use the values of α_{Si} and D_{Si}^* which correspond to a surface which is not enriched.

If the velocity of an inclusion is not very high, local equilibrium between the A and B atoms is established on the surface of a solution and in its immediate neighborhood. Therefore, the change in the concentration $\delta c_{Si}(\vec{r}_s)$ on the surface is related to a change in the concentration $\delta c_i(\vec{r}_s)$ in a region adjoining the surface. It follows from Eq. (12.6) that these changes are related by

$$\delta c_{Si} = \frac{d}{dc_i}(\chi_i' c_i)\,\delta c_i,$$

i.e.,

$$\delta c_{Si}(\vec{r}_s) = \chi(c_i)\,\delta c_i(\vec{r}_s); \quad \chi_i = \chi_i' + c_i\frac{d\chi_i'}{dc_i}, \tag{12.40}$$

$$\chi_1 = \chi_2 = \chi.$$

Influence of Diffusion Across an
Inclusion on its Velocity

An inhomogeneous change in the concentration δc_i in the bulk of the host sample at its interface with the inclusion, as well as a change in the temperature $\delta T(\vec{r}) = (1+x)\vec{r}\nabla T_\infty$ or in the stresses $\delta\sigma_{ii} = 5\sigma_1\vec{re}$, all alter the equilibrium values of the concentrations $\delta c_i^{\prime 0}$ of the A and B atoms dissolved in the inclusion [see also Eqs. (2.22) and (5.25)]:

$$\delta c_i^{\prime 0} = \frac{\partial c_i^{\prime 0}}{\partial c_i}\,\delta c_i + \frac{\partial c_i^{\prime 0}}{\partial T}\,(1+x)\,\vec{r}_S\nabla T_\infty - \frac{5}{3}\,\omega_i\sigma_1\left(\frac{\partial\mu_i^\prime}{\partial c_i^\prime}\right)^{-1}\vec{r}_S\,\vec{e}_z. \qquad (12.41)$$

Here, the derivative $\partial c_i^{\prime 0}/\partial c_i$ determines the dependence of the solubility of atoms of type i (A or B) in the inclusion on the composition of the solid solution adjoining the inclusion, and $\mu_1^\prime(c_1^\prime) \equiv \mu_A$ and $\mu_2^\prime \equiv \mu_B$ are the chemical potentials of the A and B atoms in the inclusion.

The inhomogeneity of the composition of the solid solution and the external forces produce diffusion fluxes across an inclusion, and these fluxes can be described by the general expression

$$\vec{I}_i^\prime(\vec{r}) = -N_0^\prime D_i^\prime \nabla c_i^\prime - N_0^\prime \frac{c_i^\prime \alpha_i^\prime}{T} D_i^{\prime\bullet}(1+x)\nabla T_\infty +$$
$$+ N_0^\prime \times \frac{c_i^\prime D_i^{\prime\bullet}}{f_i kT}\,ez_i^\prime\left(1+x_e\right)\vec{E}_\infty, \qquad (12.42)$$

where the primes of the various transport coefficients denote that these coefficients apply to the interior of the inclusion. Since, under steady-state conditions, $\mathrm{div}\vec{I}_i^\prime = 0,$, it follows that the distribution of the concentration in the inclusion satisfies

$$\Delta c_i^\prime = 0. \qquad (12.43)$$

At the inclusion−host interface, the normal components of the fluxes of Eq. (12.42) should be equal to the number of atoms crossing this interface. If we assume, as in Chap. I [see Eq. (2.27)], that this number is proportional to the difference between the actual change in the concentration at the interface $\delta c_i^\prime$ and the change in the equilibrium concentration $\delta c_i^{\prime 0}$, the conditions of continuity become

$$\vec{n}\vec{I_i}\,(\vec{r}) = \beta_i\left[\delta c_i'\,(\vec{r}) - \delta c_i'^0\,(\vec{r})\right] = -N_0'D_i'\vec{n}\nabla c_i' -$$
$$-N_0'\frac{c_i'\alpha_i'}{T}D_i'^\bullet(1+\varkappa)\vec{n}\nabla T_\infty + N_0'\frac{c_i'D_i'^\bullet}{f_i'kT}\,ez_i'\left(1+\varkappa_e\right)\vec{n}\vec{E}_\infty, \qquad (12.44)$$

$$(\vec{r} = \vec{r_s}),$$

where $\beta_i = \beta_i(T, c_i)$ are the transport coefficients.

It follows from Eqs. (12.41)-(12.44) that the solution of Eq. (12.43) satisfying the boundary conditions of Eq. (12.44) should be proportional to $\cos\theta$ (for $\vec{E}_\infty \parallel \nabla T_\infty$) and should be of the form

$$\delta c_i'\,(\vec{r}) = \delta c_i'\,(\vec{r_s})\frac{\vec{r}\,\vec{e_z}}{R\cos\theta}. \qquad (12.45)$$

Substituting Eq. (12.45) into (12.44), we can express the actual changes in the concentration at the interface $\delta c_i'\,(\vec{r_s})$ in terms of changes in the equilibrium concentration $\delta c_i'^0\,(\vec{r_s})$, i.e., in terms of changes in the concentration in the host sample δc_i.

The contribution of the diffusion fluxes of Eq. (12.42) across an inclusion should be added to the contribution of the surface diffusion fluxes described by Eq. (12.25). The two diffusion fluxes of Eq. (12.25) are sufficient for the determination of the velocity of a pore and the distribution of the concentrations of the components inside it (obviously, this distribution will be different from that observed in the case of pure surface diffusion). If the inclusion migrates under the influence of a temperature gradient, its velocity [12] is given by

$$\vec{v} = (1+\varkappa)\frac{a_1^T b_2 + a_2^T b_1}{c_1 b_2 + c_2 b_1}\,\nabla T_\infty. \qquad (12.46)$$

Here, we have introduced the following notation:

$$\left.\begin{array}{c} a_i^T = \dfrac{2a}{R}c_{Si}D_{Si}^\bullet\dfrac{\alpha_{Si}}{T} + \dfrac{\omega}{\omega'}\,\delta_i\left(D_i'\dfrac{\partial c_i'^0}{\partial T} + \dfrac{\alpha_i'c_i'^0}{T}D_i'^\bullet\right), \\[3mm] b_i = \dfrac{2a}{R}\chi D_{Si} + \dfrac{\omega}{\omega'}\,\delta_i D_i'\dfrac{\partial c_i'^0}{\partial c_i}, \\[3mm] \delta_i = \dfrac{\beta_i\omega' R}{\beta_i\omega' R + D_i'}. \end{array}\right\} \qquad (12.47)$$

Expressions (12.46) and (12.47) describe, generally, the complex dependence of the velocity of an inclusion on its radius and on the solid-solution composition. In particular, the addition of a small number of B atoms to a crystal consisting of A atoms can alter not only the velocity of an inclusion (as in the case of pure surface diffusion) but also its dependence on the radius.

For example, let us consider a sufficiently large inclusion which satisfies the condition

$$\beta_i \omega' R \gg D_i', \qquad \frac{a}{R} D_{S1} \ll \frac{\omega}{\omega'} D_1' c_1'^0. \tag{12.48}$$

In this case, the absorption of the A and B atoms in the inclusion is limited by the rate of diffusion in the bulk of the host sample and not by the conditions at the interface. In the case of atoms of the main component A ($c_1 \approx 1$), the diffusion across the inclusion and not the surface diffusion fluxes are of prime importance. It then follows from Eqs. (12.46) and (12.47) that, at low impurity concentrations in the solution ($c_2 \ll 1$), we obtain

$$\vec{v} = (1 + \varkappa) \frac{\omega}{\omega'} D_1' \left(\frac{\partial c_1'^0}{\partial T} + \frac{\alpha_1'}{T} \frac{c_1'^0 D_1'^\bullet}{T} \right) \times$$

$$\times \frac{\dfrac{2a}{R} D_{S2} + \dfrac{\omega}{\omega' \chi} D_2' \dfrac{\partial c_2'^0}{\partial c_2}}{\dfrac{2a}{R} D_{S2} + \dfrac{\omega}{\omega' \chi} D_2' \dfrac{\partial c_2'^0}{\partial c_2} + c_2 \dfrac{\omega}{\omega' \chi} D_1' \dfrac{\partial c_1'^0}{\partial c_1}} \nabla T_\infty. \tag{12.49}$$

When $c_2 \to 0$, the penultimate factor in Eq. (12.49) tends to unity and this equation becomes identical with the analogous formula (2.26) for one-component crystals. However, if the impurity atoms (B) dissolve weakly in the inclusion or if they diffuse slowly across it, and the surface diffusion coefficient D_{S2} is of the same (or lower) order of magnitude as D_{S1}, we find that

$$D_1' \frac{\partial c_1'^0}{\partial c_1} \gg D_2' \frac{\partial c_2'^0}{\partial c_2}, \qquad \frac{\omega}{\omega'} D_1' \frac{\partial c_1'^0}{\partial c_1} \gg \frac{a}{R} D_{S2}, \tag{12.50}$$

and at relatively low impurity concentrations in the solution $c_2 \sim c_2^*$, where

$$c_2^* \frac{\omega}{\omega' \chi} D_1' \frac{\partial c_1'^0}{\partial c_1} = \frac{\omega}{\omega' \chi} D_2' \frac{\partial c_2'^0}{\partial c_2} + \frac{2a}{R} D_{S2} \qquad (c_2^* \ll 1), \tag{12.51}$$

the impurity atoms control the diffusion in the inclusion and alter considerably the velocity of the inclusion.

If the contribution of the surface diffusion can be ignored and the migration is primarily due to diffusion fluxes across the inclusion, and if the magnitude of these fluxes is not controlled by the processes at the inclusion—host interface ($\delta_i \approx 1$), we find that Eq. (12.46) simplifies to

$$
\vec{v} = (1 + \varkappa)\frac{\omega}{\omega'}\left[\frac{D_1' D_2'\left(\dfrac{\partial c_1'^0}{\partial T}\dfrac{\partial c_2'^0}{\partial c_2} + \dfrac{\partial c_2'^0}{\partial T}\dfrac{\partial c_1'^0}{\partial c_1}\right)}{c_1 D_2'\dfrac{\partial c_2'^0}{\partial c_2} + c_2 D_1'\dfrac{\partial c_1'^0}{\partial c_1}} + \right.
$$

$$
\left. + \frac{1}{T}\frac{\alpha_1' c_1'^0 D_1'^* D_2'\dfrac{\partial c_2'^0}{\partial c_2} + \alpha_2' c_2'^0 D_2'^* D_1'\dfrac{\partial c_1'^0}{\partial c_1}}{c_1 D_2'\dfrac{\partial c_2'^0}{\partial c_2} + c_2 D_1'\dfrac{\partial c_1'^0}{\partial c_1}}\right]\nabla T_\infty. \qquad (12.51\text{a})
$$

In the range of concentrations where $1 \gg c_2 \gg c_2^*$, the velocity v is inversely proportional to c_2. However, it is also inversely proportional to the inclusion radius R or is not affected by the radius, depending on whether the diffusion fluxes of impurity atoms on the interface or across the inclusion predominate. The transition concentration c_2^* depends on the inclusion radius. If $c_2 D_1'(\partial c_1'^0/\partial c_1) \gg D_2'(\partial c_2'^0/\partial c_2)$, we find that at a fixed concentration the velocity is initially independent of the radius, which increases from very small values of R (this constant value of the velocity corresponds to a pure crystal consisting solely of A atoms). Then, when the radius reaches values at which the concentration of c_2 becomes of the order of c_2^*, the velocity begins to depend strongly on R. Finally, at very large values of R, the velocity is again constant (however, the new value is much lower than that corresponding to low values of R).

The same approach can be used in an analysis of the motion of inclusions in an inhomogeneous stress field or in an electric field, when the migration of inclusions is due to diffusion fluxes on the surface of the host or across the inclusion. In this case, the velocity of an inclusion in a solid solution subjected to the inhomogeneous stress field of Eq. (5.16) is given by the expression

$$
\vec{v} = -\frac{5}{3}\frac{a^\sigma b_2 + a_2^\sigma b_1}{c_1 b_2 + c_2 b_1}\sigma_1 \vec{e}_z, \qquad (12.52)
$$

where

$$a_i^\sigma = \frac{2a}{R} \frac{\omega_{Si}}{kTf_{Si}} c_{Si} D_{Si}^\bullet + \frac{\omega}{\omega'} \delta_i \omega_i \left(\frac{\partial \mu_i'}{\partial c_i'}\right)^{-1} \cdot D_i'. \tag{12.53}$$

If we can ignore surface diffusion fluxes, we find that the velocity of a conducting inclusion in an electric field is given by

$$\vec{v} = - (1 + \varkappa_e) \frac{1}{kT} \frac{a_1^E b_2 + a_2^E b_1}{c_1 b_2 + c_2 b_1} \cdot e\vec{E}_\infty, \tag{12.54}$$

where

$$a_i^E = \frac{\omega}{\omega'} \delta_i \frac{z_i'}{f_i} c_i'^0 D_i'^\bullet. \tag{12.55}$$

As in the case of the migration of inclusions in solid solutions under the influence of a temperature gradient, the velocity of inclusions migrating in a stress or an electric field may depend strongly on the concentration of impurity atoms (in the range $c_2 \ll 1$) and on the radius of these atoms if they dissolve weakly in the inclusion and diffuse slowly across it.

13. ROLE OF VOLUME DIFFUSION IN THE HOST

CRYSTAL AND THE MIGRATION OF

INCLUSIONS IN A CONCENTRATION GRADIENT

Migration of Inclusions in the Field

of an Inhomogeneous Concentration

Gradient

As in the case of one-component crystals, the volume diffusion fluxes in a solid solution exert a strong influence on the migration of inclusions under the action of external forces, provided the volume diffusion coefficients of the A and B atoms in the host are sufficiently large. In some respects, these volume diffusion fluxes may be more important in solid solutions than in one-component crystals and may give rise to some new effects. For example, an allowance for volume diffusion in the host substance may lead to the migration of inclusions in an inhomogeneous concentration field [12, 65]. This migration must be considered, especially as in experimental studies of diffusion processes in solutions it is often found that the inclusions act as "immobile markers" and it is necessary to determine the conditions under which they are really immobile.

If the diffusion fluxes appear only on the inclusion—host interface and across an inclusion, the inclusion cannot migrate under the influence of a concentration gradient (if there are no other thermodynamic forces). The concentrations of the A and B atoms, which are transferred initially by such diffusion fluxes from the front to the rear end of an inclusion, become equal so that the transfer of matter (and, consequently, the motion of the inclusion) stops. A constant difference between the concentrations at the front and rear ends of an inclusion can be maintained and the transfer of matter can continue in the case of balanced fluxes of the A and B atoms in an inclusion (this always applies to the migration of inclusions in solutions). The excess A or B atoms reaching a given part of the interface should be dispersed in the bulk of the host substance. This can occur only by volume diffusion in the host substance.

We mentioned in Sec. 12 that diffusion fluxes in a solid solution depend on the relationship between the characteristic lengths in which significant changes take place in the concentration (in the present case, the characteristic length is of the order of R) and the distance l between vacancy sources and sinks. We shall start by considering solid solutions with a high density of vacancy sources and sinks so that $l \ll R$. In this case, the volume diffusion fluxes in the host substance are given by Eq. (12.4) in the system of coordinates referred to the lattice.

In considering the migration of inclusions in an inhomogeneous concentration field, we shall assume that in a region whose dimensions are $L \gg R$ the concentration gradient $\nabla c_{1\infty}$ of the A atoms far from an inclusion is constant, so that when $L \gg r \gg R$, the distribution of the concentration is of the form

$$c_1(\vec{r}) = c_{10} + \vec{r}\nabla c_{1\infty}, \qquad c_2(\vec{r}) = 1 - c_1(\vec{r}) \qquad (L \gg r \gg R). \tag{13.1}$$

The diffusion fluxes of atoms far from an inclusion do not alter the nature of this distribution. However, near an inclusion, the

lines of flow of the A and B atoms are distorted and the concentra
tion given by Eq. (13.1) is disturbed.

If the concentration gradient and, consequently, the velocity of
an inclusion are not very high so that

$$v' \ll \bar{D}_i / R, \qquad (13.2)$$

we find that in a time R/v', during which an inclusion migrates a
distance of the order of R, the distance traveled by the A and B
atoms is of the order of $\sqrt{\bar{D}_i R / v'}$, which is considerably larger
than the radius R [it follows from the expressions for the velocity
v' given below that the condition (13.2) is practically always satisfied in the migration of inclusions in a concentration gradient].

Under these conditions, the distribution $c_1(\vec{r})$ can be regarded as
quasisteady-state and it can be found from the equation

$$\Delta c_1 = 0, \qquad (13.3)$$

ignoring the term $(1/D_1)(\partial c_1 / \partial t)$ in the diffusion equation.

In the region surrounding in inclusion, we must supplement the
concentration distribution of Eq. (13.1) with the solution of the
Laplace equation (13.3) which decreases with distance. This decreasing function can be expanded as a series of spherical harmonics. If the distribution $c_1(\vec{r})$ at large distances from an inclusion is of the form given by Eq. (13.1) and contains only the
first (and zeroth) harmonics and if the properties of a crystal and
its surface are isotropic, we find that the expansion of the decreasing function also contains only the first spherical harmonic,
so that near the inclusion

$$c_1(\vec{r}) = c_{10} + \vec{r}\nabla c_{1\infty} + AR^2 \frac{\vec{e}\cdot\vec{r}}{r^3}, \qquad c_2(\vec{r}) = 1 - c_1(\vec{r}), \qquad (13.4)$$

where the unit vector $\vec{e}$ is directed along $\nabla c_{1\infty}$ and the constant
A can be found from the boundary conditions on the surface of the
inclusion. If the concentration is distributed in accordance with
the above equation, all the fluxes are proportional to $\cos\theta$, the
migrating inclusion retains its spherical shape, and the problem
is self-consistent. When the Gibbs enrichment in a surface layer
is taken into account, it follows from Eqs. (12.40) and (13.4) that

the concentration is proportional to $\vec{e}\vec{n} = \cos\theta$:

$$c_{S1}\left(\vec{r}_s\right) = \chi'\left(c_{1\,0}\right)c_{1\,0} + \chi\left(Re\vec{\nabla}c_{1\infty} + A\right)\vec{e}\vec{n}. \tag{13.5}$$

The boundary condition on the surface of an inclusion, which determines the constant A, and the equation which governs the inclusion velocity $\vec{v}'$ (in a system of coordinates linked to the crystal lattice of the host crystal) are given by the system of equations (12.24). If we determine $\vec{I}_i\vec{n}$ from Eqs. (12.4) and (13.4), $\text{div}_s\vec{I}_{Si}$ from Eqs. (12.7), (13.5), and (2.19), and $\vec{I}_i'\,\vec{n}$ from Eqs. (12.41), (13.4), (12.42), and (12.44), we obtain the following version of Eq. (12.24):

$$\left.\begin{aligned}
\vec{v}'\vec{n}c_1 &= b_1\frac{A}{R}\,\vec{e}\vec{n} + a_1^C\vec{n}\nabla c_{1\infty}, \\
\vec{v}'\,\vec{n}c_2 &= -b_2\frac{A}{R}\,\vec{e}\vec{n} - a_2^C\vec{n}\nabla c_{1\infty},
\end{aligned}\right\} \tag{13.6}$$

where

$$\left.\begin{aligned}
a_i^C &= -\bar{D}_i + \frac{2a}{R}\,\chi D_{Si} + \frac{\omega}{\omega'}\,\delta_i D_i'\frac{\partial c_i^{'0}}{\partial c_i}, \\
b_i &= 2\bar{D}_i + \frac{2a}{R}\,\chi D_{Si} + \frac{\omega}{\omega'}\,\delta_i D_i'\frac{\partial c_i^{'0}}{\partial c_i},
\end{aligned}\right\} \tag{13.7}$$

and δ_i is given by Eq. (12.47).

The solution of the system of equations (13.6) yields the constant A, which represents — in accordance with Eq. (13.4) — the distortion of the distribution of the concentration near an inclusion, and the velocity of the inclusion $\vec{v}'$ relative to the crystal lattice of the host:

$$A = -\frac{c_2a_1^C + c_1a_2^C}{c_2b_1 + c_1b_2}\,Re\vec{\nabla}c_{1\infty}, \tag{13.8}$$

$$\vec{v}' = \frac{a_1^Cb_2 - a_2^Cb_1}{c_1b_2 + c_2b_1}\,\nabla c_{1\infty}. \tag{13.9}$$

According to Eqs. (13.4) and (13.8), the difference between the concentrations at the two opposite ends of the inclusion surface is of the order of $R|\nabla c_{1\infty}|$. This difference is considerably less than unity if the radius of the inclusion is small compared with the dimensions of the diffusion zone. This justifies the use of the linear theory of the migration of inclusions.

In order to determine the velocity of inclusions $\vec{v}$ in the system of coordinates linked to the host matter in a region $L \gg r \gg R$ located far from an inclusion, it is necessary to subtract from Eq. (13.9) the velocity of atoms $\vec{v}_a$ in this region [see also Eq.(12.22)], which is given by Eqs. (12.15) and (12.4):

$$\vec{v} = \vec{v}' - \vec{v}_a, \qquad \vec{v}_a = \omega\left(\vec{I}_{1\infty} + \vec{I}_{2\infty}\right) = \left(\bar{D}_2 - \bar{D}_1\right)\nabla c_{1\infty}. \tag{13.10}$$

It follows from Eqs. (13.9) and (13.7) that if the fluxes of atoms in the bulk of the host crystal vanish (for $\bar{D}_i = 0$) in a system of coordinates linked to the lattice, the migration of inclusions is impossible. However, if the diffusion takes place only in the bulk of the host substance and on the inclusion—host interface and the fluxes across the inclusion can be ignored, we find that Eq. (13.9) simplifies to

$$\vec{v}' = \frac{3a}{R}\,\chi\,\frac{D_{S1}\bar{D}_2 - D_{S2}\bar{D}_1}{\bar{D} + \dfrac{a\chi}{R}\bar{D}_S}\,\nabla c_{1\infty} \tag{13.11}$$

Here,

$$\left.\begin{array}{l} \bar{D} = c_1\bar{D}_2 + c_2\bar{D}_1, \\ \bar{D}_S = c_1 D_{S2} + c_2 D_{S1}. \end{array}\right\} \tag{13.12}$$

Inclusions can act as immobile markers only if their velocity relative to the lattice $\vec{v}'$ is small compared with the velocity $\vec{v}_a$ of the atoms relative to the lattice, or compared with the velocity of the lattice relative to the boundaries of the crystal. This condition is satisfied only in the case of fairly large inclusions and moderate temperatures, when

$$\frac{a}{R}\,D_{Si} \ll \bar{D}_i. \tag{13.13}$$

However, if

$$\frac{a}{R} D_{Si} \geq \tilde{D}_i,$$ (13.14)

we find that v' is of the same order as v_a and the inclusions migrate quite rapidly relative to the lattice, i.e., they cannot serve as immobile markers.

For example, if $\bar{D}_1 \sim \bar{D}_2 \sim 10^{-10}$ cm^2/sec, the annealing time is $\sim 10^4$ sec, the width of the diffusion zone is $\sim 10^{-3}$ cm, $|\nabla c_{1\infty}| \sim 10^3$ cm^{-1}, we find that v' $\sim 10^{-7}$ cm/sec if the condition (13.14) is satisfied and during the annealing the inclusions migrate for a distance $\sim 10^{-3}$ cm, i.e., of the order of the width of the diffusion zone.

In the range where $(a/R)D_{Si} \lesssim \bar{D}_i$, the velocity should depend strongly on the radius of the inclusion and on its nature (the latter determines the value of D_{Si}). If the radius is large so that the condition (13.13) is satisfied, the velocity of Eq. (13.11) is inversely proportional to R. When R is reduced, the velocity increases more slowly than does 1/R, and when the condition (13.13) is replaced by the opposite condition $(a/R) D_{Si} \gg \bar{D}_i$, we find that the velocity tends to a constant limit.

If the radius of the inclusions is sufficiently large so that we can ignore surface diffusion fluxes and consider only the volume diffusion in the host and in the inclusions, we find that Eqs. (13.9) and (13.7) yield the velocity $\vec{v}'$:

$$\vec{v}' = 3 \frac{\omega}{\omega'} \frac{\delta_1 D_1' \bar{D}_2 \dfrac{\partial c_1'^0}{\partial c_1} - \delta_2 D_2' \bar{D}_1 \dfrac{\partial c_2'^0}{\partial c_2}}{2\tilde{D} + \tilde{D}'} \nabla c_{1\infty}.$$ (13.15)

Here,

$$\tilde{D}' = \frac{\omega}{\omega'} \left(c_1 \delta_2 D_2' \frac{\partial c_2'^0}{\partial c_2} + c_2 \delta_1 D_1' \frac{\partial c_1'^0}{\partial c_1} \right).$$ (13.16)

This velocity is small compared with v_a and the inclusions can be regarded as immobile markers provided

$$\delta_i D_i' \left| \frac{\partial c_i'^0}{\partial c_i} \right| \ll \bar{D}_i.$$ (13.17)

If these conditions are not satisfied, the inclusions migrate at a significant velocity (which is of the same order as v_a) relative to the lattice. In this case, the velocity of the inclusions is independent of their radius but depends in a complex manner on the composition of the solid solution and on the temperature.

We shall now consider the migration of inclusions under the action of a concentration gradient in crystals with few defects so that the distances l between the vacancy sources and sinks are large compared with R. In this case, the diffusion fluxes are given by Eq. (12.1) and are related not only to the gradient of the concentration of atoms ∇c_1, but also to the gradient of the concentration of vacancies ∇c_v. We may assume that in the region where $L \gg r \gg R$, i.e., far from an inclusion, the gradient $\nabla c_v = \nabla c_{v\infty}$ is constant and the concentration of vacancies is given by the expression

$$c_v(\vec{r}) = c_{v0} + \vec{r}\nabla c_{v\infty}. \tag{13.18}$$

On the surface of an inclusion, where vacancies can appear and disappear quite freely, the vacancy concentration attains its equilibrium value c_v^0, governed by the composition of the solution, the temperature of the surface region being considered, and its curvature:

$$c_v(\vec{r}_S) = c_v^0 \equiv c_v^0 \{T(\vec{r}_S), c_i(\vec{r}_S), K(\vec{r}_S)\}. \tag{13.19}$$

If the temperature and the curvature of the surface of an inclusion are constant and only the concentration is a function of $\vec{r}_S$, the boundary condition (13.19) combined with Eq. (13.4) can be written in the form

$$c_v(\vec{r}_S) = c_{vR}^0(c_{10}) + A_v \vec{e}\vec{n}, \quad A_v = \frac{\partial c_{vR}^0}{\partial c_1}(A + R\vec{e}\nabla c_{1\infty}), \tag{13.20}$$

where $c_{vR}^0(c_{10}) \equiv c_{vR}^0$ is the equilibrium concentration of vacancies near a spherical inclusion of radius R.

If the condition (13.2) is satisfied, the condition (2.13) for the quasisteady-state distribution of vacancies is more than satisfied. Therefore, the distribution of vacancies near an inclusion is given by the equation

$$\Delta c_v = 0. \tag{13.21}$$

The solution of Eq. (13.21) satisfying the boundary conditions (13.18) and (13.20) is of the form (for $\vec{r}_s = 0$):

$$c_v(\vec{r}) = c_{v0} + (c_{vR}^0 - c_{v0})\frac{R}{r} + \vec{r}\nabla c_{v\infty}\left(1 - \frac{R^3}{r^3}\right) + A_v R^2 \frac{\vec{e}\,\vec{z}}{r^3}. \qquad (13.22)$$

For the sake of simplicity, we shall restrict our discussion to the case when the pressure on the surface of the host crystal (for example, that exerted by the gas in a bubble) is such that the concentrations c_{vR}^0 and c_{v0} are equal. If we ignore the $\sim R\omega/kT$ terms, which are usually very small [for $P \sim P_L \sim \gamma/R$ these terms are of the order of $(1\text{-}10)\,a/R$], we can also ignore the dependence $\partial c_{vR}^0/\partial c_1$ on the inclusion radius and we can assume that its derivative is the same as that in the interior of the host crystal $(\partial c_{vR}^0/\partial c_1 \approx \partial c_v^0/\partial c_1)$.

Equations (13.4) and (13.22) for the concentrations of atoms and vacancies can be used, in combination with Eq. (12.1), to determine the fluxes of the A and B atoms in the bulk of the host crystal. The diffusion fluxes at the inclusion—host interface can be determined, as in the case when $l \ll R$, from Eqs. (12.7), (13.5), and (2.19), whereas the diffusion fluxes across an inclusion can be found from Eqs. (12.41), (13.4), (12.42), and (12.44). Substituting the expressions obtained into Eq. (12.24), we find a system of equations which determines the constant A and the velocity of inclusions $\vec{v}'$:

$$\left.\begin{aligned}
\vec{v}'nc_1 &= b_1\frac{A}{R}\vec{e}\vec{n} + a_1^c\vec{n}\nabla c_{1\infty} + 3F_1\vec{n}\nabla c_{v\infty}, \\
\vec{v}'nc_2 &= -b_2\frac{A}{R}\vec{e}\vec{n} - a_2^c\vec{n}\nabla c_{1\infty} + 3F_2\vec{n}\nabla c_{v\infty}.
\end{aligned}\right\} \qquad (13.23)$$

Here, b_i is still given by the expressions in Eq. (13.7) [this can be demonstrated readily by employing the relationship (12.5) between D_i and D_i] and a_i^c is given by the formula

$$a_i^c = -D_i - 2F_i\frac{\partial c_v^0}{\partial c_i} + \frac{2a}{R}\chi D_{si} + \frac{\omega}{\omega'}\delta_i D_i'\frac{\partial c_i'^0}{\partial c_i}. \qquad (13.24)$$

The following expressions are obtained for A and $\vec{v}'$ from Eq. (13.23):

$$A = -\frac{c_2 a_1^c + c_1 a_2^c}{c_2 b_1 + c_1 b_2}\,R\vec{e}\nabla c_{1\infty} - 3\frac{c_2 F_1 - c_1 F_2}{c_2 b_1 + c_1 b_2}\,R\vec{e}\nabla c_{v\infty}, \qquad (13.25)$$

$$\vec{v}_1 = \frac{a_1^c b_2 - a_2^c b_1}{c_1 b_2 + c_2 b_1} \nabla c_{1\infty} + 3 \frac{F_1 b_2 + F_2 b_1}{c_1 b_2 + c_2 b_1} \nabla c_{v\infty}. \tag{13.26}$$

In a system of coordinates linked to the host matter far from an inclusion, the velocity $\vec{v}$ follows from Eqs. (12.3) and (12.15):

$$\vec{v} = \vec{v}' - \vec{v}_a, \qquad \vec{v}_a = (D_2 - D_1)\nabla c_{1\infty} + (F_1 + F_2)\nabla c_{v\infty}. \tag{13.27}$$

The velocity of inclusions given by Eq. (13.26) or (13.27) depends strongly on the conditions under which vacancies are created and annihilated in a crystal, i.e., on the average distance l between the vacancy sources and sinks. If l is much less than the characteristic length L in which the gradients of the concentration of atoms or vacancies can be regarded as constant (L can be the thickness of the diffusion layer or the dimension of a neighboring inclusion which strongly distorts the concentration gradient), we find that the local concentration of vacancies at a point far from an inclusion is governed by the local concentration of atoms and that the gradients of the concentrations of atoms and vacancies are related by

$$\nabla c_{v\infty} = \frac{\partial c_v^0}{\partial c_1} \nabla c_{1\infty}. \tag{13.28}$$

If we use Eq. (13.28) to express $\nabla c_{v\infty}$ in terms of $\nabla c_{1\infty}$, we find that when $l \ll L$ the expression for the velocity given by Eq. (13.26) includes only the gradient of the concentration of atoms and — according to Eqs. (13.26), (13.24), (13.7), and (12.5) — this expression is completely identical with Eq. (13.9) obtained for $l \ll R$. The values of the velocity $\vec{v}$ given by Eqs. (13.27) and (13.10) are also equal. Thus, the criterion which determines the validity of Eq. (13.9) and that of the formulas (13.11) and (13.15) which follow from it is not really the condition $l \ll R$ (which is used in the derivation) but a much less stringent condition $l \ll L$.

However, if the condition $l \ll L$ is not satisfied and l is comparable with or larger than the length L (in particular, if L is small compared with the dimensions of the grains in a polycrystalline host), we find that the gradients $\nabla c_{v\infty}$ and $\nabla c_{1\infty}$ are no longer related by Eq. (13.28) and can be independent of each other. In this case, the velocity of an inclusion $\vec{v}'$ given by Eq. (13.26) can be represented as the sum of two terms $\vec{v}_c'$ and $\vec{v}_v'$, which are proportional to the gradients of the concentrations of atoms and vacancies, respectively:

$$\vec{v}_c' = \frac{a_1^c b_2 - a_2^c b_1}{c_1 b_2 + c_2 b_1} \nabla c_{1\infty}, \qquad \vec{v}_v' = 3 \frac{F_1 b_2 + F_2 b_1}{c_1 b_2 + c_2 b_1} \nabla c_{v\infty}. \qquad (13.29)$$

If we can ignore the diffusion fluxes across an inclusion, we find that Eqs. (13.29), (13.7), and (13.24) yield the following equation for small-radius inclusions

$$\vec{v}_c' = 3 \frac{D_{S1} D_2 - D_{S2} D_1}{\bar{D}_S} \nabla c_{1\infty}, \qquad \vec{v}_v' = 3 \frac{F_1 D_{S2} + F_2 D_{S1}}{\bar{D}_S} \nabla c_{v\infty} \qquad \left(\frac{a}{R} D_{Si} \gg D_i \right),$$
$$(13.30)$$

and a different expression for large-radius inclusions

$$\vec{v}_c' = -3 \frac{F_1 D_2 + F_2 D_1}{\bar{D}} \frac{\partial c_v^0}{\partial c_1} \nabla c_{1\infty}, \qquad \vec{v}_v' = 3 \frac{F_1 D_1 + F_2 D_1}{\bar{D}} \nabla c_{v\infty} \qquad \left(\frac{a}{R} D_{Si} \ll D_i \right).$$
$$(13.31)$$

The velocity of small-radius inclusions is independent of R and is of the same order of magnitude as the velocity given by Eq. (13.11) when $l \ll L$. However, the velocities given by Eqs. (13.30) and (13.11) can differ considerably if the relationship (13.28) is not satisfied.

If $l \gg L$, the velocity $\vec{v}_c'$ of large-radius inclusions is also independent of the radius (the dependence on R appears if $a D_{Si} \sim R D_i$) but it differs from the velocity of small-radius inclusions (if $l \ll L$, we find that $v' \sim 1/R$). According to Eqs. (13.31) and (13.27), the velocities v_c' and v_v' are usually of the same order of magnitude as the velocity v_a of the matter relative to the lattice (or the velocity of the lattice relative to the boundaries of the crystal). Therefore, if the relationship (13.28) is not satisfied, the total velocity of an inclusion $\vec{v}' = \vec{v}_c' + \vec{v}_v'$, due to the volume diffusion fluxes of atoms in the host crystal does not vanish to within terms $\sim v_c' D_{Si} a (D_i R)^{-1}$ (as in the case when $l \ll L$) and again this velocity is of the order of v_a. In this case, inclusions cannot act as immobile markers.

However, it should be mentioned that usually the width of the diffusion zone is large compared with l. Therefore, the unbalance of the gradients of the concentrations of atoms and vacancies [the departure from Eq. (13.28)] is more likely when the gradients are distorted by a neighboring inclusion.

Influence of Volume Diffusion in the Host on the Migration of Inclusions in a Force Field

The influence of volume diffusion in the host crystal on the motion of inclusions in an external force field (for example, a temperature gradient) depends — as in the case of a concentration gradient — on the density of vacancy sources and sinks. This dependence is due to the fact that, as mentioned earlier, the diffusion fluxes of the A and B atoms resulting from external forces can be balanced in the host crystal by an inhomogeneous distribution of concentrations of atoms and vacancies and the expressions for the fluxes associated with the gradients of these concentrations are different for large and small values of l.

If l is small and the temperature gradient gives rise to a concentration gradient, the volume fluxes in the host substance are given by Eqs. (12.10), (12.4), (2.9), and (13.4) (if $\nabla c_{1\infty} = 0$). If in Eq. (12.24) we add these fluxes $\vec{I}_i$ to the fluxes $\vec{I}_{Si}$ and $\vec{I}_l'$ on the inclusion—host interface and across an inclusion, which have been allowed for in the derivation of Eq. (12.46), we can solve the resultant system of equations and find that the velocity of an inclusion is still given by the formula (12.46). All that changes are the expressions for the quantities a_i^T and b_i, which occur in this formula:

$$\left.\begin{aligned}
a_i^T &= -c_i D_i^{\bullet}\frac{\alpha_i}{T}\frac{1-2\varkappa}{1+\varkappa} + \frac{2a}{R}c_{Si}D_{Si}^*\frac{\alpha_{Si}}{T} + \frac{\omega}{\omega'}\delta_i\left(D_i\frac{\partial c_i'^0}{\partial T} + \frac{\alpha_i'\, c_i'^0 D_i'^{\bullet}}{T}\right), \\
b_i &= 2\bar{D}_i + \frac{2a}{R}\chi D_{Si} + \frac{\omega}{\omega'}\delta_i D_i\frac{\partial c_i'^0}{\partial c_i}.
\end{aligned}\right\} \quad (12.32)$$

The velocity of matter $\vec{v}_a$ far from an inclusion and the velocity of an inclusion relative to the host matter are now given by the formulas:

$$\left.\begin{aligned}
\vec{v}_a &= -\frac{1-2\varkappa}{T}\left(c_1\alpha_1 D_1^* + c_2\alpha_2 D_2^{\bullet}\right)\nabla T, \\
\vec{v} &= \vec{v}' - \vec{v}_a.
\end{aligned}\right\} \quad (13.33)$$

In particular, if we can neglect the diffusion across an inclusion, we find that the velocity of small-radius inclusions (in the case when $a D_{Si} \gg RD_i$) is given by Eq. (12.33) and that it is considerably larger than v_a, i.e., it is almost identical in the system of coordinates linked to the lattice and to the host matter. This velocity depends on the radius as $1/R$.

In the case of large-radius inclusions, when $RD_i \gg a D_{Si}$, the velocity $\vec{v'}$ of inclusions relative to the lattice is

$$\vec{v'} = -(1-2\varkappa)\frac{c_t\,\alpha_1 D_1^{\bullet}\,\overline{D_2} + c_2\,\alpha_2\,D_2^{\bullet}\overline{D_1}}{T\widetilde{D}}\nabla T_\infty + (1+\varkappa)\frac{2a}{R} \times$$
$$\times \frac{c_{S1}\,\alpha_{S1}\,D_{S1}^{\bullet}\overline{D_2} + c_{S2}\,\alpha_{S2}\,D_{S2}^{\bullet}\overline{D_1}}{T\widetilde{D}}\nabla T_\infty. \qquad (12.34)$$

If the thermal conductivity of an inclusion is very low and $\varkappa \approx 1/2$ [see Eq. (2.7)], we find that even when $R\overline{D_i} \gg a D_{Si}$, the second term in Eq. (13.34) is greater than the first term. As in the case of small-radius inclusions, the thermal diffusion fluxes appear primarily on the inclusion—host interface (the volume diffusion fluxes in the host are due to an inhomogeneity in the concentration, which appears in the presence of a temperature gradient) and the velocity of inclusions⁻ relative to the lattice is proportional to $1/R$. Consequently, a partial relaxation of the surface concentration gradients is now due to volume diffusion fluxes (and not surface fluxes) and all that changes is the coefficient of proportionality in the relationship $v' \propto 1/R$. However, the dependence of v' on the concentration (in particular, the possibility of a large reduction in the velocity as a result of the addition of a small number of impurities) is now governed by the relationships between the volume diffusion coefficients $\overline{D}_1$ and $\overline{D}_2$ and not by the relationships between D_{S1} and D_{S2}, as in the case of small-radius inclusions. The dependence of v' on R for $\varkappa \approx 1/2$ becomes more complex in the intermediate range of radii, where $\overline{D}_i R \sim a D_{Si}$ and the contributions of the volume and surface diffusion fluxes are comparable.

It is evident from Eqs. (13.34) and (13.33) that, in the case of large-radius inclusions with a low thermal conductivity, the velocity v' relative to the lattice is considerably less than the velocity v_a of the host matter far from an inclusion. In this case, $\vec{v}$ of an inclusion relative to the matter (this velocity is equal to the velocity relative to the boundaries of a sample if the temperature gradient ∇T_{∞} is constant throughout the sample) is approximately equal to $-\vec{v}_a$, and is independent of the inclusion radius.

In the case of large-radius inclusions with a relatively high thermal conductivity, the velocity relative to the lattice is governed by the first term in Eq. (13.34). This velocity is independent of the inclusion radius and its order of magnitude is the same as that of v_a.

Similar general dependences of the velocity on the inclusion radius and on the composition of a solid solution should apply also in the migration of inclusions in alloys subjected to an electric field.

Experimental Investigations of the Migration of Inclusions in the Field of a Concentration Gradient

The most convenient object for experimental studies of the migration of foreign inclusions in the field of a concentration gradient is the zone which is formed and expands as a result of the interdiffusion of the components forming a continuous series of solid solutions. A considerable concentration gradient is established in such a diffusion zone, i.e., the conditions are favorable for the migration of inclusions.

We shall consider the cases when the interdiffusion occurs under the condition of unequal partial diffusion coefficients, the latter which determine the magnitudes of oppositely directed fluxes of the components $j_{A \to B}$ and $j_{B \to A}$ in the diffusion zone. That component of a diffusion pair which produces the larger flux (for example, the component A) has a source of vacancies whose strength is $\sim |j_{A \to B} - j_{B \to A}|$.

A source of atoms of the same strength acts in the component B. Naturally, we do not mean that new atoms are generated but simply that A atoms arrive in the component B in numbers which exceed the available vacancies in B. Since the sources of vacancies and atoms are spatially separated, it is also necessary to have spatially separated sinks. Dislocations can act as sinks for atoms. The motion of dislocations is accompanied by the absorp-

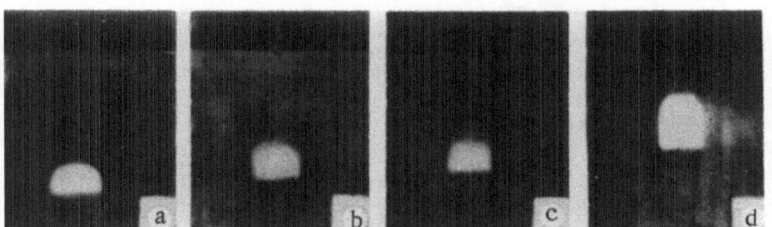

Fig. 44. Downward migration of a pore in a diffusion zone in a KCl–KBr system, relative to the initial contact plane [82].

tion of atoms and it gives rise to new atomic planes in B. For vacancies, the sinks may be in the form of dislocations or microcracks, where the absorbed vacancies produce pores. These pores behave as foreign inclusions, which should migrate under the influence of, say, a temperature gradient.

The migration of pores in the diffusion zone was investigated experimentally by Geguzin et al. [82]. They used diffusion pairs made of KCl and KBr crystals, in which the migration of a single pore was studied in the diffusion zone. This could be done because of the optical transparency of the KBr crystals in which nucleation, growth and migration of pores were observed.

The sequences of events presented in Fig. 44 illustrates the migration of a pore in the diffusion zone in a KBr single crystal. During isothermal annealing, the velocity of the pores decreased (Fig. 45) because they traveled across into a diffusion zone with a lower value of $|\nabla c|$.

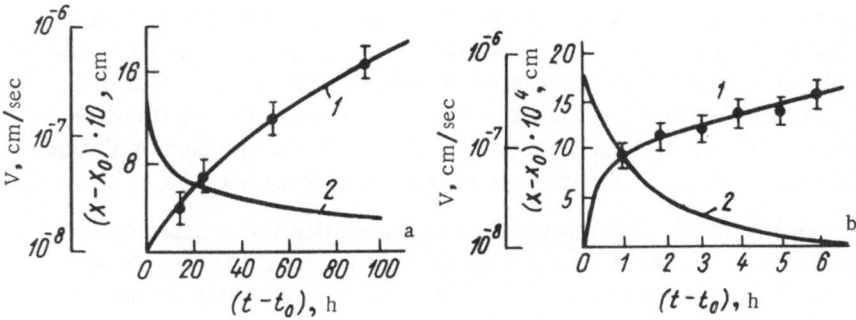

Fig. 45. Time dependences of the displacement (1) and velocity (2) of pores in diffusion zones of the KCl–KBr (a) and Cu–Ni (b) systems [82].

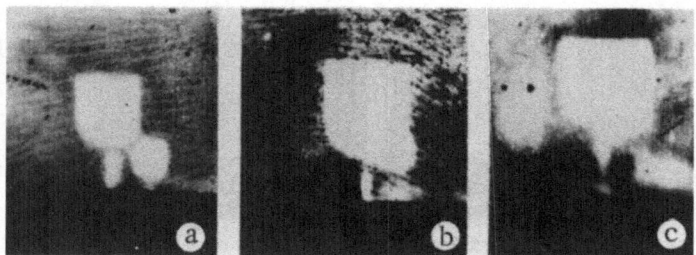

Fig. 46. Collision of pores formed in a diffusion zone in the
KCl—KBr system (a, b, and c are successive stages of the collision) [82].

The direct consequence of the dependence of the velocity of inclusions on ∇c was the coalescence of the colliding pores because
the pores located closer to the diffusion contact plane, i.e., the region with a high value of ∇c, migrated fast and caught up with the
pores in the region with a lower value of $|\nabla c|$. This coalescence
mechanism is illustrated by micrographs in Fig. 46.

The coalescence of the pores in the diffusion zone occurred
not only because of the collisions between pores moving at different velocities but also because of the collisions which occurred
during the growth of the pores which absorbed excess vacancies
(Fig. 47).

The motion of pores in the diffusion zone of the Cu—Ni system
was investigated in three-layer samples of the Ni—Cu—Ni type
[82]. The migration of pores was accompanied by coalescence, as
shown in Fig. 48. It is evident from this figure that the distance
between two rows of pores, located in the copper parallel to the
plane of contact between the copper and the nickel, decreased
during isothermal annealing. The time dependence of the displacement, which was defined as half the reduction in the distance
between two rows of pores, is shown clearly in Fig. 48.

Bailly et al. [106] studied experimentally the migration of
cylindrical channels created artificially in the CdTe—HgTe system. Such migration was observed during high-temperature annealing of the diffusion zone. Scratches were made on the surface
of one of the components of the diffusion pair: after high-temperature welding of the CdTe—HgTe pairs, these scratches changed
to cylindrical channels located in the scratched crystal.

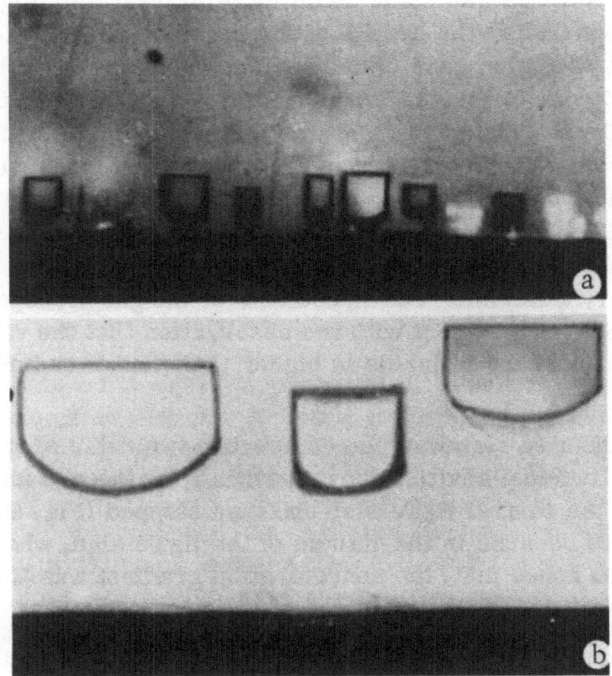

Fig. 47. Collision-induced enlargement of pores migrating in a diffusion zone (a and b are successive
stages) [82].

Some important observations were made in the experiments
with the CdTe−HgTe system. It was established that, irrespective of which of the components was scratched, the resultant channel always migrated in HgTe. Under isothermal conditions, the
displacement of the channel was proportional to $t^{1/2}$. This law was
in agreement with the hypothesis that the displacement of a channel was due to the propagation of a diffusion front.

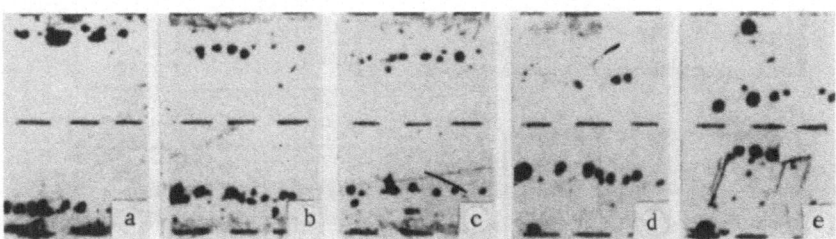

Fig. 48. Convergence of pores in a three-layer Ni−Cu−Ni sample (a-e
are successive stages) [82].

When a scratch was made on an HgTe crystal, the migrating cavity did not distort the profile of the constant-concentration surfaces in the diffusion zone (these surfaces remained parallel to the original contact plane). When a scratch was made on the surface of CdTe, the diffusion front was distorted as shown in Fig. 49. This observation demonstrated indirectly that the transport of matter from the front to the rear of a migrating cylinder was controlled by the diffusion either in the occluded gas or on the surface of the cylinder. In both cases, the front of the cavity was enriched with mercury telluride. The hypothesis of the gas transport mechanism was in agreement with the observation that the vapor pressure of mercury telluride is higher than that of cadmium telluride.

When an HgTe slab was placed between two CdTe single crystals, the cylindrical cavities located initially at the two interfaces moved into the central HgTe slab and then stopped (Fig. 50). This was expected because in the middle of the HgTe slab, where the two diffusion zones met, the concentration gradient was $\nabla c = 0$ and there was no force on the inclusion.

The dominant role played by the value of $|\nabla c|$ in the motion of cavities in a diffusion zone was supported also by the following experiment. Bailly et al. [106] studied the motion of cylindrical channels in a three-layer sample of the CdTe−HgTe−HgTe type (Fig. 51). These channels were initially located on the two interfaces. A cavity formed at the CdTe−HgTe interface started to migrate right from the beginning of the diffusion annealing treatment. The cavity located at the HgTe−HgTe interface migrated

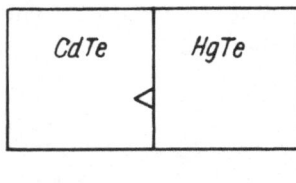

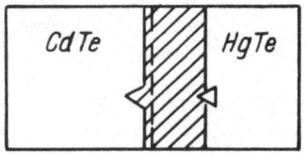

Fig. 49. Migration of channels in a diffusion zone of the CdTe−HgTe system [106].

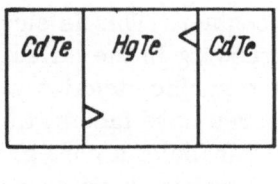

Fig. 50. Migration experiments in three-layer CdTe−HgTe−CdTe samples [106].

only when the other cavity approached it. When the two cavities reached the same distance from the CdTe−HgTe interface they continued to move at the same velocity because they were in the region of identical values of ∇c.

The velocity of pores in a diffusion zone can be estimated using Eq. (13.11) and assuming that $a\chi \tilde{D}_s \gg R\tilde{D}$ and that the transport of matter across the host crystal can be ignored. Then, if the values of $\bar{D}_1$ and $\bar{D}_2$ (or D_{S1} and D_{S2}) differ considerably, we find that $v' \propto \tilde{D}\nabla c_\infty$. In a diffusion zone, we find that $| \nabla c | \propto (\tilde{D}t)^{-1/2}$; hence, it follows that $v' \propto (\tilde{D}/t)^{1/2}$. In the experiments carried out on KCl−KBr pairs at 660°C, the relevant parameters were $\tilde{D} \sim 10^{-10}$ cm^2/sec, $t \sim 10^4$ sec, whereas in the Cu−Ni system at 950°C, $\tilde{D} \sim 10^{-9}$ cm^2/sec. In both cases, $v' \sim 10^{-7}$ cm/sec, i.e., it was close to the experimentally observed value.

The migration of a pore as a whole may be inhibited by a flux of vacancies directed toward the pore from the boundary of a dif-

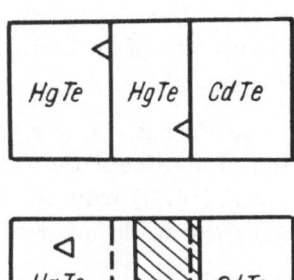

Fig. 51. Migration experiments in three-layer HgTe−HgTe−CdTe samples [106].

fusion contact. This is supported by the dynamic shape of the pores growing in the diffusion zones in alkali halide crystals [82]. Since the surface tension in these crystals is strongly anisotropic, these pores have facets, i.e., they are "negative crystals." The dynamic shape of a pore growing in a diffusion zone differs from the equilibrium shape in that a part of the surface facing the diffusion contact plane is elongated toward this plane (Fig. 47). This feature is due to the fact that the flux of vacancies toward the pore exceeds the flux necessary for equilibrium faceting.

It is worth noting that the degree of departure of the dynamic from the equilibrium shape, which can be represented conveniently by the ratio of the sagitta of the convex part of the surface to the linear dimensions of the pore x/R, should decrease with diminishing pore size. At some critical value, R^*, the ratio x/R^* approaches zero. This occurs when $D_v R^* \sim D_S a$ and it can be used to estimate the value of $D_S a$.

14. NONLINEAR EFFECTS IN THE MIGRATION

OF PORES IN SOLID SOLUTIONS

In the preceding sections, we considered the migration of inclusions in solid solutions subject to the condition

$$v' \ll \overline{D}_i / R \qquad (14.1)$$

i.e., when atoms diffuse over a long distance in the time in which an inclusion migrates a distance R. In this case, the distribution of the concentration of atoms is determined by the steady-state solution of the diffusion equation, and the velocity of an inclusion is a linear function of the external driving forces.

However, since the volume diffusion coefficients $\overline{D}_i$ are frequently small, in some cases the criterion of applicability of the linear theory to solid solutions, given by Eq. (14.1), may be violated and the opposite condition may apply even in the case of relatively weak driving forces:

$$v' \gg \overline{D}_i / R. \qquad (14.2)$$

For example, if $D_{Si} \sim 10^{-6}$ cm^2/sec, $\bar{D}_i \sim 10^{-13}$ cm^2/sec, $R \sim 10^4 a \sim 3 \times 10^{-4}$ cm, $\alpha_{Si} \sim 10$, $T \sim 10^3$ °K, $|\nabla T_\infty| \sim 10^3$ deg/cm, it follows from Eq. (12.33) that $v' \sim 10^{-9}$ cm/sec and $v'R > D_i$, i.e., the condition (14.1) is not obeyed.

When Eq. (14.1) is not satisfied, the distribution of the concentration of atoms around an inclusion in a coordinate system linked to the lattice is no longer in a steady state (in this case, the diffusion equation must include the term $\partial c_i / \partial t$) but it depends on the velocity of the inclusion. Therefore, the diffusion fluxes and, consequently, the velocity of inclusions are found to depend nonlinearly on the external forces.

Moreover, if an inclusion is a micropore and an allowance is made for the nonlinear effects, it is found that the diffusion fluxes of atoms cease to be proportional to cos θ, even in the isotropic case. Therefore, the pore is distorted and the distortion depends quadratically, or in a more complex manner, on its velocity (or on the applied forces). This distortion is somewhat different from that associated with the anisotropy of the properties of the pore's surface (Sec. 7).

Another cause of nonlinear effects is the dependence of the transport coefficients on the temperature and composition. In the nonlinear theory, this dependence leads to an inhomogeneity of diffusion fluxes, even when the external forces are homogeneous. Consequently, once again a pore is distorted during its migration. This effect is not specific to solid solutions and should also occur in one-component crystals. However, it is manifested only when the driving forces acting on inclusions are very large.

The above nonlinear effects associated with the migration of inclusions in solid solutions will be analyzed by considering the motion of micropores in a temperature gradient or a concentration gradient [65]. We shall limit our analysis to the range of temperatures in which atoms hardly evaporate from the surface of a pore and in which we can ignore the diffusion fluxes across the pore. We shall thus consider only the volume diffusion in the host crystal and the diffusion along the pore–host interface. We shall discuss the specific case of crystals with a high density of vacancy sources and sinks, i.e., the case when the condition $l \ll R$ is satisfied.

Distribution of the Concentration of

Atoms around a Rapidly Moving Pore

If a pore is migrating rapidly, the distribution of the concentrations of the A and B atoms in the bulk of the host is not governed by the Laplace equation but by the complete diffusion equation:

$$\frac{\partial N_i}{\partial t} + \operatorname{div} \vec{I}_i = 0. \tag{14.3}$$

In crystals with a high density of vacancy sources and sinks, the fluxes of atoms $\vec{I}_i$, due to temperature or concentration gradients are given by Eqs. (12.4) and (12.10) in a system of coordinates linked to the crystal lattice. Near a distorted pore we must also allow for the presence of additional diffusion fluxes

$$\vec{I}_i = N_0 \, c_i \, \frac{\omega_i \, D_i^{\bullet}}{3kT} \, \nabla \sigma_{ll}, \tag{14.4}$$

which are due to the inhomogeneous field of internal stresses generated by the Laplace pressure. The quantities ω_i in Eq. (14.4) are of the order of the atomic volume and include the factors $1/f_i$.

The concentrations of atoms in imperfect crystals ($l \ll R$) change as a result of the absorption or annihilation of vacancies (if $\operatorname{div} \vec{I}_v = -\operatorname{div} (\vec{I}_1 + \vec{I}_2) \neq 0$). It follows from Sec. 12 that the lattice near a rapidly moving pore is deformed in a region $r \sim R$ (if a pore moves slowly, $\operatorname{div} \vec{I}_v = 0$ and this deformation is absent). Since different parts of the lattice in the region $r \sim R$ move relative to one another, it is more convenient to discuss the diffusion fluxes, the velocity of a pore $\vec{v}'$, and the rate of flow of the lattice $\vec{w}'(r)$ in a system of coordinates linked to those parts of the lattice which are far from a pore in the region where $r \gg R$. If the values of ∇T_∞ and $\nabla c_{1\infty}$ are constant, it follows from the linear theory that this distant region is not deformed and is immobile relative to the ends of a crystal (Sec. 12); however, if ∇T_∞ and $\nabla c_{1\infty}$ vary smoothly over distances $\sim L \gg R$, we find that the one-dimensional deformation and motion of this region relative to the ends of the crystal are given by Eq. (12.21).

As a result of the deformation of the lattice at the rate $\vec{w}'(r)$, we find that, in addition to the fluxes of atoms given by Eqs. (12.10) and (12.4) and resulting from temperature and concentration gradients, there are additional fluxes

$$\vec{I}_i = N_i(\vec{r})\,\vec{w}'(\vec{r}),   \tag{14.5}$$

associated with the motion of a given region of the lattice as a whole.

If we substitute Eqs.(12.4), (12.10), (14.4), and (14.5) into Eq. (14.3) and if we bear in mind that the temperature distribution can be regarded as steady-state and that it obeys the equation $\Delta T = 0$ (because the thermal diffusivity is much larger than the diffusion coefficient), we obtain the following temporal diffusion equation:

$$\frac{\partial N_i}{\partial t} = -\operatorname{div}\vec{I}_i = N_0\,\bar{D}_i\,\Delta c_i + N_0\,\nabla D_i\,\nabla c_i + N_0\,\nabla\left(\frac{c_i\,\alpha_i\,D_i^*}{T}\right)\nabla T -$$

$$- N_0 c_i\frac{\omega_i\,D_i^*}{3kT}\,\Delta\sigma_{ll} - \frac{N_0}{3k}\,\nabla\left(\frac{c_i\,\omega_i\,D_i^*}{T}\right)\nabla\sigma_{ll} - \operatorname{div}(N_i\vec{w}').   \tag{14.6}$$

Here, N_i and N_0 refer to a unit volume of the deformed and undeformed parts of a crystal, respectively.

For the sake of simplicity, we shall assume that far from a pore the gradients ∇T_∞ and $\nabla c_{i\infty}$ are initially constant over the whole crystal (i.e., $\Delta c_i = 0$ for $r \gg R$). In the nonlinear (with respect to ∇c and ∇T) theory, the deformation of a crystal and the change in the concentration are once again determined by the dependences of the transport coefficients on the coordinates [Eq. (14.6)] but a significant inhomogeneity of the concentration does not arise if the time is relatively short.

If we confine ourselves to the terms which are quadratic in ∇T_∞ and $\nabla c_{1\infty}$, we find that, with the exception of the region next to the pore and certain regions near the boundary of a crystal, $\partial N_{1\infty}/\partial t = -\partial N_{2\infty}/\partial t$ and $\operatorname{div}\vec{w}'_\infty$ are constant in time and space. According to Eq. (14.6), these quantities are given by the expressions

$$\left.\begin{aligned}\frac{\partial N_{i\infty}}{\partial t} &= N_0 k_i - \operatorname{div}\left(N_{i\infty}\vec{w}'_\infty\right),\quad \operatorname{div}\vec{w}'_\infty = k_1 + k_2,\\[2mm] k_i &= \nabla\bar{D}_{i\infty}\,\nabla c_{i\infty} + \nabla\frac{c_{i\infty}\,\alpha_{i\infty}\,D_{i\infty}^*}{T}\,\nabla T_\infty.\end{aligned}\right\}   \tag{14.7}$$

Here, all the quantities represent regions far from a pore, where $c_i \equiv c_{i\infty}$ and $T \equiv T_\infty$ depend linearly on the coordinates.

The diffusion equations can conveniently be considered in a system of coordinates linked to a pore moving at a velocity $\vec{v}'$. After a time interval equal to the volume diffusion relaxation time

$$\tau \sim R^2/\tilde{D} \qquad (14.8)$$

the distribution of the concentration in the new system of coordinates reaches a quasisteady state and varies slowly with time, in accordance with some simple law. [In a system of coordinates linked to the lattice, a rapidly moving pore dragging a region of inhomogeneous composition produces a much greater change in the concentration.]

This quasisteady-state distribution of the concentration varies with time because of a uniform "background" change in the concentration throughout the crystal at a rate given by Eq. (14.7), and because a pore moving in a crystal of variable composition may pass through regions with different concentrations. The corresponding change in the volume concentration per unit time is $\vec{v}' \nabla N_{i\infty} = N_0 \vec{v}' \nabla c_{i\infty}$. Bearing in mind that

$$N_i = c_i (N_1 + N_2), \qquad \frac{\partial N_i}{\partial t} = (N_1 + N_2)\frac{\partial c_i}{\partial t} - c_i \operatorname{div} (\vec{I}_1 + \vec{I}_2),$$

i.e., that

$$(N_1 + N_2)\frac{\partial c_1}{\partial t} = c_1 \operatorname{div} \vec{I}_2 - c_2 \operatorname{div} \vec{I}_1, \qquad (14.9)$$

we can write the diffusion equations in the system of coordinates linked to the pore:

$$\tilde{D}\Delta c_1 + \frac{N_1 + N_2}{N_0}\left(\vec{v}'\nabla c_1 - \vec{v}'\nabla c_{1\infty}\right) + c_1 c_2 \frac{\omega_2 D_2^\bullet - \omega_1 D_1^\bullet}{3kT}\Delta\sigma_{ll} +$$

$$+ (c_1 \nabla \tilde{D}_2 + c_2 \nabla \tilde{D}_1)\nabla c_1 + \left[c_2 \nabla\left(\frac{c_1 \alpha_1 D_1^\bullet}{T}\right) - c_1 \nabla\left(\frac{c_2 \alpha_2 D_2^\bullet}{T}\right)\right]\nabla T -$$

$$- \frac{1}{3k}\left[c_2 \nabla\left(\frac{c_1 \omega_1 D_1^\bullet}{T}\right) - c_1 \nabla\left(\frac{c_2 \omega_2 D_2^\bullet}{T}\right)\right]\nabla\sigma_{ll} + \frac{1}{N_0}(c_1 \nabla N_2 - c_2 \nabla N_1)\vec{w}' -$$

$$- (c_1 k_2 - c_2 k_1) = 0 \quad (c_1 + c_2 = 1). \qquad (14.10)$$

The boundary conditions for Eq. (14.6) or (14.10) are the conditions at infinity given by Eq. (13.1) and the condition of continuity of the number of atoms on the surface of a pore given by Eq. (12.24), in which we must now substitute $\vec{I}'_t = 0$. If surface fluxes are considered, the latter condition must include the fluxes (12.11) and (12.7) in the field of temperature and concentration gradients as well as the fluxes

$$\vec{I}_{Si} = 2N_0\, c_{Si}\, \omega'_{Si}\, \frac{D_{Si}\, a}{f_{Si}\, kT}\, \gamma \nabla_S K \tag{14.11}$$

[see Eq. (7.7)] associated with the inhomogeneity of the curvature of the pore's surface (ω'_{Si} are of the order of the atomic volume), as well as the surface fluxes (12.13) in an inhomogeneous field of internal stresses around a distorted pore.

The solutions of the diffusion equations given above will be quite different for slowly moving pores, which satisfy the condition (14.1), and for rapidly moving pores, when the opposite condition (14.2) is obeyed.

In the case of low velocities, Eq.(14.10) can be solved, subject to the boundary conditions (13.1) and (12.24), by the method of successive approximations, in which it is assumed that ∇T, $\nabla c_{i\infty}$ and $\vec{v}'$ are all proportional to some small parameter ξ. In the linear theory, the parts of the lattice near a pore are not displaced relative to one another so that $\vec{w}'(\vec{r}) \propto \xi^2$. In the first approximation, we can reject terms of the order of ξ^2 and we may simplify Eq. (14.10) by substituting $\vec{v}'=0$, $\vec{w}'=0$, $\Delta\sigma_{ll}=0$, $\nabla\vec{D}_i=0$, $\nabla\left(\dfrac{c_i\alpha_i D_i^\bullet}{T}\right)=0$, $\nabla\left(\dfrac{c_i\omega_i\vec{D}_i}{T}\right)=0$. In the boundary conditions (13.1) and (12.24), we can replace c_i with c_{i_0}, substitute $\vec{w}'=0$, $\nabla\sigma_{ll}=0$, and ignore the dependences of all the transport coefficients on the coordinates.

In this linear approximation, the distribution of atoms is given by the solution (13.4) of the Laplace equation, and the pore retains its spherical shape during migration. The velocity of the pore and the constant A in Eq. (13.4) are now given for formulas analogous to Eqs. (13.8) and (13.9):

$$A = (1+\varkappa)\frac{c_1\, a_2^T - c_2\, a_1^T}{c_2\, b_1 + c_1\, b_2}\, Re\, \vec{\nabla}T_{\cdots} - \frac{c_2\, a_1^c + c_1\, a_2^c}{c_2\, b_1 + c_1\, b_2}\, Re\, \vec{\nabla}c_{1\infty}, \tag{14.12}$$

$$\vec{v}' = (1 + \varkappa) \frac{a_1^T b_2 + a_2^T b_1}{c_1 b_2 + c_2 b_1} \nabla T_\infty + \frac{a_1^c b_2 - a_2^c b_1}{c_1 b_2 + c_2 b_1} \nabla c_{1\infty}. \qquad (14.13)$$

Here, a_i^T, a_i^c, and b_i are given by Eqs. (13.7) and (13.32), in which we must substitute $D_i' = 0$ and $D_i'^* = 0$.

It follows from Eqs. (13.4) and (14.12) that the correction to the concentration is proportional to the small parameter ξ. It is clear from Eq. (13.4) that the angular dependences of the changes in the concentration is given by the factor $\cos \theta$.

The next approximations with respect to the small parameter ξ yield the higher-order corrections to the distribution of atoms (the angular dependences of these corrections are governed by higher spherical harmonics) and the expression for the elastic or plastic deformation fields. This allows us to determine the correction to the velocity of a pore and to find its distortion. However, if the condition (14.1) is satisfied, the correction to the velocity and the relative distortion of the pore are very small. Much larger effects may be expected in the case of high velocities, when the condition (14.2) is satisfied. Therefore, we shall now consider specifically the case of high velocities.

At high velocities, the distribution of atoms near a pore changes drastically. The surface diffusion fluxes of the A and B atoms give rise to an inhomogeneous distribution of the concentrations on the surface of a pore. However, in contrast to low velocities, when this inhomogeneity is partly relieved by the diffusion fluxes, at high velocities there is hardly any relaxation of the inhomogeneity in the bulk of the host crystal because, during the time (of the order of $R^2/\tilde{D}$) needed to establish volume fluxes in the region $r \sim R$, the pore travels a large distance away from this region. During the period that a rapidly moving pore spends in this region, volume diffusion can occur only in a thin layer near the pore surface (the thickness of this layer is $\tilde{D}/v'$, which is small compared with the radius R).

In discussing rapidly moving pores, we shall assume that the surface diffusion coefficients of the A and B atoms are sufficiently large so that the following conditions are satisfied:

$$aD_{Si} \gg R\bar{D}_i \left| \frac{\partial \ln \bar{D}_i}{\partial c_i} \right|, \qquad aD_{Si} \gg R\bar{D}_i \left| \frac{\partial \ln \bar{D}_i}{\partial \ln T} \right|. \qquad (14.14)$$

This means that we can satisfy the criterion

$$v' \ll \frac{aD_{Si}}{R^2} \qquad (14.15)$$

as well as the criterion (14.2). It should be noted that a departure from the conditions expressed in Eq. (14.14) [when the conditions of Eq. (14.2) are satisfied] or in Eq. (14.15) is possible only under the action of extremely strong driving forces.

It is evident from Eq. (14.13) that the condition (14.2) can be satisfied only in a temperature gradient but not in a concentration gradient (because $R \mid \nabla c_{1\infty} \mid < 1$). Therefore, we shall assume that $\nabla c_{1\infty} = 0$ and we shall consider only the forces associated with ∇T_∞.

Even when the velocity of a pore is high, so that the condition (14.2) is satisfied, the results which will be given later show that a change in the concentration near a pore $\delta c_i(\vec{r})$ may be small. If we confine ourselves to the case when

$$|\delta c_i(\vec{r})| \left| \frac{\partial \ln \overline{D}_i}{\partial c_i} \right| \ll 1, \qquad \frac{R \mid \nabla T_\infty \mid}{T} \left| \frac{\partial \ln \overline{D}_i}{\partial \ln T} \right| \ll 1, \qquad (14.16)$$

we can simplify considerably the diffusion equation (14.10). In this case and when the conditions of Eq. (14.14) are satisfied, we can drop the terms containing the products of the gradients (in particular, k_i) as well as $\vec{v}' \nabla c_i$, and we can assume that $(N_1 + N_2)/N_0 \approx 1$ and ignore the term $(c_1 \nabla N_2 - c_2 \nabla N_1) \vec{w}'$. Equation (14.10) then becomes

$$\overline{D} \Delta c_1 + \vec{v}' \nabla c_1 + c_1 c_2 \frac{\omega_2 D_2^* - \omega_1 D_1^*}{3kT} \Delta \sigma_{ll} = 0. \qquad (14.17)$$

The total number of vacancies which disappear per unit volume and per unit time (if the continuity is preserved, this leads to deformation of the lattice) is now given by

$$-\frac{\partial N_v}{\partial t} = -\operatorname{div}(\vec{I}_1 + \vec{I}_2) = N_0(\overline{D}_1 - \overline{D}_2)\Delta c_1 - N_0 \frac{c_1 \omega_1 D_1^* + c_2 \omega_2 D_2^*}{3kT} \Delta \sigma_{ll},$$

$$(14.18)$$

where $\vec{I}_i$ are the fluxes in the system of coordinates linked with the part of the lattice under consideration [the above expression is derived using Eqs. (12.4) and (14.4)].

The internal stress σ_{ll} , which occur in Eq. (14.17), and the rate of deformation $\vec{w}'$, depend on whether plastic or elastic deformation occurs near a pore.

If the stresses exceed the flow limit and the deformation becomes plastic, we can ignore the term containing $\Delta\sigma_{ll}$ in Eq. (14.17). If we assume that the distances l between the vacancy sources and sinks are small compared with the thickness of the diffusion layer next to a pore, $\sim\tilde{D}/v'$, and that the continuity of the medium is preserved during diffusion (i.e., that pores are not formed), we find that the rate of plastic deformation can be calculated from the equation

$$\operatorname{div}\vec{w}' = -\frac{1}{N_0}\frac{\partial N_v}{\partial t} = (\overline{D}_1 - \overline{D}_2)\,\Delta c_1 = \frac{\overline{D}_2 - \overline{D}_1}{\tilde{D}}\vec{v}'\,\nabla c_1. \qquad (14.19)$$

However, if the stresses do not exceed the flow limit and are elastic, the rate of change of the stresses and the rate of deformation $\vec{w}' = \dot{\vec{u}}$ ($\vec{u}$ is the displacement vector) in an elastic isotropic continuum can be found from the equation of the theory of elasticity

$$\dot{\sigma}_{ij} = K\omega\,\dot{N}_v\,\delta_{ij} + K\dot{u}_{ll}\,\delta_{ij} + 2G\left(\dot{u}_{ij} - \frac{1}{3}\dot{u}_{ll}\delta_{ij}\right), \qquad (14.20)$$

where $\dot{N}_v \equiv \partial N_v/\partial t$ is given by Eq. (14.18); K and G are the bulk and shear moduli, respectively.

We shall first consider the case of plastic deformation when we can substitute $\Delta\sigma_{ll} = 0$ in Eq. (14.17). The solution of this equation is quite different for the front and rear halves of a pore. If the condition (14.2) is satisfied, a significant change in the concentration near the front half of a rapidly moving pore can occur (in the time available) only in a thin layer whose thickness $\tilde{D}/v'$ is considerably less than the pore radius. In this region, the problem is one-dimensional and the solution of Eq.(14.17) is of the form

$$c_1\,(\vec{r}) = c_{10} + \delta c_1\,(\vec{r}), \qquad \delta c_1\,(\vec{r}) = \varphi\,(\vec{r}_s)\exp\left(-\frac{\vec{v}'\,\vec{n}}{\tilde{D}}x\right) \qquad (14.21)$$

$$\left(\text{for}\;\; \vec{v}'\,\vec{n} > 0\right), \qquad x = (\vec{r} - \vec{r}_s)\,\vec{n}.$$

Here, $\varphi(\vec{r}_s)$ or $\delta c_1(\vec{r}_s)$ is a function of the point $\vec{r}_s$ on the surface of a pore, which should be determined from the boundary condition applicable to $\vec{r} = \vec{r}_s$.

If we substitute Eq. (14.21) into Eq. (14.19) and integrate the resultant expression for the derivative of $\vec{w}'(r)$ in a thin layer near the surface of a pore, we obtain an expression for the rate of plastic deformation:

$$\vec{w}'\,(\vec{r}) = \vec{w}'_\infty + \frac{\overline{D}_2 - \overline{D}_1}{\overline{D}}\,\vec{v}'\,\vec{n}\,\delta c_1\,(\vec{r})\,\vec{n}. \tag{14.22}$$

Behind a moving pore, Eq. (14.21) gives an exponentially rising solution (for $\vec{v}'\vec{n} < 0$, which is evidently invalid. In the region which becomes filled after the passage of a pore, the concentration varied relatively slowly over distances of ~R. We shall not give explicit expressions for the concentration near the rear half of a pore but we shall point out that because of the slow changes in the concentration, the diffusion fluxes and the rate of plastic deformation are considerably smaller than those for the front half and, therefore, they can be ignored.

We shall now discuss the distribution of the atomic concentrations around a rapidly moving pore under elastic deformation conditions. Such deformation may be due to diffusional growth or the disappearance of dislocation loops near a pore (the density of these loops is such that $\tilde{D}/v' \gg l$). In a stressed thin surface layer near the front half of a pore, only the components $\dot{u}_{nn}$ do not vanish in the derivative of the strain tensor. It follows from the boundary condition $\sigma_{nn} = 0$ at $\vec{r} = \vec{r}_s$ (the stresses σ_{nn} generated by the Laplace pressure and by the gas pressure in a pore can be allowed for separately and included in an additive manner) and from the equilibrium conditions that in this layer $\sigma_{nn} = 0$ for any value of $\vec{r}$. Therefore, it follows from Eq. (14.20) that, in a system of coordinates at rest, we obtain

$$\dot{\sigma}_{ll} = \frac{2\,(1-2v)}{1-v}\,K\omega\dot{N}_v, \qquad \dot{u}_{nn} = -\frac{1+v}{3\,(1-v)}\,\omega\,\dot{N}_v. \tag{14.23}$$

In a system of coordinates linked to a pore, we have $\dot{\sigma}_{ll} = 0$ and therefore, if we use Eq. (14.18), we obtain the following equation for σ_{ll} in the layer under consideration:

$$D_\sigma \Delta\sigma_{ll} + \vec{v}' \nabla\sigma_{ll} + \frac{2(1-2\nu)}{1-\nu} K (\bar{D}_2 - \bar{D}_1)\Delta c_1 = 0,$$

$$D_\sigma = \frac{2(1-2\nu)}{1-\nu} K \frac{c_1\omega_1 D_1^* + c_2\omega_2 D_2^*}{3kT}. \qquad (14.24)$$

The derivative $\dot{N}_V$ represents the rate of change in the concentration N_d of atoms in defects (for example, in dislocation loops) which generate stresses. We may assume that these defects disappear as a result of diffusion at the pore's surface, i.e., that $N_d = 0$ at $\vec{r} = \vec{r}_S$. Since in the one-dimensional case considered here we have $\sigma_{ll} \propto N_d$ [see Eq. (14.23)], we find that the boundary condition for the stresses is

$$\sigma_{ll} = 0 \quad \text{for } \vec{r} \cdot \vec{r}_S. \qquad (14.25)$$

If we solve the system of equations (14.17) and (14.24) subject to the boundary condition (14.25) and subject to $\delta c(\vec{r}_S) = \varphi(\vec{r}_S)$, we obtain

$$\delta c_1(\vec{r}) = \frac{1}{\vec{v}' \vec{n}(p_2 - p_1)} \Big[p_2 (\vec{v}' \vec{n} - D_\sigma p_1) e^{-p_1 x} -$$
$$- p_1(\vec{v}' \vec{n} - D_\sigma p_2) e^{-p_2 x}\Big] \varphi(\vec{r}_S),$$
$$\sigma_{ll} = \frac{2(1-2\nu)}{1-\nu} K(\bar{D}_2 - \bar{D}_1) \frac{p_1 p_2}{\vec{v}' \vec{n}(p_2 - p_1)} \big[e^{-p_1 x} - e^{-p_2 x}\big] \varphi(\vec{r}_S) \qquad (14.26)$$
$$(\text{for } \vec{v}' \vec{n} \cdot 0),$$

where

$$p_{1,2} = \frac{(\tilde{D} + D_\sigma)\vec{v}' \vec{n} \pm \sqrt{(\tilde{D} + D_\sigma)^2(\vec{v}' \vec{n})^2 - 4\zeta \bar{D}_1 \bar{D}_2 (\vec{v}' \vec{n})^2}}{2\zeta \bar{D}_1 \bar{D}_2},$$
$$\zeta = \frac{2}{3} \frac{(1-2\nu)}{(1-\nu)} \frac{K(c_1\omega_1 + c_2\omega_2)}{kT}. \qquad (14.27)$$

Since, according to Eq. (14.23), we have $\dot{u}_{nn} \propto \dot\sigma_{ll} = -\vec{v}'\nabla\sigma_{ll}$ and since $\sigma_{ll} = 0$ on the pore's surface, it follows that the rate of deformation of the lattice $\vec{w}'(\vec{r}_S) = \dot{u}_n(\vec{r}_S)\vec{n}$ vanishes near the surface.

In the vicinity of the rear half of the pore, we find that, as in the case of plastic deformation, the concentration and stress distributions are much smoother and we can ignore the volume fluxes of atoms resulting from these distributions.

Velocity and Distortion of a Rapidly Moving Pore

The velocity and curvature of a pore are found, together with the surface concentration $\varphi(\vec{r}_S)$, from the continuity conditions of Eq. (12.24). Substituting in Eq. (12.24) the fluxes given by the solutions of the volume problem [Eqs. (14.21), (14.22), or (14.26)], we obtain the equation

$$c_i \vec{v}' \vec{n} + c_{Si} \frac{\alpha_{Si} D^*_{Si}}{T} a\Delta_S T + D_{Si} a\Delta_S c_{Si} - 2c_{Si} \omega'_{Si} \times$$

$$\times \frac{D_{Si} a}{f_{Si}kT} \gamma\Delta_S K - c_{Si} \omega_{Si} \frac{D^*_{Si} a}{3f_{Si}kT} \Delta_S \sigma_u + \frac{c_i \alpha_i}{T} D^*_i \vec{n} \nabla T -$$

$$- \vec{v}' \vec{n} \delta c_i (\vec{r}_S) \theta.(\vec{e}\,\vec{n}) = 0. \qquad (14.28)$$

Here, $\theta(x) = 1$ for $x > 0$ and $\theta(x) = 0$ for $x < 0$. The term containing $\Delta_S \sigma_u$ represents the surface fluxes in the field of stresses generated by the inhomogeneous distortion (this term is absent if the stresses exceed the limit of plasticity).

Equation (14.28) can be simplified considerably if the relative change in the curvature $|\delta K|/K_0$ is small compared with the maximum change in the concentration δc_m on the surface. The main effect of the distortion can then be allowed for by retaining in Eq. (14.28) the terms with $\Delta_S K$ and $\Delta_S \sigma_u$; the difference between $\vec{n}(\vec{r}_S)$ and $\vec{m} = \vec{r}_S/R$ and the difference between the operator Δ_S on the distorted surface and the Laplacian on the surface of a sphere give rise to corrections of higher orders of smallness that can be ignored. As in Sec. 7, it is convenient to study distortion by expanding all the quantities in terms of the Legendre polynomials $P_n(\cos\theta) = P_n(\vec{e}\vec{n}) \equiv P_n$.

When the conditions of Eq. (14.14) are satisfied the last term in Eq. (14.28) is small compared with $a D_{Si} \Delta_S c_{Si}$ and this equation can be solved by the method of successive approximations. In the first approximation, we reject the last term and find that $K = K_0 = R^{-1} = $ const and that

$$\left.\begin{aligned}
\delta c_1 &= A' \cos \theta; \quad A' = \frac{R^2 (1 + \varkappa)}{2 a \chi \widetilde{D}_S} \left(c_{10} a_2^T - c_{20} a_1^T \right) \vec{e} \, \nabla T_\infty, \\
\vec{v}' &= \frac{(1 + \varkappa)}{\widetilde{D}_S} \left(a_1^T D_{S2} + a_2^T D_{S1} \right) \nabla T_\infty,
\end{aligned}\right\} \tag{14.29}$$

where a_i^T, and b_i are given by Eq. (13.32) with $D_i' = 0$.

The relative difference between the velocities given by Eqs. (12.46), (13.32), and (14.29) is

$$\frac{R}{a\chi} \frac{\widetilde{D}_S}{\widetilde{D}_S + \dfrac{R}{a\chi} \overline{D}} \left[\frac{a_1^T \overline{D}_2 + a_2^T \overline{D}_1}{a_1^T D_{S2} + a_2^T D_{S1}} - \frac{\widetilde{D}}{\widetilde{D}_S} \right]. \tag{14.30}$$

In the case considered here, the condition (14.4) is satisfied and the above difference is usually small. However, in the general case, $R\widetilde{D}$ can be of the order of $a\widetilde{D}_S$ and considerable deviations from the linear dependence of $\vec{v}'$ on ∇T_∞ can be expected in the range of very high driving forces and high velocities $v \sim \widetilde{D}/R$. It can be seen from Eqs. (14.30) and (13.32) that the nonlinearity may be appreciable also at lower velocities if the surface diffusion coefficient for one of the components is much greater than that for the other and the volume diffusion coefficients of both components are of the same order of magnitude.

In the second approximation, we must include the last term in Eq. (14.28), replacing δc with Eq. (14.29). The expansion of this strongly asymmetrical term in the Legendre polynomials is of the form

$$v' \cos \theta \delta c_1 \left(\vec{r}_S \right) \theta \left(\cos \theta \right) = \frac{1}{6} v' A' \left[1 + 2 P_2 + \sum_{n=0}^{\infty} (-1)^{n+1} \right.$$

$$\left. \times \frac{n^2 + 3n - 5}{2^{2n+2} (n!)^2 (n+1)(n+2)(2n-1)} P_{2n+1} \right]. \tag{14.31}$$

If we consider Eq. (14.28) for the individual harmonics and use the relationship (7.19) between the expanded expressions for the curvature and the internal stresses, we can find the coefficients K_n

in the expansion of the curvature in terms of the Legendre polynomials:

$$
\left.
\begin{aligned}
K_2 &= \frac{1}{36}\, A'\, v'\, R^2\, \frac{kT}{a\gamma\,\widetilde{\omega}}\, \frac{D_{S2}-D_{S1}}{D_{S1}D_{S2}},\quad K_{2n}=0 \text{ for } n>1,\; K_1=0, \\[2mm]
K_{2n+1} &= \frac{1}{12}\, \frac{(n^2+3n-5)\,(-1)^{n+1}}{2^{2n+2}\, n\,(n+1)^2\,(n+2)\,(2n-1)\,(n!)^2}\, A'\, v'\, R^2\, \times \\[2mm]
&\quad \times\, \frac{kT}{a\gamma\,\widetilde{\omega}}\, \frac{D_{S2}-D_{S1}}{D_{S1}\,D_{S2}} \quad \text{for } n\geqslant 1,
\end{aligned}
\right\}
\tag{14.32}
$$

where $\widetilde{\omega} = c_{S1}\omega'_{S1} + c_{S2}\omega'_{S2} + \Gamma_n\,(c_{S1}\omega_{S1} + c_{S2}\omega_{S2})$ for elastic deformation and $\omega = c_{S1}\omega\,\xi_1 + c_{S2}\omega\,\xi_2$ for plastic deformation. In practice, all the terms in the expansion If $K(\theta)$ can be ignored, apart from $K_2 P_2\cdot(\cos\theta)$.

It is evident from Eq. (14.32) that the relative distortion of a pore is of the order of

$$
\frac{|K_2|}{K_0} \sim 10^{-2}\left(\frac{v'\,R^2}{aD_S}\right)^2 \frac{R}{a}.
\tag{14.33}
$$

We have made here an allowance for the fact that the ratio $kTa \times (\gamma\widetilde{\omega})^{-1}$ amounts to a few tenths.

In the derivation of Eq. (14.32), we assumed that $|\delta K|/K_0 \ll \delta c_m = A' \ll 1$. It follows from Eqs. (14.33) and (14.29) that this assumption is correct if

$$
v' \lesssim 10^2\, \frac{aD_S}{R^2}\, \frac{a}{R},
\tag{14.34}
$$

which may be more stringent than the condition (14.15).

If the condition (14.34) is not obeyed, the expression for $K(\theta)$ and its dependence on the velocity of a pore are different. The order of the change in the curvature can still be determined using Eq. (14.28) [65]. We then find that the ratio $|\delta K|/K_0$ is given by

$$
6\,\frac{|\delta K|}{K_0} \sim \frac{|\delta \vec{r}_S|}{R} \sim A' \sim \frac{R^2 v'}{aD_S} \quad \text{for } 10^2\,\frac{a^2 D_S}{R^3} \gtrsim v' \lessgtr \frac{aD_S}{R^2}.
\tag{14.35}
$$

It is clear from Eqs. (14.33) and (14.35) that initially the value of K_2 increases proportionally to $(v')^2$ and then proportionally to v'. The quantity $|\vec{\delta r_s}|/R$ is of the order of unity for high driving forces, i.e., for an appreciable distortion of a pore, if a pore is of large radius and $R\tilde{D}$ is small compared with $a\tilde{D}_S$.

The nonlinear mechanism of distortion of a pore may act simultaneously with the mechanisms considered in Sec. 7, i.e., those due to the anisotropy of the characteristics of a pore or its host or due to the inhomogeneity of the external forces. These mechanisms usually result in a strong distortion of a pore at much lower values of the driving forces than those which are needed for the nonlinear effects. The various causes of the distortion may be divided in accordance with the dependence of the distortion on the velocity.

15. MIGRATION OF INCLUSIONS IN IONIC CRYSTALS

The migration of inclusions in ionic crystals resembles that in binary solid solutions because two types of ion are disturbed by a moving inclusion. However, the migration of inclusions in ionic crystals has certain features that distinguish it from the migration in metallic solutions.

These features are associated primarily with the fact that ionic crystals are ordered systems. The formation of defects in such crystals cannot be regarded as consisting of statistically independent events because even in an imperfect crystal certain rigorous relationships must be satisfied between the numbers of sites of the first and second kinds.

Moreover, defects in ionic crystals are usually electrically charged and they interact strongly with one another. Even in the absence of external forces near surfaces, such as inclusion–host boundaries, an inhomogeneous distribution of the defects in an ionic crystal may have a strong influence on the properties of the surface, particularly on the transport coefficients. Any redistribution of charged defects – for example, that caused by an external force field – generates electric fields. In contrast to solid solutions, in which diffusion fluxes of the A and B atoms are balanced by a redistribution of the concentration near a moving inclusion, electric fields play an important role in the balancing of the fluxes of ions of different types in ionic crystals. The difference between the signs of the ionic charges and the eléctrical neutrality of each unit cell also give rise to specific features in the motion of inclusions in ionic crystals under the action of external electric fields.

One of the most important aspects that must be considered in the migration of inclusions in ionic crystals is the inhomogeneous equilibrium distribution of the defects near the inclusion—host interface in the absence of external forces.

Distribution of Defects near External or Internal Surfaces in Ionic Crystals

The condition of electrical neutrality of each unit cell demands that the concentration of defects of opposite sign should be equal in the interior of a bulk ionic crystal. However, this condition may not be satisfied in a thin layer near an external or internal surface in an ionic crystal. The departure from neutrality is due to different values of the "work function" of ions of different types, which are transferred into a phase adjacent to the ionic crystal or on its surface. Therefore, different amounts of positive and negative ions are dissolved in the second phase or in the surface layer and, at the same time, a charged layer forms in the crystal near the surface in question. The new layer is quite thick, of the order of the Debye screening radius r_D. The charge in this layer alters the potential in the ionic crystal in such a way that the value of the work functions of different types of ion are rapidly equalized with depth from the surface. Therefore, at depths considerably greater than the Debye screening radius, the ionic crystal is neutral. On the other hand, it is obvious that the charge in the contact layer, whose thickness is of the order of r_D, must be compensated (to avoid the appearance of an enormous electrostatic energy) by charges of opposite sign localized in a contact layer in the second phase or on the external boundary of the ionic crystal.

The change in the concentration of charged defects in contact layers has been discussed by many workers (the first of such discussions is given in [107]), who have dealt with the following cases: a free boundary of an ionic crystal [9, 108-110] (a consistent solution of this problem is given in [9, 110]); an interface with an electrolyte [111]; a boundary with a metal, a covalent crystal, or a second ionic crystal [112].

The special nature of the distribution of defects near an internal or external surface in an ionic crystal can be demonstrated by considering the relatively simple case of a crystal of the A^+B^- type, in which the only defects are charged vacancies located at lattice sites (the more general case of an arbitrary stoichiometric composition and defects of different types is considered in [112]). We shall discuss an element of volume of this crystal which contains N unit cells. In this volume, the lattice sites of the first type are occupied by N_1 of the A ions and n_1 vacancies, whereas the sites of the second kind are occupied by N_2 of the B ions and n_2 vacancies. Since the volume in question is subject to the electric field generated by the charges of ions elsewhere in the crystal, the field produces a potential V.

The Gibbs free energy of the region subject to a potential V can be represented in the form

$$\Phi = \Phi_0 + U(N_1 + N_2) + (u_1 - eV)\, n_1 + (u_2 + eV)\, n_2 - kT \prod_{i=1,2} \frac{(N_i + n_i)!}{N_i!\, n_i!}.$$
(15.1)

Here, Φ_0 is a constant; $-e$ is the electronic charge; 2U is the binding energy of the AB "molecule."

In the configurational approximation the quantities u_1 and u_2 are constants which determine the energy of the formation of vacancies (in a more rigorous theory, they would have to be regarded as weak functions of the temperature). The last term in Eq. (15.1) represents the configurational part of the entropy associated with different arrangements of the vacancies in the lattice sites.

If we minimize Eq. (15.1) in a determination of the vacancy concentration, we must make allowance for the fact that n_1 and n_2 are not independent but must satisfy an additional condition of equality of the sites of the first and second kind:

$$N = N_1 + n_1 = N_2 + n_2.$$
(15.2)

This condition can conveniently be satisfied by the method of Lagrange multipliers, which we shall use to find the minimum of the expression $\Phi - \lambda(N_1 + n_1) + \lambda(N_2 + n_2)$. Differentiating this ex-

pression with respect to n_1 and n_2, we obtain the following formulas for the equilibrium concentrations of vacancies in an ionic
crystal:

$$
\left.
\begin{aligned}
c_1 &= \frac{n_1}{N} = \exp\left(-\frac{u_1 - eV - \lambda}{kT}\right), \\[2mm]
c_2 &= \frac{n_2}{N} = \exp\left(-\frac{u_2 + eV + \lambda}{kT}\right).
\end{aligned}
\right\}
\tag{15.3}
$$

If we substitute the vacancy concentrations from Eq. (15.3)
into Eq. (15.2), we can express the Lagrange multiplier λ in terms
of the numbers of ions N_1 and N_2. However, the values of N_1 and
N_2 are governed by the conditions of equilibrium with the second
phase, i.e., by the chemical potentials μ_1 and μ_2 of the A and B
ions (in equilibrium, these potentials are the same in the ionic
crystal and in the second phase).[†] If we differentiate Eq. (15.1)
with respect to N_i and bear in mind that n_i also depends on N_i,
we can easily express λ directly in terms of μ_1 and μ_2 (for details see [112]):

$$
\left.
\begin{aligned}
\mu_1 &= U - \lambda, \\
\mu_2 &= U + \lambda, \\
\lambda &= \frac{1}{2}(\mu_2 - \mu_1).
\end{aligned}
\right\}
\tag{15.4}
$$

Hence, we see that the multiplier λ has the same value throughout
the crystal.

Although Eqs. (15.3) and (15.4) determine completely the concentrations of vacancies in an ionic crystal, in particular, the
concentrations near its surfaces, we shall find it more convenient
to rewrite these concentrations in terms of the concentrations of
defects in the bulk of a crystal, which can be regarded as known
(at least in principle). This can be done by finding the distribution
of the potential $V(\vec{r})$ and the potential difference between the boundary and the bulk of the ionic crystal.

[†] The chemical potentials of the A and B ions are related to the chemical potentials
of the corresponding A and B atoms in the second phase (for example, a metal) by
the expressions $\mu_1 = \mu_1^* - \mu_e$, $\mu_2 = \mu_2^* + \mu_e$, where μ_e is the electrochemical potential of an electron.

   The distribution of the potential is given by the Poisson equation

$$\Delta V = -\frac{4\pi}{\varepsilon} e N_0 (c_2 - c_1). \tag{15.5}$$

Here, ε is the permittivity; N_0 is the number of unit cells per unit volume; and the vacancy concentrations c_1 and c_2 are functions of the potential given by Eq. (15.3).

   The generality of our treatment will not be affected if we assume that the potential $V(0)$ vanishes on the boundary of an ionic crystal. In the interior of the crystal, the potential assumes a constant value V_∞ and the volume (bulk) concentrations of vacancies of different signs $c_1(\infty)$ and $c_2(\infty)$ should be equal (this is necessary to ensure that the crystal as a whole remains electrically neutral and that a very large amount of electrostatic energy is not generated). It follows from $c_1(\infty) = c_2(\infty)$, and from Eq. (15.3) that the contact difference between the potentials V_∞ is

$$V_\infty = -\frac{\lambda}{e} + \frac{u_1 - u_2}{2e} = \frac{1}{2e} (\mu_1 - \mu_2 + u_1 - u_2). \tag{15.6}$$

The differences between the chemical potentials $(\mu_1 - \mu_2)$ and between the energies of the formation of vacancies $(u_1 - u_2)$ are usually of the order of a few tenths of an electron volt. Therefore, the potential V_∞ is of the order of a few tenths of a volt, i.e., it may be quite large. The ratio eV_∞/kT is much larger than unity and — in accordance with Eq. (15.3) — the concentrations of defects near the boundary and in the interior of an ionic crystal may differ quite considerably.

   In the case of a planar boundary of an ionic crystal we must bear in mind that at infinity $V = V_\infty$ and $\nabla V = 0$, whereas on the boundary $V = 0$. If we introduce a dimensionless potential $y = e(V_\infty - V)/kT$, we find that the Poisson equation (15.5) and the boundary conditions can be rewritten in the form

$$\left. \begin{array}{l} \dfrac{d^2 y}{dx^2} = r_D^{-2} \sinh y, \qquad y(\infty) = 0, \qquad y'(\infty) = 0, \\[2ex] y(0) \equiv y_0 = \dfrac{eV_\infty}{kT}, \qquad r_D^{-2} = \dfrac{8\pi e^2}{\varepsilon kT} N_0 c_1(\infty). \end{array} \right\} \tag{15.7}$$

The solution of Eq. (15.7) is of the form

$$y' = -2r_D^{-1} \sinh \frac{y}{2}, \qquad \tanh \frac{y}{4} = \tanh \frac{y_0}{4} \exp\left(-\frac{x}{r_D}\right), \qquad (15.8)$$

it decreases rapidly away from the boundary when the distance x becomes large compared with the Debye radius r_D.

If we know the distribution of the potential V(x) or of the dimensionless potential y(x), we can express the vacancy concentrations in different parts of an ionic crystal in terms of their concentration $c(\infty) = c_i(\infty)$ in the interior of the crystal:

$$c_1(x) = c(\infty)e^{-y}; \quad c_2(x) = c(\infty)e^{y}. \qquad (15.9)$$

The charge that appears in a layer of thickness $\sim r_D$, located near the boundary of an ionic crystal, is compensated by a charge of the same magnitude but opposite sign, which is located in the boundary layer of the second phase (the role of the charge on the free surface of an ionic crystal will be discussed later). If the second phase is metallic, the charge generated in that phase is located on the surface and its density is

$$\sigma = -\frac{\varepsilon kT}{2\pi e r_D} \sinh \frac{eV_\infty}{2kT}. \qquad (15.9a)$$

In this case, the potential in the metal is constant and equal to zero.

If the second phase is a nonmetal (for example, a covalent crystal), the compensating charge in that phase is distributed in a layer whose thickness is of the order of the Debye radius r_{D1} of the second phase. In this layer, the potential gradually increases from V = 0 at the boundary to $V = V_{1\infty}$ in the interior of the covalent crystal and is described by a formula resembling Eq. (15.8). The condition for the continuity of the normal components of the electric induction at the interface $\varepsilon \left.\frac{dV}{dx}\right|_{x=+0} = \varepsilon_1 \left.\frac{dV}{dx}\right|_{x=-0}$ (ε_1 is the permittivity of the second phase) yields a relationship between V_∞ and $V_{1\infty}$:

$$\frac{\varepsilon}{r_D} \sinh \frac{eV_\infty}{2kT} = -\frac{\varepsilon_1}{r_{D1}} \sinh \frac{eV_{1\infty}}{2kT}. \qquad (15.10)$$

The difference $(V_\infty - V_{1\infty})$ can be expressed in terms of the equilibrium concentrations c_A and c_B of the A and B atoms in the interior of the covalent crystal, which is in equilibrium with the ionic crystal under consideration. For example, if c_A and c_B are small, we find that in the interior of the covalent crystal

$$\mu_1 = \mu_1^0 + kT \ln c_A + eV_{1\infty}, \qquad \mu_2 = \mu_2^0 + kT \ln c_B - eV_{1\infty}, \qquad (15.11)$$

so that Eq. (15.6) becomes

$$V_\infty - V_{1\infty} = \frac{1}{2e} \left(u_1 - u_2 - 2\delta\mu \right), \qquad (15.12)$$

where

$$2\delta\mu = \mu_2^0 - \mu_1^0 + kT \ln \frac{c_B}{c_A}.$$

Here, $\delta\mu$ is determined solely by the volume properties of the covalent crystal and not by the properties of the contact layer.

If the second phase is metallic, we find that $V_{1\infty} = 0$ and the potential drop $(V_\infty - V_{1\infty})$ occurs in the ionic crystal. As the conductivity of the second phase is gradually reduced, the thickness r_{D1} increases, an increasing proportion of the potential drop occurs in the second phase, and V_∞ becomes smaller. Finally, when the second phase becomes an insulator (this applies to the case of an ionic crystal provided the atoms in the vapor of the same crystal are practically all neutral), we find that $r_{D1} \to \infty$ and the potential drop in the ionic crystal V_∞ tends to vanish so that there is no redistribution of the defects near the boundary of this crystal.

However, these results have been derived ignoring the surface charge. As $r_{D1} \to \infty$, which is true of a free boundary, the surface charge may become the principal factor governing the value of V_∞. In particular, the surface charge may appear as the result of changes in the energies of the formation of vacancies (u_1, u_2) near the boundary in a layer of thickness equal to several lattice constants [9, 110]. It follows from Eq. (15.3) that an allowance for the dependence of the energies u_i on the coordinates gives rise to a charge on the surface of an ionic crystal and the density of this charge is

$$\sigma_i = edN_0 c_1(\infty) \sum_{i=1,2} (-1)^i \exp\left(-\frac{e_i V_\infty}{kT} \right) \times$$
$$\times < \exp \frac{u_i(\infty) - u_i(x) + e_i V(x)}{kT} - 1 >. \qquad (15.13)$$

Here, $e_1 = e$, $e_2 = -e$, and the averaging is carried out over a surface layer of thickness d [in which $u_i(x)$ differs from $u_i(\infty) \equiv u_i$].

An allowance for the surface charge σ_i on an ionic crystal and for the surface charge σ_c on a covalent crystal, as well as for the electronic surface charges of density σ_e, alters the boundary conditions of Eq. (15.10) to

$$\varepsilon \left. \frac{dV}{dx} \right|_{x=+0} - \varepsilon_1 \left. \frac{dV}{dx} \right|_{x=-0} + 4\pi\sigma = 0$$

or

$$\frac{\varepsilon}{r_D} \sinh \frac{eV_\infty}{2kT} + \frac{\varepsilon_1}{r_{D1}} \sinh \frac{eV_{1\infty}}{2kT} = -2\pi \frac{e\sigma}{kT}, \qquad (15.14)$$

where $\sigma = \sigma_i + \sigma_c + \sigma_e$.

In particular, if r_{D1} is very large, V_∞ is determined solely by the surface charge (for example, in the case of a free boundary of an ionic crystal, where $\sigma = \sigma_i$):

$$\sinh \frac{eV_\infty}{2kT} = -2\pi \frac{r_D e}{kTe} \sigma. \qquad (15.15)$$

It follows from Eq. (15.15) that the value of $|V_\infty|$ can be much smaller than that in the case of a boundary with a metallic phase. An allowance for the dependence of u_i on x (and for the gap between the phases) simply modifies the value of y_0 in the solution given by Eq. (15.8) but does not alter the basic results.

The contact potential difference V_∞ and the Debye radius r_D may change considerably if impurity ions C are added to an ionic crystal AB [9, 110, 112]. The influence of impurity ions may be particularly strong in the case of a free surface. These ions may be localized at interstices on the surface of the ionic crystal. Among the various impurity ions (including monovalent ions) which are always found in ionic crystals, there may be some for which the energy u_C' of an ion C in a surface interstice is much less than the corresponding energy u_C in the interior of the ionic crystal. As a result of internal absorption, the number $N_s c_C'$ of these ions in surface interstices may be comparable with the total number of surface ions N_s even if $c_C(\infty)$ is negligibly small [if $u_C - u_C' = 3.2 \times 10^{-7}$ nJ (2 eV) and $kT = 0.16 \times 10^{-7}$ nJ (0.1 eV), we find that $c_C' \sim 4 \times 10^8 c_C(\infty)$]. In this case, the surface charge σ_i and the potential V_∞ will be considerably larger than in the case of an impurity-free crystal.

If the boundary of an ionic crystal is not planar but spherical (for example, the boundary surrounding a spherical inclusion), the potential can be found by solving the Poisson equation (15.5) in the three-dimensional case. If the Debye radius is small compared with the radius of the sphere ($r_D \ll R$ and $r_{D1} \ll R$), we may assume approximately that the boundary is planar and still use the solution (15.8) in which x must be replaced with ($r - R$). In the opposite case, when $R \ll r_D$, the solution will be quite different. If the condition

$$R^2 \ll \frac{y_0}{\sinh y_0} \, r_D^2 \qquad (15.16)$$

is satisfied, we can show that the potential around a metal sphere in an ionic crystal is described by the approximate formula

$$y = y_0 \frac{R}{r} \exp\left(-\frac{r}{r_D}\right). \qquad (15.17)$$

As r increases, the potential increases first in accordance with the Coulomb law and then exponentially.

A redistribution of vacancies around the inclusions in ionic crystals has a strong influence on the diffusion coefficients and, consequently, on the migration of inclusions, which is governed by diffusion. In order to avoid unnecessary complications of the formulas, we shall consider first the migration of inclusions for which $|V_\infty|$ is relatively small and this influence can be ignored. Then we shall consider specially the influence of contact effects on the migration of inclusions under the influence of diffusion fluxes.

Migration of Inclusions in a Homogeneous Ionic Crystal Subjected to a Temperature Gradient

We shall discuss the migration of inclusions in ionic crystals of the A^+B^- type, in which the vacancy diffusion is the predominant mechanism [9, 11, 113] (the role of diffusion fluxes associated with interstitial ions is considered in [113]).

In such crystals, the volume diffusion fluxes $\vec{I}_1$ and $\vec{I}_2$ of the A and B ions are equal and opposite to the fluxes of the vacancies at the corresponding lattice sites. Since the chemical potential gradients of the vacancies are equal but opposite in sign to the corresponding gradients of the ions (in diffusion jumps, a vacancy and an ion move in opposite directions), we obtain the following expression for the fluxes:

$$\vec{I}_i = -\frac{N_0 D_i^*}{kTf_i} \nabla\mu_i - N_0 D_i^* L_i \nabla T. \tag{15.18}$$

In this formula, D_i^* represents the self-diffusion coefficients of the ions and the last term represents the fluxes associated directly with a temperature gradient (and not with a chemical potential gradient), for example, the fluxes due to the "phonon wind." The products $L_i D_i^*$ determine the corresponding transport coefficients.

Similarly, the expressions for the surface diffusion fluxes at the inclusion—host interface in an ionic crystal can be written in the form

$$\vec{I}_{Si} = -\frac{N_0 D_{Si}^* a}{kTf_{Si}} \nabla_S \mu_i - N_0 D_{Si}^* a L_{Si} \nabla_S T. \tag{15.19}$$

The fluxes of the A and B ions (or atoms) across an inclusion can also be expressed in terms of the gradients of the chemical potentials of the ions (in the case of low concentrations of ions):

$$\vec{I}_i' = -\frac{D_i'^* n_i'}{kTf_i'} \nabla\mu_i - D_i'^* n_i' L_i' \nabla T, \tag{15.20}$$

where $n_i' = N_0' c_i'$ are the volume concentrations of the i-th ions (atoms) in the inclusion.

In the presence of a temperature gradient, the chemical potentials of the ions have a discontinuity at the inclusion—host interface in the ionic crystal. Then the crystal is dissolved in the inclusion or it grows at the expense of the inclusion. If the inclusion has a sufficiently rough surface (Sec. 2), the number $\dot{N}_{AB}$ of pairs of the AB ions dissolved per unit time over a unit area of the interface can be regarded as proportional to the difference

$\Delta(\mu_1 + \mu_2)$ between the total chemical potentials of a pair of atoms in the crystal and in the inclusion:

$$N_{AB} = N_0 \frac{\beta'}{kT} \Delta(\mu_1 + \mu_2), \qquad \beta' = \beta'_0 \exp\left(-\frac{Q'}{kT}\right). \tag{15.21}$$

Here, β' is a transport coefficient.

If the external forces are not very strong so that the condition (14.1) is satisfied, the chemical potential and the defect distributions can be regarded as stationary. If the change in the vacancy concentration in the contact layer can be ignored, the self-diffusion coefficients D_i^* (which are proportional to c_i) are independent of the coordinates and the crystal is homogeneous (the role of the contact effects will be considered later). In the linear theory, the case of weak external forces is represented by $\mathrm{div}\,\vec{I}_i = 0$ and $\Delta T = 0$; therefore, it follows from Eq. (15.18) that the chemical potentials in a homogeneous crystal should obey the Laplace equation

$$\Delta \mu_i = 0. \tag{15.22}$$

In the case of a crystal subjected to a temperature gradient and an external electric field, the boundary conditions for the chemical potentials at infinity can be written in the form

$$\nabla \mu_i(\infty) = \frac{\partial \mu_i}{\partial T} \nabla T_\infty + e_i \nabla V_\infty, \tag{15.23}$$

where $\nabla V_\infty = -\vec{E}_\infty$, and e_i is the charge of the i-th ion. The conditions of continuity given by Eq. (12.24) should be satisfied by the inclusion—host interface in ionic crystals as well as in binary solutions. These equations give one of the boundary conditions for the inclusion—host interface (they also determine the velocity of the inclusion).

Moreover, if the exchange of the A and B ions between the crystal and the inclusion is sufficiently rapid, i.e., if the coefficient β' in Eq. (15.21) is sufficiently large, we can ignore the discontinuities $\Delta \mu_i$ of the chemical potentials at the inclusion—host interface and assume that these potentials vary continuously. Such continuity of the potentials will be assumed in our subsequent discussion.

It follows from Eqs. (12.24) and (15.23) that, in the isotropic linear problem under consideration, a change in the chemical potential $\delta\mu_i$ associated with external forces and with the diffusional migration of an inclusion is proportional to cos θ. The corresponding solution of Eq. (15.22) satisfying the above boundary conditions can be written in the following form with the aid of Eq. (2.6):

$$
\left.
\begin{aligned}
\delta\mu_i &= \frac{\partial\mu_i}{\partial T}\,\vec{r}\nabla T_\infty + e_i\,\vec{r}\nabla V_\infty + \frac{\partial U}{\partial T}\,\varkappa\frac{R^3}{r^3}\,\vec{r}\nabla T_\infty + \\
&\quad + (-1)^i\,\frac{R^3}{r^3}\,\vec{B}\,\vec{r} \qquad (r \geqslant R); \\
\delta\mu_i &= \left(\frac{\partial\mu_i}{\partial T} + \varkappa\frac{\partial U}{\partial T}\right)\vec{r}\nabla T_\infty + e_i\,\vec{r}\nabla V_\infty + \\
&\quad + (-1)^i\,\vec{B}\,\vec{r} \qquad (r \leqslant R).
\end{aligned}
\right\}
\tag{15.24}
$$

Here, $\partial U/\partial T$ is determined by the temperature dependence of the chemical potential of a pair of ions AB [see Eq. (15.4)] and the constant $\vec{B}$ must be found from the boundary condition (12.24) in which we must substitute $c_i = 1$.

Equations (15.18)-(15.20) and (15.24) can be used to determine the diffusion fluxes across the host crystal and the inclusion, as well as on the inclusion—host interface. Next, Eq. (12.24) can be used to eliminate the constant $\vec{B}$ and to determine the velocity of inclusions $\vec{v}'$ relative to the lattice of the host crystal.

We shall first consider the migration of inclusions under the action of a temperature gradient when only the diffusion fluxes in the interior of the host crystal and on the inclusion—host interface are important. Equation (15.24) must then be modified by the substitution $\nabla V_\infty = 0$ and we must assume that $\vec{I}_i' = 0$. Equations (12.24), (15.18), (15.19), and (15.24) yield

$$
\vec{v}' = (1 + \varkappa)\,\frac{a_1^T b_2 + a_2^T b_1}{b_1 + b_2}\,\nabla T_\infty,
\tag{15.25}
$$

where

$$
\left.
\begin{aligned}
a_i^T &= \frac{D_i^*}{(1+\varkappa)f_i\,kT}\left[\left(2\varkappa\frac{\partial U}{\partial T} - \frac{\partial\mu_i}{\partial T}\right) + L_i\,kTf_i\,(2\varkappa - 1)\right] + \\
&\quad + \frac{2D_{Si}^*}{(1+\varkappa)f_{Si}\,kT}\,\frac{a}{R}\left[\left(\varkappa\frac{\partial U}{\partial T} + \frac{\partial\mu_i}{\partial T}\right) + L_{Si}\,kTf_{Si}\,(\varkappa + 1)\right], \\
b_i &= \frac{1}{f_i}\,D_i^* + \frac{1}{f_{Si}}\,\frac{a}{R}\,D_{Si}^*.
\end{aligned}
\right\}
\tag{15.26}
$$

In the case of small-radius inclusions, when the contribution of the surface diffusion fluxes is considerably greater than that of the volume fluxes, it follows from Eqs. (15.25) and (15.26) that

$$\vec{v}' = \vec{v} = 2(1 + \varkappa)\, \frac{a}{R}\, \frac{\alpha_{S\,eff}}{T} D_{S\,eff}\, \nabla T_\infty, \tag{15.27}$$

where

$$D_{S\,eff} = \frac{D_{S1}^* D_{S2}^*}{D_{S1}^*\, f_{S2} + D_{S2}^*\, f_{S1}},$$

$$\frac{\alpha_{S\,eff}}{T} = \frac{2}{kT}\, \frac{\partial U}{\partial T} + (L_{S1} f_{S1} + L_{S2} f_{S2}). \tag{15.28}$$

In the case of large-radius inclusions, when the volume diffusion fluxes predominate, the velocity of an inclusion relative to the lattice of the host crystal follows from Eqs. (15.25) and (15.26)

$$\vec{v}' = -(1 - 2\varkappa)\, \frac{\alpha_{eff}}{T} D_{eff} \nabla T_\infty, \tag{15.29}$$

where

$$D_{eff} = \frac{D_1^* D_2^*}{D_1^* f_1 + D_2^* f_2}, \qquad \frac{\alpha_{eff}}{T} = \frac{2}{kT}\, \frac{\partial U}{\partial T} + (L_1 f_1 + L_2 f_2). \tag{15.30}$$

The velocities given by Eqs. (15.27) and (15.29) are of the same form as the velocities of inclusions in one-component crystals [see Eqs. (2.21) and (2.16)] and these velocities have the same dependences on the radius. The only difference is that the velocities of inclusions in ionic crystals are governed by the effective diffusion coefficients D_{eff} or $D_{S\,eff}$. It follows from Eqs. (15.28) and (15.30) that when the mobilities of the A and B ions differ considerably, we find that — as in the case of solid solutions — the effective diffusion coefficients are approximately equal to the coefficients of the less mobile ions, which then control the migration of the inclusions. This result applies also to the diffusion of ions between the interstices.

If the volume and surface (interface) diffusion fluxes are of comparable importance, the dependence of the velocity of inclusions on their radii and on the temperature become more complex. In particular, special features are exhibited by these dependences if the ratios of the diffusion coefficients of the A and B ions differ strongly for the surface and volume diffusion.

In the most general case, when the fluxes $\vec{I}'_i$ of ions across the inclusion are also of importance, we can determine these fluxes from Eqs. (15.20) and (15.24) on the assumption that the exchange of ions across the interface is quite rapid. It then follows from Eq. (12.24) that the velocity of an inclusion $\vec{v}'$ is again given by a formula which resembles Eq. (15.25). However, this formula must be modified by the substitutions

$$a^T_i \to \widetilde{a}^T_i = a^T_i + \frac{D'^{*}_i c'_i}{(1+\varkappa) f'_i \, kT} \left[\frac{\partial \mu_i}{\partial T} + \varkappa \frac{\partial U}{\partial T} + (1+\varkappa) L'_i f'_i \, kT \right] \frac{\omega}{\omega'},$$

$$b_i \to \widetilde{b}_i = b_i + \frac{1}{2} \frac{D'^{*}_i}{f'_i} c'_i \frac{\omega}{\omega'}. \qquad (15.31)$$

In particular, if $D^*_i \ll D'^{*}_i$, $(a/R) D^*_{Si} \ll D'^{*}_i$, we find that the migration of inclusions is governed primarily by the diffusion fluxes across the inclusions and the velocity is given by

$$\vec{v} = \vec{v}' = (1+\varkappa) \frac{\omega}{\omega'} \frac{\alpha'_{eff}}{T} D'_{eff} \nabla T_\infty, \qquad (15.32)$$

where

$$D'_{eff} = \frac{D'^{*}_1 D'^{*}_2 c'_1 c'_2}{D'^{*}_1 c'_1 f'_2 + D'^{*}_2 c'_2 f'_1}, \quad \frac{\alpha'_{eff}}{T} = \frac{2}{kT} \frac{\partial U}{\partial T} + (L'_1 f'_1 + L'_2 f'_2). \qquad (15.33)$$

In this case, the velocity of the inclusions is governed by the concentrations and mobilities of the A and B ions (atoms) in the inclusions.

If the diffusion coefficients D'^{*}_i are very large and the transport of ions across an inclusion is limited not by the rate of diffusion but by the rate of dissolution of these ions in the inclusion, we can assume that the chemical potentials inside the inclusion are constant. The discontinuities $\Delta \mu_i$ are then equal to the change in the chemical potential $\delta \mu_i$ on the surface of a crystal, as given by Eq. (15.24). We shall now consider the case when the contribution of the volume and surface (interface) diffusion fluxes in an ionic crystal can be ignored: if we assume that $N_0 \vec{v} \vec{n} = \dot{N}_{AB}$, we find that Eqs. (15.21) and (15.24) yield

$$\vec{v} = \vec{v}' = 2\frac{\beta'}{kT}(1 + \varkappa)\frac{\partial U}{\partial T}R\nabla T_{\ldots}. \tag{15.34}$$

In this case, the velocity $\vec{v}$ is proportional to the inclusion radius and depends on the transport coefficient β', which determines the rate of dissolution of the host ions in the inclusion.

Migration of Inclusions in a Homogeneous Ionic Crystal Subjected to an External Electric Field

The forces exerted on ions in a metal by an electric field and by electrons dragged by the field give rise to a flux of ions and to the migration of inclusions under the influence of volume or sur-face diffusion (Sec. 3). The situation is quite different in ionic crystals because the migration of an inclusion demands that an equal number of ions with opposite charges is transferred to or from the interface. If the diffusion fluxes of just one type (either surface or volume) predominate, we find that these fluxes must be equal for the A and B ions $(\vec{I}_{S1} = \vec{I}_{S2}$ or $\vec{I}_1 = \vec{I}_2)$ and that an ex-ternal field produces directional diffusion fluxes not of individual ions but of ion pairs (A^+B^- "molecules"). However, the total charge or such "molecules" is zero and, in contrast to a tem-perature gradient, such diffusion fluxes are not produced by an ex-ternal field so that an inclusion cannot migrate.

However, if an allowance is made for the simultaneous pres-ence of fluxes of different types (volume and surface fluxes or fluxes across the host and inclusion), we find that the migration of inclusions in an ionic crystal is possible under the influence of an external electric field [114, 113]. In this case, the volume dif-fusion in the ionic crystal may be responsible for the removal of ions of one type from the region occupied by a moving inclusion, whereas the surface (interface) diffusion or the diffusion across the inclusion may be responsible for the removal of ions of the other type. The fluxes $\vec{I}_1$ and $\vec{I}_2$, $\vec{I}_{S1}$ and $\vec{I}_{S2}$, or $\vec{I}_1'$ and $\vec{I}_2'$ are no longer equal and the directional diffusion in an external field does not represent the motion of individual charged ions but of A^+B^- neutral "molecules." Such diffusion can occur in an external electric field and can give rise to the migration of inclusions. Since this requires the balancing of diffusion fluxes of different types, the migration of inclusions in an ionic crystal subjected to an electric field should differ from the migration of inclusions in a metal subjected to an electric field (Sec. 5), and from the migration of inclusions in an ionic crystal subjected to a temperature gradient.

We can determine the velocity of inclusions in an electric field from the expressions for the diffusion fluxes given by Eqs. (15.18)–(15.20) and (15.24), where we must substitute $\nabla T_\infty = 0$. If the expressions obtained in this way are used in Eq. (12.24) and the constant $\vec{B}$ is eliminated, it is found that the velocity $\vec{v}'$ of inclusions relative to the lattice of the ionic crystal becomes

$$
\vec{v}' = \left[3\, \frac{a}{R}\, \frac{D_1^* D_{S2}^* f_1^{-1} f_{S2}^{-1} - D_2^* D_{S1}^* f_2^{-1} f_{S1}^{-1}}{\tilde{b}_1 + \tilde{b}_2} +\right.
$$
$$
\left. + \frac{3}{2}\, \frac{\omega}{\omega'}\, \frac{D_2^{'*} c_2' D_1^* f_2^{-1} f_1^{-1} - D_1^{'*} c_1' D_2^* f_1^{-1} f_2^{-1}}{\tilde{b}_1 + \tilde{b}_2} \right] \frac{e\vec{E}_\infty}{kT}. \tag{15.35}
$$

Equation (15.35) simplifies if $D_i^{'*} \ll (a/R)\, D_{Si}^*$ and if the diffusion fluxes across the inclusion can be ignored. In this case [114], we find that

$$
\vec{v}' = 3\, \frac{a}{R}\, \frac{D_1^* D_{S2}^* f_1^{-1} f_{S2}^{-1} - D_2^* D_{S1}^* f_2^{-1} f_{S1}^{-1}}{D_1^* f_1^{-1} + D_2^* f_2^{-1} + \dfrac{a}{R}\,(D_{S1}^* f_{S1}^{-1} + D_{S2}^* f_{S2}^{-1})}\, \frac{e\vec{E}_\infty}{kT}. \tag{15.36}
$$

It follows from our discussion that the velocity of inclusions given by Eq. (15.36) vanishes if $D_i^* = 0$ or $D_{Si}^* = 0$, i.e., if there is no volume or surface (interface) diffusion in the ionic crystal. The migration of inclusions is related to the difference between the mobilities of ions of different types and becomes impossible if $D_1^* f_1^{-1} = D_2^* f_2^{-1}$ and $D_{S1}^* f_{S1}^{-1} = D_{S2}^* f_{S2}^{-1}$.

In the case of inclusions of sufficiently large radius, so that $(a/R)\, D_{Si}^* \ll D_i^*$, Eq. (15.36) becomes

$$
\vec{v}' = 3\, \frac{a}{R}\, \frac{D_1^* D_{S2}^* f_1^{-1} f_{S2}^{-1} - D_2^* D_{S1}^* f_2^{-1} f_{S1}^{-1}}{D_1^* f_1^{-1} + D_2^* f_2^{-1}}\, \frac{e\vec{E}_\infty}{kT}, \tag{15.37}
$$

i.e., the velocity of an inclusion becomes inversely proportional to its radius. This unusual dependence of the velocity of large inclusions on their radius is due to the fact that the lines of flow of ions of each type must have regions lying on the inclusion–host interface, where the resistance to motion is inversely proportional to R. Conversely, the velocity of small-radius inclusions $[(a/R) D_{Si}^* \gg D_i^*]$ in an ionic crystal subjected to an electric field is independent of R:

$$\vec{v}' = 3 \frac{D_1^* D_{S2}^* f_1^{-1} f_{S2}^{-1} - D_2^* D_{S1}^* f_2^{-1} f_{S1}^{-1}}{D_{S1}^* f_{S1}^{-1} + D_{S2}^* f_{S2}^{-1}} \frac{e \vec{E}_\infty}{kT}. \tag{15.38}$$

If the condition $c_i' D_i'^* \gg (a/R) D_{Si}^*$ is satisfied, i.e., if the diffusion fluxes across an inclusion are much more important than the surface fluxes, we find that Eq. (15.35) becomes [113]:

$$\vec{v}' = 3 \frac{\omega}{\omega'} \frac{D_2'^* c_2' D_1^* f_2'^{-1} f_1^{-1} - D_1'^* c_1' D_2^* f_1'^{-1} f_2^{-1}}{2 \left(D_1^* f_1^{-1} + D_2^* f_2^{-1} \right) + \frac{\omega}{\omega'} \left(D_1'^* c_1' f_1'^{-1} + D_2'^* c_2' f_2'^{-1} \right)} \frac{e \vec{E}_\infty}{kT}. \tag{15.39}$$

In this case, the velocity of large inclusions in an electric field is independent of their radius and is governed solely by the coefficients of volume diffusion in the ionic crystal and in the inclusion. This is due to the fact that the closure of the diffusion fluxes of the A and B ions, starting at infinity, occurs not at the inclusion–host interface, but inside the inclusion.

The diffusion flux of charged defects, which arises in an external electric field, changes the distribution of the defects near an inclusion and the change is proportional to the external field. A distorted distribution of vacancies is characterized by a dipole moment $\vec{P}$ of the region surrounding the inclusion [114]. This moment depends strongly on the relationship between the volume and the surface diffusion coefficients and on the value of R, so that when $D_i \gg (a/R) D_{Si}$, we find that

$$\vec{P} = - \frac{\varepsilon}{2} R^3 \vec{E}_\infty \tag{15.40}$$

(ε is the permittivity), whereas if $D_i \ll (a/R) D_{Si}$, we obtain

$$\vec{P} = \varepsilon R^3 \vec{E}_\infty. \tag{15.41}$$

Influence of Contact Effects on the

Migration of Inclusions

We have shown that a layer of thickness $\sim r_D$ appears near
the boundary between an ionic crystal and a second phase or near
the free surface of such a crystal. In this layer, the concentration of charged vacancies is different from the volume concentration and the contact potential difference V_∞ appears in this layer.
The distribution of vacancies in a surface layer of this type is
given by the formulas in Eq. (15.9). Since the contact potential V_∞
can be of the order of several tenths of a volt and since $y_0 = eV_\infty/kT$
can range from a few units to 10, the concentration of vacancies of
one type near the surface of an inclusion may be much higher than
elsewhere in the crystal, whereas the concentration of vacancies
of the other type may be much lower.

The self-diffusion coefficients of ions D_i^* are proportional to
the vacancy concentrations. Therefore, in general, these coefficients should depend (like the vacancy concentrations) on the coordinates, and near the surface of an inclusion they should differ
considerably from the values $D_i^*(\infty)$ that obtain in the interior of the
crystal:

$$D_1^* \, (\vec{r}) = D_1^* \, (\infty) \, e^{-y(\vec{r})} \, , \qquad D_2^* \, (\vec{r}) = D_2^* \, (\infty) \, e^{y \, (\vec{r})} \, . \tag{15.42}$$

This circumstance may have a strong influence on the diffusion
fluxes of ions near an inclusion and on the velocity of this inclusion [113]. Similarly, the surface diffusion coefficients at the inclusion—host interface are proportional to the concentrations of
surface defects and they may contain exponential factor $\exp(\pm y_0)$.
Therefore, these coefficients depend strongly on the nature of the
inclusion, which affects the value of y_0.

The influence of the contact potential on the volume diffusion
fluxes near an inclusion in an ionic crystal depends strongly on the
ratio between the inclusion radius R and the Debye radius r_D. If
the Debye radius is small so that

$$R \gg r_D, \tag{15.43}$$

we find that the thickness of the surface layer with a modified
vacancy concentration is much less than the dimensions of a region $r \sim R$ in which the diffusion fluxes responsible for the mi-

gration of an inclusion are generated. Therefore small parts of the layer can be regarded as planar.

The presence of a layer with a lower concentration of vacancies at sites of the first or second kind gives rise to a contact resistance, which impedes the flow of vacancies migrating under the influence of the difference between the chemical potentials. This difference is proportional to the external forces responsible for the motion of the inclusion and such a difference is small in the linear theory. A calculation of the contact resistance for a small difference between the chemical potentials shows [115] that, in the case of vacancies which replace ions with $e_i V_\infty > 0$, the resistance ρ of a unit area of the contact layer is

$$\rho = 2r_D\, \sigma^{-1}(\infty)\, \exp \frac{|y_0|}{2}, \qquad \sigma(\infty) = \frac{e^2\, D_i\,(\infty)}{f_i\, kT}. \tag{15.44}$$

In the case of vacancies which replace ions of the opposite sign, for which $e_i V_\infty < 0$, the contact layer has "antiblocking" properties and the change in its resistance can be ignored (provided the layer is thin).

If $|y_0|$ is not too large so that the condition

$$2r_D \exp \frac{|y_0|}{2} \ll R \tag{15.45}$$

is satisfied, the resistance of the contact layer is small compared with the resistance $\sim \sigma^{-1}(\infty)R$ of the remainder of the line of flow of vacancies near an inclusion (the length of this line is $\sim R$), the contact resistance can be ignored, and the expressions deduced for the velocity of an inclusion in a homogeneous crystal can be used. However, when the opposite inequality

$$2r_D \exp \frac{|y_0|}{2} \gg R \tag{15.46}$$

is satisfied, the resistance of the contact layer plays the dominant role and the expression for $\vec{v}'$ changes considerably.

We shall illustrate the influence of the contact resistance by considering the motion of inclusions in a perfect ionic crystal subjected to a temperature gradient [113]. We shall assume that in this crystal vacancies are formed easily only at the inclusion−host interface. It is shown in [113] that an allowance for the contact resistance in the formulas for the diffusion fluxes across the host crystal and for the inclusion velocity reduces to the substitution

$$D_i^* \to \frac{D_i^*(\infty)}{1 + 4\dfrac{r_D}{R}\exp\dfrac{|y_0|}{2}} \tag{15.47}$$

for those i-type ions for which $e_i V_\infty > 0$ [for the other ions, we still have $D_i^* = D_i^*(\infty)$].

If the volume diffusion fluxes in the host crystal make a significant contribution to the velocity of the inclusions, it follows from Eq. (15.47) that when $(r_D/R)\exp(|y_0|/2) > 1$ the substitution specified by this equation alters considerably the velocity and its temperature dependence (the value of $|y_0|/2$ decreases with increasing temperature, i.e., in the presence of contact effects, the activation energy becomes larger). In the case of pores, the value of $|y_0|$ may be large because of the adsorption of impurity ions. In this case, the degree of adsorption and the contact potential difference V_∞ are functions of the temperature, which complicates additionally the temperature dependence of the velocity of the pores.

If the Debye radius is so large (compared with the inclusion radius) that instead of the condition (15.43) we find that the opposite condition (15.16) is satisfied, i.e., that

$$r_D \sqrt{|y_0|}\exp\left(-\frac{|y_0|}{2}\right) \gg R, \tag{15.48}$$

the distribution of the potential around a metallic inclusion is described by Eq. (15.17), and in the region $r \sim R$, which is used in discussing the motion of an inclusion, we find that $y \cong y_0 R/r$. In this case, the problem cannot be reduced to the one-dimensional case and it is more difficult to determine the contact resistance ρ. However, if we bear in mind that when $|y_0| \gg 1$ the main contribution to the resistance of a sequence of regions arranged in series along the line of flow corresponding to $e_i V_\infty > 0$ is made by a thin almost planar high-resistance region at the surface of the inclusion, we can readily demonstrate that, in this case, the contact resistance ρ for ions with $e_i V_\infty > 0$ can be written in the form

$$\rho = \frac{R}{|y_0|}e^{|y_0|}\sigma^{-1}(\infty) \qquad \left(|y_0| \gg 1,\ e_i V_\infty > 0\right). \tag{15.49}$$

For these ions, an allowance for the contact effects reduces to the following substitution in Eqs. (15.18)-(15.33)

$$D_i^* \to \frac{|y_0|}{2} D_i^* (\infty) \, e^{-|y_0|} \qquad (e_i V_\infty > 0, \; |y_0| \gtrsim 1). \tag{15.50}$$

In the case of ions with $e_i V_\infty < 0$, the concentration of vacancies near inclusions is $\exp |y_0|$ times higher, particularly near the surface of an inclusion in a layer of thickness $\sim R/|y_0|$, where $y \approx y_0$. The lines of flow are closed in this low-resistance layer. Therefore, for the ions in question, the effective diffusion coefficient is $\sim e^{|y_0|}/|y_0|$ times larger, and in order-of-magnitude estimates D_i^* should be replaced with

$$D_i^* \to \sim \frac{1}{|y_0|} D_i^* (\infty) \, e^{|y_0|} \qquad (e_i V_\infty < 0, \; |y_0| \gtrsim 1). \tag{15.51}$$

It is evident from Eqs. (15.50) and (15.51) that when the condition (15.48) is satisfied, the contact effects have an even stronger influence on the velocity of inclusions than in the case corresponding to the conditions (15.43) and (15.46). Basically, in the present case the effective diffusion coefficients D_i change by the factors $\exp |y_0|$ or $\exp(-|y_0|)$, compared with their values in the interior of an ionic crystal.

An allowance for the contact effects complicates the dependence of the velocity of inclusions in an ionic crystal on their radius. For example, if the conditions (15.43) and (15.46) are satisfied, it follows from Eq. (15.47) that the effective diffusion coefficient of ions with $e_i V_\infty > 0$ migrating in an ionic crystal is proportional to R. The dependence on R becomes even more complex if we bear in mind that there are three regions $R \ll (|y_0|)^{1/2} \times \exp(-|y_0|/2) r_D$, $r_D \ll R \ll r_D \exp(|y_0|/2)$, and $R \gg r_D \times \exp(|y_0|/2)$, in which the effective values of D_i^* are quite different [for example, if $e_i V_\infty > 0$, these coefficients may have the values $D_i^* (\infty)(|y_0|/2) \exp(-|y_0|)$, $D_i^* (\infty)(R/4r_D) \exp(-|y_0|/2)$, and $D_i^* (\infty)$].

Experimental Investigations of the

Migration of Inclusions in Ionic

Crystals

Ionic crystals, particularly alkali halides, are very convenient for investigations of the migration of macroscopic inclusions because the transparency of these crystals makes it possible to use optical methods and to follow continuously the motion of a single inclusion.

Since the vacancies in the anion and cation sublattices in an ionic crystal carry effective charges, an external electric field should affect the migration of inclusions in such crystals. The present subsection is concerned with the results of four studies of the influence of an electric field on the migration of various types of inclusion. These investigations were concerned with the migration of gas-filled bubbles and metallic inclusions, as well as with vacancy breakdown, which is a phenomenon closely related to the motion of microcavities (microbubbles) in defective regions in a crystal.

Migration of Bubbles in an Electric Field. It follows from general considerations that the effects due to the influence of an external electric field on closed gas-filled cavities (bubbles) may be linear or nonlinear functions of the field. The linear effects include the migration of vacancies, their collisions, coalescence, etc. On the other hand, the deformation of inclusions is a nonlinear effect.

The migration of gas-filled bubbles in NaCl and LiF single crystals was investigated using samples of two kinds containing bubbles of different origin [116, 117]. In some of these samples, cracks were produced artificially in the (100) plane. When these cracks were healed by the application of high temperatures and external compressive stresses, cavities partly filled with air and partly with the vapor of the host substance formed at the tips of the cracks. Prolonged annealing at temperatures close to the

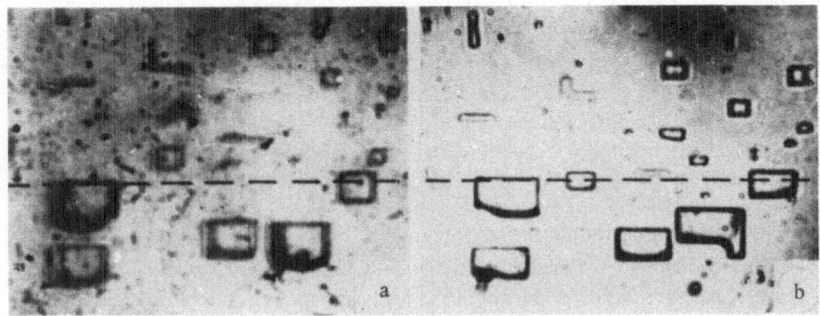

Fig. 52. Motion of gas-filled bubbles (a and b) show successive stages in an NaCl single crystal subjected to an electric field [116].

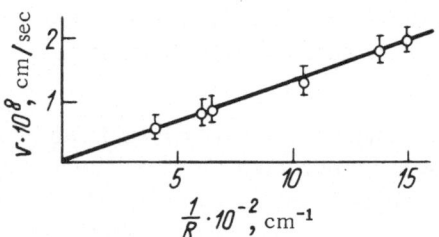

Fig. 53. Velocity of gas-filled bubbles, migrating in an electric field, plotted as a function of the reciprocal of their radius [116].

melting point healed these cracks completely, leaving behind trains of cavities (bubbles) oriented along the directions of original cracks. In other samples, the bubbles appeared during growth. The annealing was carried out in the temperature range 600-800°C in fields $E = 50$-200 V/cm. [†]

It was found that the displacements of bubbles in samples of the first type obeyed the law $v \propto 1/R$. Figures 52 and 53 show a typical sequence of photographs illustrating the dependence of the velocity of bubbles on their radii (Fig. 52) and the dependence $v \propto 1/R$ (Fig. 53). In the range of temperatures and the values of E employed in these experiments, the velocity of the bubbles was $\sim 10^{-7}$-10^{-8} cm/sec.

The migration of bubbles which appeared during the growth of NaCl and LiF crystals was characterized by a velocity which was practically independent of their radii, which were in the range $(1$-$2) \times 10^{-4}$ cm (Figs. 54 and 55).

The results obtained can be explained in a natural manner by means of Eqs. (15.36) and (15.37). The dependence $v \propto 1/R$ applies

---

[†] In view of the unreliability of the various markers used in crystals (such as edges, scratches, etc.), the absolute displacements of bubbles were found in [116, 117] by a method which could be used in those cases when the velocity of an inclusion was inversely proportional to its radius: $v \propto 1/R$. The displacements of different bubbles in an ensemble were determined in a system of coordinates linked to one of the bubbles. Use was made of the relationship $x_{ik} = x_i - x_k = C(1/R_i - 1/R_k)$, where x_{ik} is the displacement of the i-th bubble in a system of coordinates linked to the k-th bubble; x_i and x_k are, respectively, the displacements of the i-th and k-th bubbles in the laboratory system of coordinates. The value of x_k could be found by extrapolation of the straight line $x_{ik} = \varphi(1/R_i)$ to zero value of $1/R_i$. This corresponded to the displacement of an infinitely large bubble (i.e., of the laboratory system of coordinates) relative to the k-th bubble or, conversely, to the absolute displacement of the k-th bubble in the laboratory system of coordinates.

Fig. 54. Migration, in an electric field, of empty pores formed
during the growth of an LiF single crystal [117].

in the case when $(a/R)D_{Si}^* < D_i^*$, i.e., when the migration is controlled (in the case of balanced surface and volume fluxes) by the
surface diffusion. Such diffusion is affected strongly by the state
of the inclusion—host interface. This may be due to the influence
of impurities on the mobilities of ions, on the distribution of the
potential, and on the concentration of surface defects (see preceding subsection). A considerable influence on the migration of
bubbles can also be exerted by insoluble surface impurities which
can hinder the free motion of atomic steps and kinks and practically suppress directional surface diffusion fluxes, i.e., the
directional transport of matter by surface diffusion [118]. The
contamination of bubble surfaces may be responsible for the different laws [Eqs. (15.37) and (15.38)] obeyed by the velocities of

Fig. 55. Dependence of the velocity on
the reciprocal of the radius for "contaminated" (1) and "pure" (2) pores in an LiF
single crystal [117].

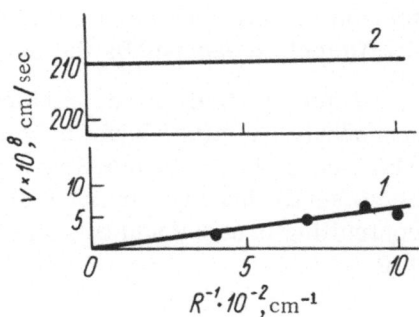

bubbles formed at the tips of healed cracks and during the growth of single crystals: the former bubbles were in contact with air and the atmosphere in the annealing oven.

The bubbles formed during the growth of single crystals had much cleaner surfaces and obeyed the inequality $(a/R)D^*_{Si} > D^*_i$ so that the transport of matter was limited by the diffusion fluxes in the interior of the host crystal. In this case the velocity is given by Eq. (15.38) and it should be independent of the bubble radius.

The surface of a bubble migrating in an ionic crystal may become contaminated during its motion and this may affect the migration velocity. We found [116, 117] that the velocity of bubbles in an NaCl single crystal subjected to an electric field decreased as a result of its irradiation with x rays. This was probably due to the fact that radiolysis of the bubble surface produced metallic sodium which contaminated the surface.

It is worth stressing the following point in connection with the different behavior of "clean" and "contaminated" bubbles. The formulas for the velocity of bubbles have been derived on the assumption that the transport of matter along the bubble—host interface is governed by the coefficient D_S, which can be found by the labeled atom method. However, the surface transport of matter is not always governed by this coefficient because a directional flux of atoms along a surface must give rise to the motion of atomic layers and steps in the opposite direction. This motion is due to the fact that geometrical surface defects are sources of diffusing atoms. If the mobility of atomic steps and layers is hindered by impurities, the directional transport of matter is also hindered, irrespective of the value of D_S measured by the mass transport method. Therefore, the surface diffusion coefficients measured by the labeled atom technique are always larger than the coefficients measured by the mass transport method [21].

It follows that, on the surfaces of "contaminated" bubbles, the transport of matter should be described by an effective coefficient which depends on the mobility of the steps. This coefficient may be so small that the surface transport of matter will be the rate-controlling process and this will give rise to the dependence $v \propto 1/R$.

The influence of impurities, which hinder the motion of steps, on the kinetics of surface mass transfer was demonstrated in experiments concerned with the healing of a scratch on the surface of copper coated with a thin film of molybdenum [118]. During high-temperature annealing, the molybdenum, which did not wet the copper surface, collected in the form of microscopic particles at the geometrical defects on the surface. The rate of healing of a scratch decreased with the density of the distribution of molybdenum particles on the surface of the copper (Chap. III).

A study of the migration of bubbles in an electric field [122] revealed that the directions of motion of individual bubbles changed during high-temperature annealing in a field of constant direction. These "anomalous" bubbles were usually located far from the main body of the bubbles, which moved without a change in direction. Formally, a change in the direction of migration of a bubble represented such a change in the volume and surface self-diffusion coefficients of anions and cations, which altered the sign of the differences of the combinations of these coefficients in the numerators of Eqs. (15.36)-(15.39). This could occur as a result of changes in the temperature or under isothermal conditions. In the latter case, it could be due to a change in the chemical composition of the surface of a bubble as a result of the diffusional exchange of atoms between this surface and the bulk of the host crystal (the local chemical composition of the host near an anomalous bubble could be different from the average composition).

An anomalously high velocity was exhibited by bubbles generated in KBr single crystals in the following way.[†] A nickel wire, $\sim 10^{-2}$ cm in diameter, was fused into a KBr single crystal. Next, this wire was exploded by passing a current pulse through it. The thermal and mechanical shock waves generated by the explosion produced "clouds" of nickel particles and disperse (1-2 μ) bubbles. When a crystal containing a cloud of bubbles was annealed in an electric field whose vector was oriented at right-angles to the axis of the explosion channel, a bubble migrated, forming a "jet" behind it (Fig. 56). The shadow behind the empty channel represented a current flowing in the electrically conducting medium. In these experiments (T = 630°C, E = 200

[†] G. A. Gurevich, Diploma Thesis [in Russian], Khar'kov University (1967).

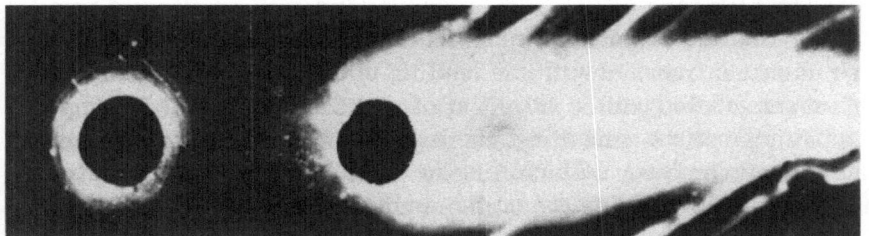

Fig. 56. Migration, in an electric field, of microcavities formed around an explosion channel.

V/cm, t = 7 h), the velocity of the bubbles was 4×10^{-6} cm/sec. This very high velocity was due to the small dimensions of the bubbles [see, for example, Eq. (15.37)] and the considerable damage to the crystal around the explosion channel (the vacancy concentrations and, therefore, the volume diffusion coefficients D_i^* were higher near this channel).

The bubbles formed during the growth of a crystal sometimes leave a trail of fine faceted pores of different interference colors (Fig. 57). This effect was thermodynamically unexpected because it increased the free surface and, therefore, the surface energy of the system. However, the formation of micropores may relax the local stresses in the host crystal.

A similar effect was the expansion of a localized porous region [119]: high-temperature isothermal annealing produced a multitude of fine pores around this region. These pores formed as a result of the condensation of vacancies, which diffused from the porous region where vacancy supersaturation was established by the strong distortion of the pore surfaces. Vacancies condensed into pores on various defects and inhomogeneities in the nonporous region. This process increased the free surface and it occurred because

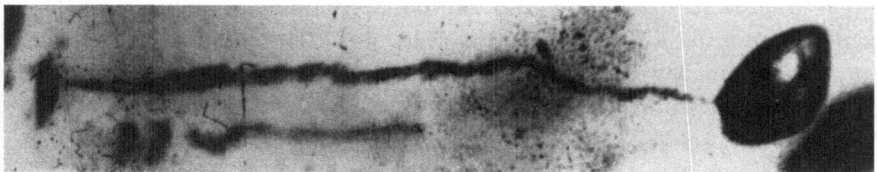

Fig. 57. Trail of pores left behind by a moving cavity (Geguzin et al.).

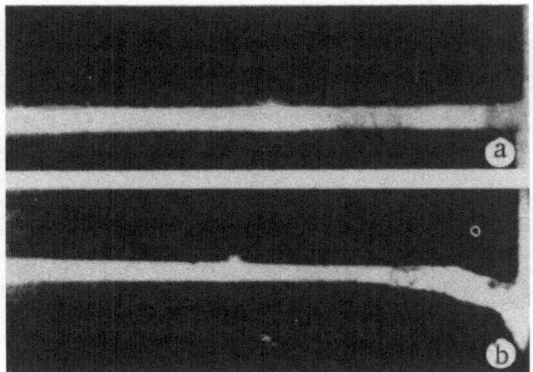

Fig. 58. Bending of a needle-shaped cavity in an
LiF single crystal subjected to an electric field (a
and b show successive stages) (Geguzin et al.).

a rise in the number of pores relaxed the stresses localized near
growing pores.

An external electric field applied to an ionic crystal may give
rise to diffusion-induced deformation of a single-crystal cavities,
which are sometimes known as negative crystals. The observed
types of deformation of negative crystals resemble the deformation of normal crystals (bending, elongation, etc.).

The diffusion-induced deformation of negative crystals in an
electric field was observed clearly in LiF single crystals containing needle-shaped cavities (negative whiskers). These cavities were generated by irradiation with a flux of thermal neutrons (total dose 10^{18} cm^{-2}) followed by high-temperature annealing, which healed the radiation defects and produced vacancies.
These vacancies condensed and produced needle-shaped cavities.
Some of these cavities were pinned by various obstacles at one or
both ends. Those which were pinned at one end were used in experiments in which the electric field was oriented along the cavity. Under the influence of the field, the cavities became longer
and the free ends emitted fine pores which traveled along the field.

Bending of needle-shaped cavities was observed when the field
was applied at right angles to their axes (Fig. 58). In a given
field, a cavity was bent to a curvature $1/R^*$ and, for a given distance between the pinning points L the bending could be described

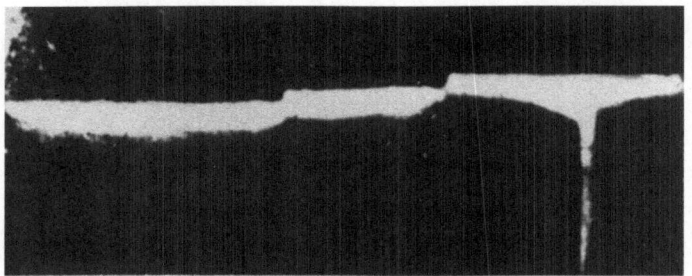

Fig. 59. Fracture of a needle-like cavity (Geguzin et al.).

by the maximum displacement (sagitta) $\sim L^2/2R^*$. In some cases, an analog of the brittle fracture of negative whiskers was observed: different parts of the cavity moved relative to one another and acquired a faceting in accordance with the anisotropy of the surface tension γ (Fig. 59).

The physical origin of the steady-state curvature of a needle-like cavity bent by an external field can be explained as follows. When a cavity is bent, it produces surface and volume diffusion fluxes in the host crystal. In equilibrium, the steady-state curvature of the bent cavity corresponds to the balance of the fluxes generated by the bending and by the external field.

The solution of the problem of steady-state (equilibrium) curvature of a cylindrical cavity bent by a field yields the relationship

$$\frac{1}{R^*} = \frac{1}{a^2\gamma} q^* E, \tag{15.52}$$

where

$$q^* = e \frac{D_2^* D_{S1}^* - D_1^* D_{S2}^*}{\left(D_2^* + \dfrac{a}{R} D_{S2}^*\right)\left(D_1^* + \dfrac{a}{R} D_{S1}^*\right)} \tag{15.53}$$

represents the effective charge of an ion on the surface of the cavity (it is assumed that $f_i = f_{Si} = 1$, $L_i = L_{Si} = 0$).

The charge q* is of dynamic origin: it is due to the diffusion fluxes which are maintained by the field.[†] The value and sign of the effective charge are governed by the relationship between the surface and volume self-diffusion coefficients in the cation and anion sublattices. It follows from Eqs. (15.36) and (15.52) that a pinned cavity bends in the same direction in which it would move if it were free.

A gas-filled pore migrating in an external electric field may become deformed. We have mentioned (Sec. 7) that in the linear theory the distortion of the shape of a pore migrating in a homogeneous field can be related to the anisotropy of the surface diffusion coefficient. Since the migration of pores in ionic crystals under the influence of an electric field is governed by volume and surface diffusion fluxes, this distortion mechanism may be important (in contrast to the migration of pores in a temperature gradient) for small and large pores.

A considerable distortion of the pores in ionic crystals in relatively weak electric fields may also be due to three nonlinear effects. The first of these effects arise because the expression for the chemical potential of a pair of ions (A^+ and B^-) contains a term which is proportional to E^2 and $\cos^2 \theta$. The curvature is similar to that observed for pores in ferromagnets or in pyroelectrics (Sec. 7): the order of magnitude of the distortion is $|\delta R|/R_0 \sim (\varepsilon R/\gamma)E^2$. However, this effect appears only in relatively strong fields ($E \sim 10^5$ V/cm for $R \sim 10^{-3}$ cm).

The second nonlinear effect is observed when the condition $v' \ll \overline{D}_i/R$ of Eq. (14.1) is not obeyed so that the distribution of

---

[†]Having postulated the existence of a charge q*, we can easily derive a relationship such as Eq. (15.52) on the basis of the following elementary considerations. The steady-state curvature is determined by the equality of two forces: the bending force F_E due to the field E and the opposing force F_γ governed by the surface tension γ. Since the surface density of ions is $n_S \sim 1/a^2$ and the surface area of a cylindrical cavity of length L is $S = 2\pi RL$, it follows that $F_E \sim q^* n_S SE \sim q^* \dfrac{2\pi RL}{a^2}E$. The opposing force is $F_\gamma \sim \left.\dfrac{\partial w}{\partial x}\right|_{x=x_0}$, where $w \sim 2\gamma \dfrac{\pi R x^2}{L}$ is the increase in the surface energy of the cavity when the sagitta is x_0. It follows from $|F_E| = |F_\gamma|$ that $x_0 \sim (L^2/\gamma a^2)q^* E$. Since $R^* \sim L^2/x_0$, the relationship obtained for x_0 is equivalent to Eq. (15.52).

defects around a pore is no longer stationary. This distribution depends on the field and is not proportional to cos θ, as has been the case in weak fields. As in solid solutions (Sec. 14), this effect should result in the distortion of pores moving at velocities v $\gtrless$ D/R.

The third effect causing nonlinear distortion of pores is associated with the influence of an electric field on the transport coefficients, i.e., on the diffusion coefficients of ions. In ionic crystals, this influence is associated particularly with a redistribution of the concentration of vacancies near a moving pore. This effect is estimated in [121] for the case of a weak nonlinearity in the distortion of pores in an electric field when it is sufficient to consider only the terms linear and quadratic in the field. It is shown in [121] that the expressions for the initial velocities of different parts of a spherical pore contain not only the linear terms $v(\vec{r}_s) \sim$ E cos θ, which do not cause distortion, but also the quadratic terms

$$\vec{n}\, \delta \vec{v}\, (\vec{r}_s) = A_2\, E_\infty^2 \cos^2 \theta. \tag{15.54}$$

This dependence of the angle θ between $\vec{r}_s$ and the direction of the field means that the front and rear surfaces of a pore move either in the direction toward the host or away from it (depending on the sign of A_2). This distorts the originally spherical pore by elongation or flattening.

The expression for the constant A_2 obtained in [121] can be simplified in the following two limiting cases. If $RD_i^* \gg aD_{Si}^*$, we find that

$$A_2 = -\frac{27}{4}\, \frac{D_1^* D_{S2}^* + D_2^* D_{Si}^*}{D_1^* + D_2^*}\, c_v \left(\frac{e}{kT}\right)^2 r_D \frac{a}{R}. \tag{15.55}$$

In this case, the pore becomes flattened along the electric field. If $RD_i^* \ll a\, D_{Si}^*$, we find that

$$A_2 = 27\, \frac{\left(D_1^* + D_2^*\right) D_{S1}^* D_{S2}^*}{\left(D_{S1}^* + D_{S2}^*\right)^2}\, c_v \left(\frac{e}{kT}\right)^2 r_D. \tag{15.56}$$

In this case, the pore becomes elongated along the field.

Equations (15.54)-(15.56) give only the direction and initial rate of deformation of a pore. Since the deformation of a constant-volume pore is accompanied by an increase in its surface, the force opposing the deformation increases as the deformation increases. This means that the rate of deformation will gradually decrease and at some stage no further deformation will occur.

The principal qualitative characteristics of the deformation of gas-filled bubbles in alkali halide crystals were established in [122]. They can be summarized as follows.

1. The deformation of bubbles can be represented quantitatively by the dimensionless ratio $R_{\parallel}/R_{\perp}$ ($R_{\parallel}$ and $R_{\perp}$ are the dimensions of the bubble parallel and perpendicular to the applied electric field). This ratio increases with time, reaching some limiting value for a given electric field. This is in agreement with the hindering effect of the surface energy, which increases as the degree of deformation increases.

Figure 60 shows the results of studies of the deformation of bubbles in an electric field. The bubbles were deformed by elongation either at right angles to the applied field (in an NaCl single crystal) or along this field (in an LiF single crystal).

2. The deformation of bubbles by the field was a reversible process. The original isomeric shape of the bubbles ($R_{\parallel} \approx R_{\perp}$) was reestablished by high-temperature annealing after the application of the field. The direction of deformation of a bubble could be altered by changing the direction of the applied field (Fig. 61).

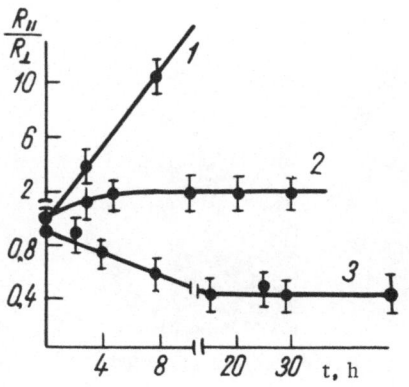

Fig. 60. Deformation of bubbles by the application of an electric field to ionic single crystals [122]: 1) LiF; 2) NaCl + $10^{-2}\%$ Cu; 3) NaCl [122].

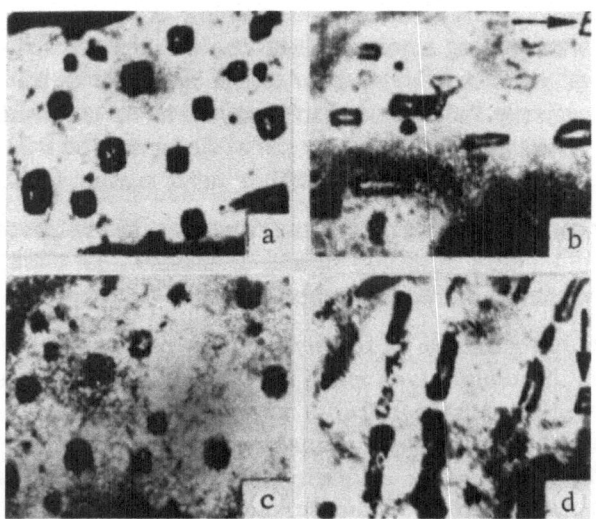

Fig. 61. Reversibility of the deformation of bubbles
in an LiF single crystal (a-d represent successive stages)
[122]. ×300.

Some bubbles in LiF single crystals split into several parts
in the final stages of deformation along the field (Fig. 62). This
process was due to the loss of stability by deformed bubbles.

The experimentally observed deformation of bubbles was
most probably due to the second and third of the nonlinear effects
discussed earlier. In particular, the second of the effects should
be active in the cases considered ($R \sim 10^{-3}$ cm, $v \sim 10^{-7}$ cm/sec)
if the diffusion coefficient, which depends strongly on the presence of impurities, is $D \leq 10^{-10}$ cm^2/sec.

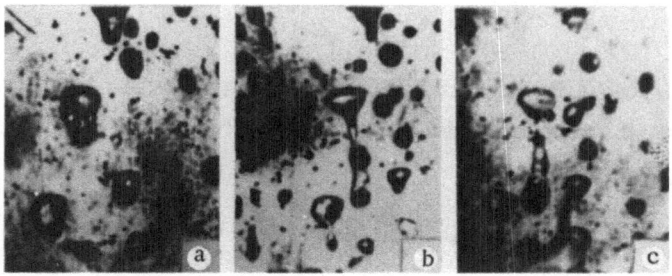

Fig. 62. Breakup of bubbles in an LiF single crystal subjected
to an electric field (a-c represent successive stages). ×300.

Migration of Metallic Inclusions in an Electric Field. If a metallic inclusion consists of A atoms, corresponding to the cations in an ionic crystal of the A^+B^- type, and if B atoms have a considerable solubility and mobility in the metal A, we find that the velocity $v = v_s$ of large spherical inclusions in an electric field is described by Eq. (15.39). In this case, the velocity of the inclusions is independent of their radius. If we apply the same reasoning as in the derivation of Eq. (15.39), we can readily obtain the corresponding expressions for the velocities $v_{c\perp}$ and v_p of cylindrical and plate-like inclusions oriented at right angles to the field. All three velocities are related by the simple expression

$$v_s : v_{c\perp} : v_p = 3 : 2 : 1. \tag{15.57}$$

It is more difficult to determine the velocity $v_{c\parallel}$ of a cylindrical inclusion of finite length L oriented parallel to the field because an allowance must be made for the edge effects. However, it is obvious that, as the current through the inclusion becomes larger, the velocity should increase with ascending values of L/R.

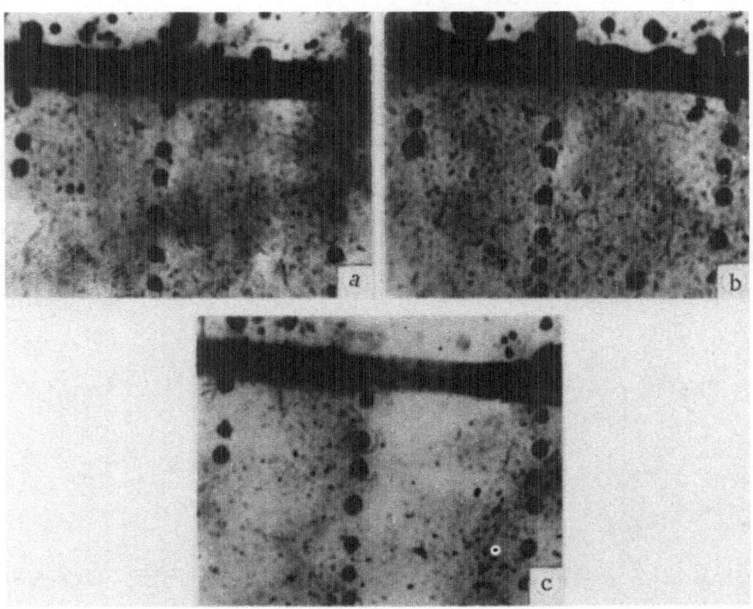

Fig. 63. Motion of lithium inclusions in LiF single crystals in an electric field. The inclusions are migrating toward the stem of a dendrite (a–c represent successive stages). ×300.

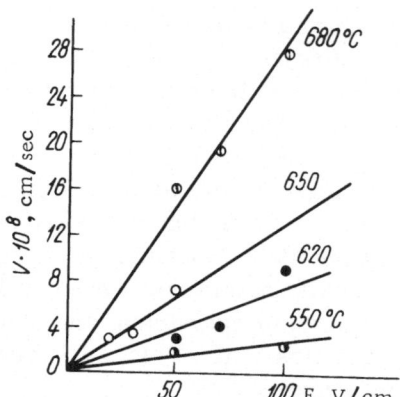

Fig. 64. Velocity of lithium inclusions in LiF single crystals plotted as a function of the field intensity [83].

The motion of a metallic inclusion in an ionic crystal subjected to an external field was investigated experimentally by Geguzin et al. [83], who studied Li inclusions in an LiF single crystal. The lithium inclusions were generated in the same way as described in Sec. 2. Many of these inclusions were spherical but some were cylindrical (L/R ≈ 2-4).

The migration of lithium inclusions was investigated in fields E = 30-100 V/cm. The field was oriented at right angles to the stem of a lithium dendrite. The experiments were carried out at 550-680°C, which is higher than the melting point of lithium (T_{mp} = 186°C).

Lithium particles of 3-30 μ size migrated, in accordance with Eq. (15.39), at a velocity which was independent of their size. Figure 63 shows a typical sequence of photographs of an ensemble

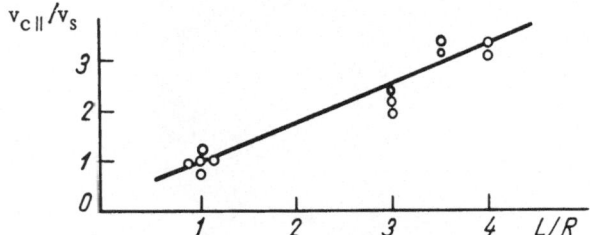

Fig. 65. Dependence of the velocity of cylindrical lithium inclusions in LiF single crystals on the ratio of the length of the inclusion to its radius [83].

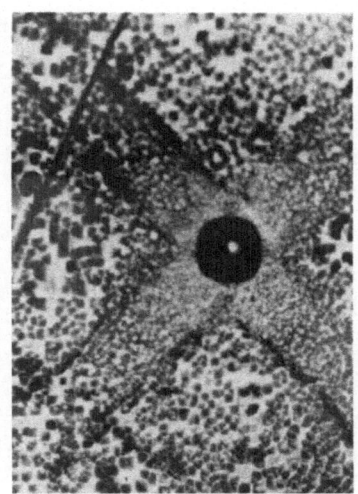

Fig. 66. Distribution of dislocations in a lithium inclusion in an LiF single crystal [83]. ×270.

of inclusions moving toward the stem of the dendrite. The linear dependence of the velocity on the field, predicted by Eq. (15.39), is plotted in Fig. 64.

The reality of the shape effect was demonstrated by Geguzin et al. [83] in a study of the motion of cylindrical inclusions with different values of the ratio $v_{c\|}/v_s$. This ratio increased linearly when L/R was increased from 1 to 4 (Fig. 65).

It follows from the experimental results that $v_{c\|} \propto v_s L/R$; this means that coalescence should occur during the motion of an ensemble of originally parallel cylindrical metallic inclusions with different values of the ratio L/R. The inclusions with larger values of L/R overtook the less elongated inclusions and coalesced with them. A cylindrical inclusion resulting from such coalescence was found to migrate at a velocity greater than that of either of the original inclusions. This situation was opposite to that found in the coalescence of spherical inclusions migrating at a velocity $v \propto 1/R$ (a particle resulting from the coalescence of two inclusions migrated more slowly than either of the original inclusions) but was analogous to the situation in the case of gas-filled bubbles migrating at a velocity $v \propto R$.

The velocity of spherical liquid-lithium inclusions in an LiF single crystal increased rapidly with rising temperature. However, this dependence could not be used to deduce the mass transfer mechanism which governed the velocity of the inclusions. This was due to the fact that the region of the crystal in the direct vicinity of these inclusions was very distorted (Fig. 66) and, therefore, the values of the diffusion coefficient governing the mass transfer would differ considerably from the equilibrium values.

Role of Contact Effects. The distribution of defects near the boundary of an ionic crystal depends strongly on the nature of the phase which is in contact with this crystal (see the first subsection in the present section).

The properties of the Debye layer which appears near the surface of an ionic crystal, the thickness of this layer, and the distribution of anion and cation vacancies in the layer, all depend in a complex manner on the physical properties of the phase which is in contact with the crystal. Obviously, the kinetics of the diffusional transfer of matter along the inclusion—host interface must depend strongly on the properties of such a surface layer. This should be manifested in the migration of an inclusion in an ionic crystal, particularly when this crystal is subjected to an external electric field.

The influence of contact effects on the motion of ionic crystalline inclusions in an ionic crystal was found in experiments in which spherical SiO_2 particles ($\sim 50\ \mu$) migrated in an NaCl single crystal [170].

The SiO_2 particles migrated toward the negative terminal at the velocity $v \propto 1/R$, i.e., their migration was governed by surface (interface) diffusion. Under similar conditions, gas-filled bubbles in NaCl migrated in the opposite direction at a velocity $\sim 10^3$ times smaller than the velocity of SiO_2 inclusions. A comparison of these observations with Eq. (15.37), on the assumption that the correlation factors differed little from unity, suggested that the bubbles obeyed the inequality $D_1^{\bullet} D_{S2} f_1^{-1} f_{S2}^{-1} < D_2^{\bullet} D_{S1}^{\bullet} f_2^{-1} f_{S1}^{-1}$, whereas the crystalline inclusions were subject to $D_1^{\bullet} D_{S2} f_1^{-1} f_{S2}^{-1} > D_2^{\bullet} D_{S1}^{\bullet} f_2^{-1} f_{S1}^{-1}$.

These inequalities between the products of the volume and surface self-diffusion coefficients indicated that, at the interface between the SiO_2 inclusions and the NaCl host, the boundary self-diffusion coefficient of the anions was considerably higher than the self-diffusion coefficient of the anions in a surface layer of NaCl in contact with the air.

These experiments showed qualitatively that the structure of a surface layer in an ionic crystal depends strongly on the medium which is in contact with this crystal, and that this structure affects the magnitude and direction of the velocity of the inclusions in the crystal.

Vacancy Breakdown. Vacancy breakdown is a phenomenon specific to ionic crystals: it appears as a result of the migration of cavities in an electric field.

Vacancy breakdown occurs in the following manner. If an ionic crystal contains a localized source of excess vacancies, a chain of pores is formed during the annealing of this crystal in an external electric field. The direction of the chain coincides with that of the field, or is close to it. Sometimes, a chain of closely spaced pores merges into a needle-shaped cavity. Since a chain of fine pores resembles an electric breakdown channel, this phenomenon has been called vacancy breakdown. Such breakdown has been observed and investigated by many workers [120, 123-125].

A localized source of excess vacancies in an ionic crystal can be produced in many ways. In particular, such a source appears in a crystal with an inhomogeneous distribution of defects or in the diffusion zone in a system in which mutual diffusion results from unequal partial diffusion coefficients of the components.

Vacancy breakdown was observed in one-component systems in the following experiments. Films of NaCl, KCl, and KBr were evaporated on the (100) natural cleavage or the (110) artificially induced faces of NaCl, KCl, and KBr single crystals. The evaporation conditions were such that the films contained many defects and were milky in color. Identical samples were assembled in pairs in such a way that they enclosed the evaporated films, i.e., three-layer samples of the equilibrium−defective−equilibrium type were formed. The central defective films acted as localized sources of excess vacancies. The defects in these films were healed by annealing in an external electric field (T = 680-700°C, E = 50-100 V/cm). Such annealing produced chains of pores

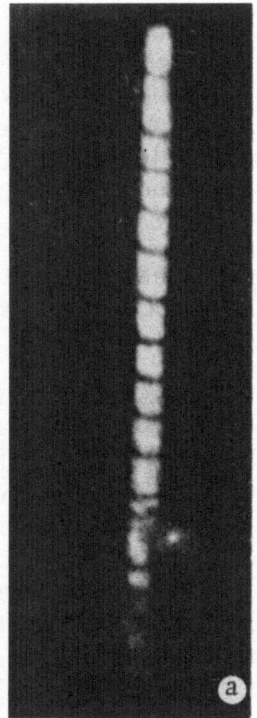

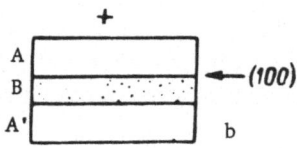

Fig. 67. Typical chain observed in vacancy breakdown as
a result of self-diffusion (a) and experimental setup used in
observations of breakdown (b) [123].

in the equilibrium parts of the crystal, i.e., vacancy breakdown
was observed.

A typical chain of pores observed in vacancy breakdown is
shown in Fig. 67.

Vacancy breakdown was also observed in mutual diffusion experiments in KCl−KBr and NaCl−NaBr systems. The application
of an external field of ~100 V/cm to three-layer samples of the
$K_1-K_2-K_1$ type (K_1 and K_2 are different crystals) resulted in the
asymmetric formation of pores so that the volumes of the pores
in one of the crystals (K_1) became larger compared with the other
crystal (K_2). Needle-shaped cavities and chains of pores of the
vacancy breakdown type were observed in the K_1 crystal (Fig. 68).

The mechanism of vacancy breakdown can be described as follows. The coalescence of excess anion and cation vacancies in the vacancy generation zone produces neutral microscopic complexes. The formation of these complexes is due to the need to annihilate the continuously generated excess vacancies and to the fact that the formation of vacancy complexes carrying charges of the same type would have resulted in a large increase in the electrostatic energy of the system. These neutral complexes of anion and cation vacancies have linear dimensions of the order of several atomic spacings $(R \sim 10^{-7}$ cm). In an external electric field, these complexes migrate as a whole, like macroscopic pores, and this

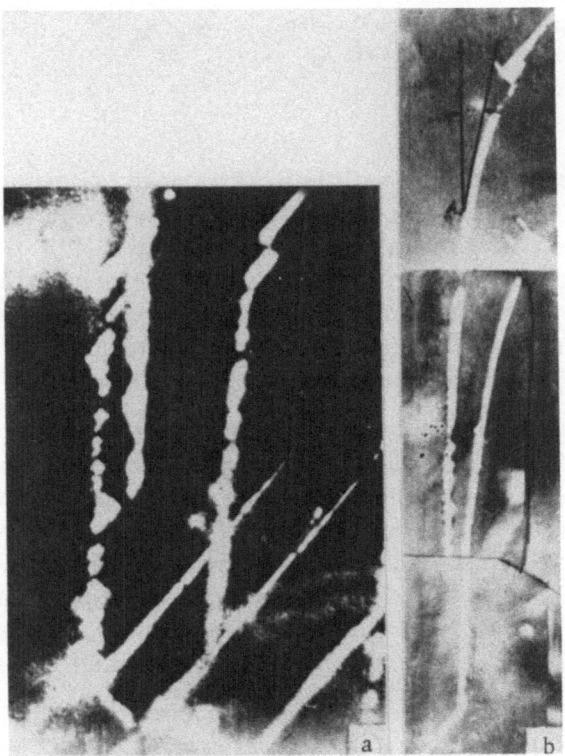

Fig. 68. Needle-shaped cavities resulting from vacancy breakdown in the diffusion zone in the KCl−KBr system [120]: a, b) different parts of the same sample. ×800.

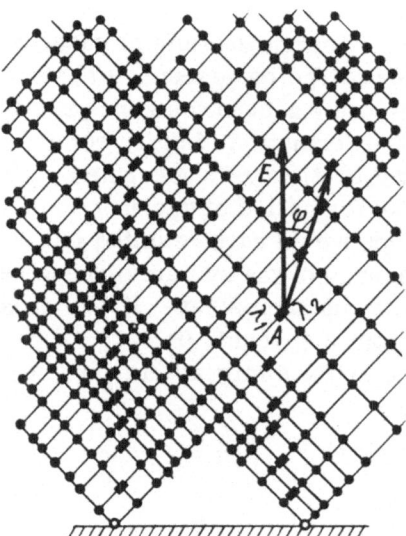

Fig. 69. Formation of bent chains of pores due to local differences in the dislocation density in intersecting slip systems in which the distances between dislocations are λ_1 and λ_2 [120].

migration occurs under conditions of balanced volume and surface self-diffusion fluxes. These neutral vacancy complexes are highly mobile. If we use the results reported in [116] on the velocity of macroscopic pores v_R ($R \sim 10^{-4}$ cm) in an external field, we can estimate the velocity of the complexes v_c from $v_c \approx v_R \times (R/R_c)$.[†] Under the conditions in the experiments described in the published work, we should have $v_c \sim 10^{-5}$ cm/sec.

It is worth pointing out that, in the case of cavities whose linear dimensions are only slightly larger than the atomic spacings, $R \approx (3-5)a$, the mechanisms of volume and surface diffusion may differ somewhat from those in the case of larger cavities because the number of atoms which can participate in diffusion is limited. However, the action of the fluxes remains the same as in the case of cavities of radii $R \gg a$: these fluxes prevent, in a balanced way, the appearance of a migration-inhibiting field in the cavity.

Neutral microscopic complexes can travel not only in a defect-free lattice but also along the dislocation lines. The latter process is more likely because kinks in dislocation lines are possible sources of neutral complexes.

[†] This estimate is valid only if the velocity varies with the radius in accordance with the law $v \propto 1/R$.

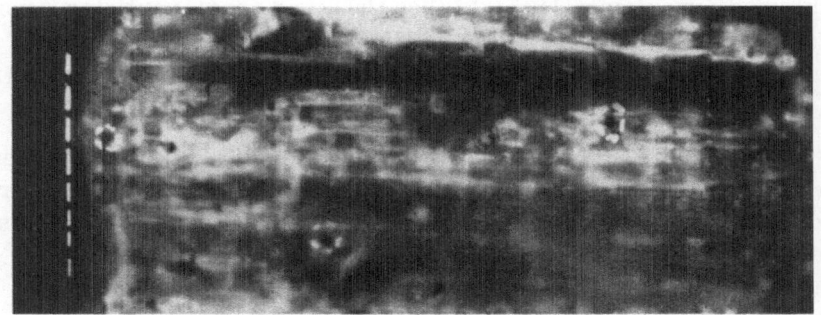

Fig. 70. Migration of Cr_2O_3 particles in the vacancy breakdown zone in a KBr single crystal [125]. ×180.

The neutral complexes migrating along dislocation lines may accumulate at the points of intersection of these lines, forming macroscopic pores visible under an optical microscope. The observed "bending" of pore chains, i.e., their deviation from the direction of the applied field, is explained by the fact that the local densities of dislocations in an intersecting slip system can have various values. This is illustrated schematically in Fig. 69, which shows a two-dimensional model.

Second-phase inclusions also migrate at anomalously high velocities along the vacancy breakdown channels. It has been es-

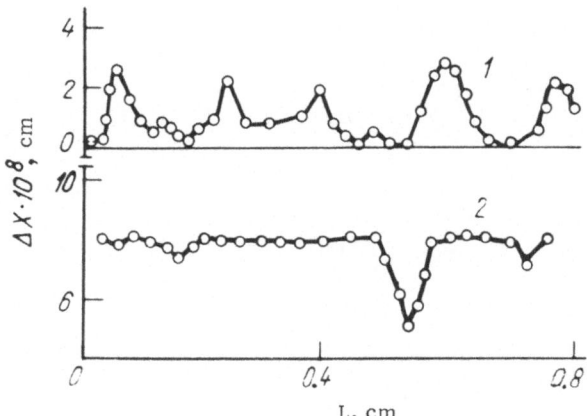

Fig. 71. Distribution of the displacements of Cr_2O_3 particles in the vacancy breakdown zone in a KBr single crystal [125]: 1) undeformed crystal; 2) deformed crystal.

tablished experimentally [125] that CuO_2 and Cr_2O_3 particles (linear size $\sim 10^{-3}$ cm) migrate in the vacancy breakdown region in KBr crystals at a velocity which is an order of magnitude higher than the velocity in the defect-free parts of such crystals.

The results of the experiments described in [125] can be seen in Figs. 70 and 71. Figure 70 shows the distribution of Cr_2O_3 particles in the vacancy breakdown zone. Originally, these particles were located at the interface between a KBr single crystal and a defective evaporated film (represented by the dashed line in the figure). Figure 71 shows the distribution of the same particles along the migration front in annealed (1) and specially deformed (2) crystals. The results of these experiments demonstrated that the migration of inclusions in the vacancy breakdown zone is highly sensitive to the structure of this zone.

The anomalously high mobility of inclusions in a defective region in a crystal is evidently a common phenomenon which is independent of the origin of the force responsible for the migration of inclusions. In dispersion-hardened structural materials, such regions may be nuclei of local softening, where the inclusions hindering the motion of the dislocations have a high mobility.

CHAPTER III

Influence of the Migration of Inclusions on High-Temperature Processes in Solids

The diffusive migration of inclusions in solids can affect many important processes at high temperatures. This is because at these temperatures the diffusional mobility of atoms is sufficient to produce considerable displacements of inclusions in a reasonable time.

The basic idea behind the development of dispersion-hardened alloys and composites is the combination of a plastic host and disperse inclusions which act as obstacles hindering the motion of dislocations.

If the temperature is low and the stresses are high, the presence of disperse inclusions increases the flow stress of dispersion-hardened materials to a value above the flow stress of the host substance. At high temperatures and at stresses below the flow limit, the mechanism by which inclusions harden a material is different.

The kinetics of the high-temperature creep of dispersion-hardened alloys and composites containing foreign inclusions may be determined by the concentration and the velocity of migration of these inclusions. This happens because the effective velocity of dislocations in stress fields depends on the number and type of inclusions as well as on the mechanism and the velocity of motion of these inclusions. Thus, the heat resistance of a material intended for prolonged use at high temperatures may depend strongly on the diffusional migration of the inclusions which harden this material.

In many technological processes aimed to produce a material with definite mechanical, magnetic and other properties, the most important parameters are the rate of recrystallization growth of the material's grains, the optimal range of temperatures at which recrystallization annealing is carried out, and the maximum size of the grains which are formed as a result of secondary recrystallization. All these parameters of the recrystallization process in dispersion-hardened materials depend on the presence of foreign inclusions and on the velocity of their migration under the influence of the forces exerted on them by the moving grain boundaries. The motion of these boundaries, like the motion of dislocations, is hindered by inclusions.

This hindering effect is manifested in many structural alloys and composites containing foreign inclusions. For example, it is reported in [126] that the growth of grains in tungsten is hindered by thorium oxide inclusions. It is also reported [127] that the presence of insoluble inclusions reduces considerably the maximum size of the grains forming as a result of secondary recrystallization in α-brass (of the 70−30 type) under isothermal annealing conditions. The results of numerous investigations of the recrystallization of porous bodies prepared by the pressing of powders shows clearly that the average size of the grains produced by recrystallization decreases with increasing porosity [128].

Thus, the kinetics of two important processes − the high-temperature creep and the recrystallization in materials containing disperse inclusions − may be determined by the velocity of inclusions.

The velocity of dislocation−inclusion and boundary−inclusion systems plays an important role in creep and recrystallization processes. However, there are some important processes and phenomena in which the velocity of free inclusions is a key factor (these inclusions may move under the influence of external forces or as a result of the Brownian motion). In particular, empty and gas-filled pores may coalesce as a result of collisions. In the case of foreign inclusions or pores, such coalescence may accompany diffusional coalescence, in which the enlargement of an inclusion is due to a directional diffusion flux of atoms or vacancies. However, in the case of bubbles filled with gases insoluble in the host material, the collision mechanism is the only possible one. The coalescence of gas-filled bubbles colliding during their diffusional migration is an important effect in the mechanism by which fuel elements swell in nuclear reactors.

In the present chapter, we shall consider some of the important processes whose kinetics depends on the presence and motion of foreign inclusions.

16. DISPERSE FOREIGN INCLUSIONS IN REAL CRYSTALS

We shall start by considering some experimental data obtained in studies of the distributions of disperse foreign inclusions in crystals and of the influence of such inclusions on some processes and properties of dispersion-hardened alloys. A review of such data is essential for the understanding of those high-temperature properties of dispersion-hardened materials which are due to the diffusional migration of inclusions.

Inclusions and Dislocations

The results of many experimental investigations show clearly that inclusions of various types (second-phase precipitates, cavities formed as a result of the condensation of vacancies, particles of oxides created in internal oxidation processes, gas-filled bubbles) are usually found on grain or block boundaries or around single dislocations and dislocation networks.

The preferential location of inclusions on dislocations can be demonstrated in optically transparent crystals in which this preference is exploited in the decoration of dislocations. Hedges and Mitchell [129] demonstrated that colloidal silver, which appears in AgBr single crystals during photographic development, is usually precipitated at dislocations. Copper precipitates preferentially on dislocations in silicon single crystals [130]. Dislocations in NaCl single crystals can be decorated by the additive coloration with metallic sodium [131] or by the introduction of AgCl, $BaCl_2$, $CaCl_2$, and similar chlorides [132, 133]. These two methods reveal dislocations by the precipitation of Ag, Ba, and Ca particles. Figure 72 shows photographs of parts of an NaCl single crystal in which dislocations are decorated by silver precipitates [133].

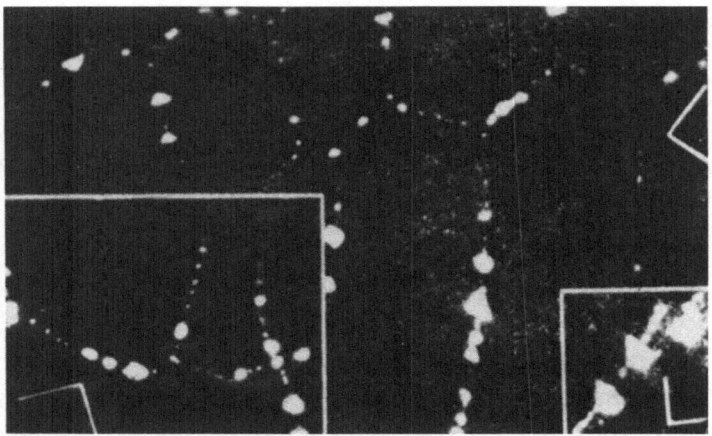

Fig. 72. Silver precipitates decorating dislocations in NaCl single crystals [133].

Bubbles and cavities, the latter resulting from the condensation of excess vacancies, also settle on dislocations [86]. Excess vacancies in crystals may be generated by irradiation with neutrons or other high-energy particles, or by the annealing of crystals prepared under nonequilibrium conditions (for example in an electrolytic bath using a high current density) or subjected to considerable plastic deformation. Excess vacancies also appear in a diffusion zone if the values of the partial diffusion coefficients of a diffusion pair are different. Supersaturation with vacancies may also result from quenching from a high temperature.

Excess vacancies escaping from a solid-solution lattice form cavities directly on dislocations or they condense on various structure inhomogeneities (particularly microcracks). The cavities formed in this way are obstacles which impede the motion of grain boundaries and dislocations.

The precipitation of second-phase particles on dislocations in supersaturated solid solutions may occur within Cottrell clouds, in which the concentration of impurity atoms c* is considerably higher than the average volume concentration c. Such precipitation may be regarded as the final stage of the formation of Cottrell clouds in supersaturated solutions.

The order of magnitude of time necessary for the formation of a Cottrell cloud can be estimated in a simple manner, as suggested in [134, 135]. The velocity of diffusion of impurity atoms v_i under the action of the force of interaction with a dislocation $f_i = -\nabla w_i$ (w_i is the energy of the interaction) follows from the Einstein relationship

$$v_i = \frac{D_i}{kT} f_i, \qquad (16.1)$$

where D_i is the diffusion coefficient of impurity atoms. This velocity is derived ignoring the diffusion fluxes resulting from changes in the impurity concentration. The dependences of w_i and f_i on the distance r of a given atom from a dislocation can be written in the following form (ignoring the angular dependences)

$$w_i(r) \sim w_{i0} \left(\frac{b}{r} \right)^n, \qquad f_i \sim \frac{n w_{i0}}{r} \left(\frac{b}{r} \right)^n. \qquad (16.2)$$

Here, w_{i0} is of the same order as the binding energy of an impurity atom in a Cottrell cloud (for $r \approx b$); n = 1 if the interaction is due to the difference between the atomic radii of the impurity and the host, and n = 2 if the dominant effect is the difference between the elastic moduli.

If we assume that all the impurity atoms in a cylinder of radius $r_0 = \int_0^t v_i dt$, reach a dislocation in a time t, i.e., that $dr_0/dt = v_i$, it follows from Eqs. (16.1) and (16.2) that

$$r_0^{n+2} = n(n+2) \frac{D_i |w_{i0}| b^n}{kT} t. \qquad (16.3)$$

All the atoms in this cylinder (end area πr_0^2) are thus transferred to a cylinder surrounding a dislocation (end area $\sim b^2$). Hence, it follows that

$$\frac{c^* - c}{c} \sim \frac{\pi r_0^2}{b^2} \sim \pi \left[\frac{n(n+2) D_i |w_{i0}| t}{kT b^2} \right]^{\frac{2}{n+2}}. \qquad (16.4)$$

Since $c^* \sim 1 \gg c$, the formation time of a Cottrell cloud τ_C is of the order of

$$\tau_C \sim \left(\frac{c^*}{\pi c} \right)^{\frac{n+2}{2}} \frac{kT b^2}{n(n+2) D_i |w_{i0}|}. \qquad (16.5)$$

Within a Cottrell cloud, $c^* \sim 1$; impurity atoms forming a supersaturated solution may precipitate preferentially on kinks in a dislocation line. These are the precipitates which are used in various dislocation decoration methods. Obviously, the mobility of dislocations is reduced by the presence of a cloud of impurity atoms or precipitates. The constant τ_C is thus a characteristic time in which a supersaturated solid solution hardens during its aging under isothermal conditions.

The formation of cavities on dislocations as a result of the condensation of vacancies from a lattice supersaturated with vacancies is complicated by the fact that dislocations can absorb vacancies [135]. Cavities formed on dislocations act as vacancy sinks, and, consequently, their dimensions increase with time. Moreover, the size of cavities may change as a result of their coalescence with neighboring cavities [147].

Obviously, dislocations are not the only sinks for excess vacancies, and cavities may form also on various defects far from dislocations. The relative importance of dislocations and other defects as vacancy sinks depends on the temperature and on the dislocation density. At low temperatures and low dislocation densities, when the diffusion of vacancies to dislocations is relatively slow, the role of other defects may be considerable. The cavities formed at these defects cannot harden a crystal, at least during the initial stage of deformation before these cavities encounter dislocations.

The binding force between a dislocation and an inclusion may be very strong and the simple detachment of a moving dislocation from inclusions is not a typical process in the deformation of heterogeneous systems. This force can be estimated in the same way as in the interaction between a particle and a grain boundary (Sec. 11). A spherical inclusion of radius R located on a dislocation reduces the length of the latter by $\sim$R and, consequently, reduces the energy by $\sim Gb^2R$. This reduction in the energy occurs when an inclusion approaching a dislocation travels a distance $\sim$R toward the dislocation. Therefore, the maximum force binding a dislocation to an inclusion can be estimated approximately from

$$F_m \propto \frac{Gb^2R}{R} \propto Gb^2. \tag{16.6}$$

Thus, the force F_m is independent of the inclusion radius.

An external stress can detach a dislocation from an inclusion. We can estimate the stress capable of such detachment by assuming that a segment of a dislocation line of length l_d (l_d is the distance between two inclusions) is acted upon by a force $F' \sim \tau_l b l_d$. A dislocation can be detached from inclusions if the force F' exceeds the maximum binding force of Eq. (16.6). Consequently, the external stress should be of the order of

$$\tau_l^* \sim G \frac{b}{l_d}. \tag{16.7}$$

If $G \approx 100$ kJ/cm^3 (10^{12} ergs/cm^3) and $l_d \sim 10^{-5}$ cm, we find that $\tau_l \sim 10$ J/cm^3 (10^9 ergs/cm^3) = 10 kN/n^2 (10^3 kgf/cm^2). This value is considerably greater than the stresses encountered in the service life of heat-resistant alloys at high temperatures, i.e., under realistic conditions dislocations cannot be detached from inclusions. It should be noted that τ_l depends on the shape of the inclusions. We can easily show that, in the case of a cylindrical inclusion of length L, the approximate formula for τ_l, analogous to Eq. (16.7), is

$$\tau_l^* \sim G \frac{b}{l_d} \frac{L}{R}. \tag{16.8}$$

In real heterogeneous materials, the nature of the interaction between dislocations and inclusions is governed, to a considerable degree, by the structure of the inclusion—host interface. Obviously, this structure must depend on the size of the inclusions and the method by which they are formed or introduced into the host. If inclusions are small and are formed by a diffusion mechanism, the inclusion—host boundaries may be coherent. In the diffusionless nucleation, the coherence of the boundaries is a natural attribute of the process, whereas in the diffusional nucleation coherent boundaries are formed only if the precipitate and the host satisfy the geometrical matching conditions.

A coherent inclusion—host interface has a low energy. However, an inclusion itself may store considerable energy in the form of elastic strain. A departure from coherence is usually accompanied by the relaxation of the elastic stresses and by an increase in the surface energy of the inclusion—host interface. Since the elastic strain energy in an inclusion is $w_0 \propto R^3$ and the surface energy is $w_S \propto R^2$, there should be a critical value of the inclusion radius R^* at which a departure from coherence will be favored by the energy considerations: if $R > R*$, the inclusion—host interface will be incoherent.

The orders of magnitude of w_0 and w_S are given by the self-evident relationships

$$w_0 \propto 3GR^3 \varepsilon^2,$$
$$w_S \propto 4\pi R^2 \gamma,$$

where 3ε is the relative change in the atomic volume due to the formation of an inclusion. If we assume that $w_0 \approx w_S$, we can estimate the value of $R*$:

$$R* \sim \frac{\gamma}{\varepsilon^2 G}. \tag{16.9}$$

If $\gamma \sim 10$ mJ/cm^2 (10^3 ergs/cm^2), $\varepsilon \approx 0.03$, $G \approx 100$ kJ/cm^3 (10^{12} ergs/cm^3), it follows from Eq. (16.9) that $R* \sim 10^{-6}$ cm.

This estimate of $R*$ is obtained on the assumption that the inclusions are spherical. A different value is obtained for inclusions of other shapes, such as disks or needles.

The comment on the departure from coherence applies also to inclusions which form in the bulk of a host crystal, for example, during its internal oxidation or the precipitation (aging) of a supersaturated solution. It should be noted that precipitates forming in the matrix of an alloy during the aging process may form boundaries of different degrees of coherence with the matrix. The Guinier—Preston zones which form during the initial aging stages have coherent boundaries. In Al—Cu alloys, these zones are platelets oriented parallel to the (100) plane. The thickness of these platelets is $\sim 10^{-7}$ cm and the characteristic linear size in the (100) plane is $\sim 5 \times 10^{-6}$ cm. The precipitates formed during the later stages have partly coherent boundaries with the matrix. In Al—Cu alloys, these precipitates are particles of the θ' phase, whereas in Ag—Mg—Zn alloys they are particles of the M' phase (MgZn$_2$). The precipitates formed in "overaged" alloys have incoherent boundaries.

The nature of the interaction of dislocations with precipitates depends strongly on the properties of the boundaries of the precipitates. Nicholson et al. [136] carried out an electron-microscopic study of the interaction between moving dislocations and

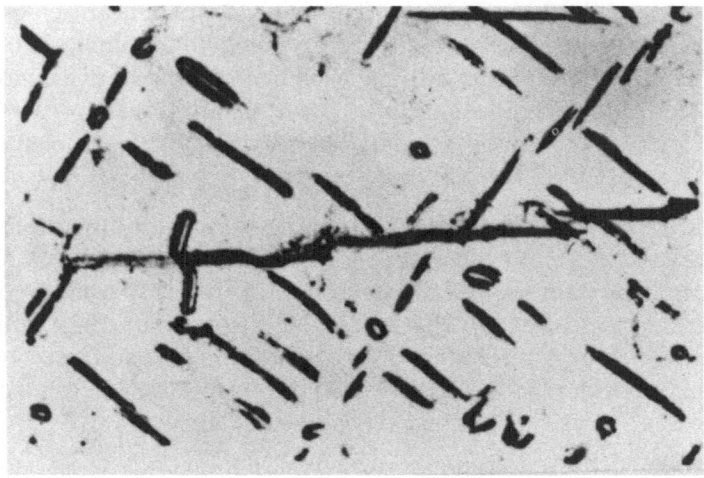

Fig. 73. Bending of a coherent inclusion as a result of intersection by a slip band [136].

the precipitates formed at the different stages of aging of aluminum alloys (Al−4%Cu, Al−10%Zn, Al−7.5%Zn−2.5%Mg). They
established that the inclusions with coherent or partly coherent
boundaries hindered the motion of dislocations in the slip plane.
However, at high stresses, the inclusions could be bypassed by
dislocations. Figure 73 shows an electron micrograph of the Al−

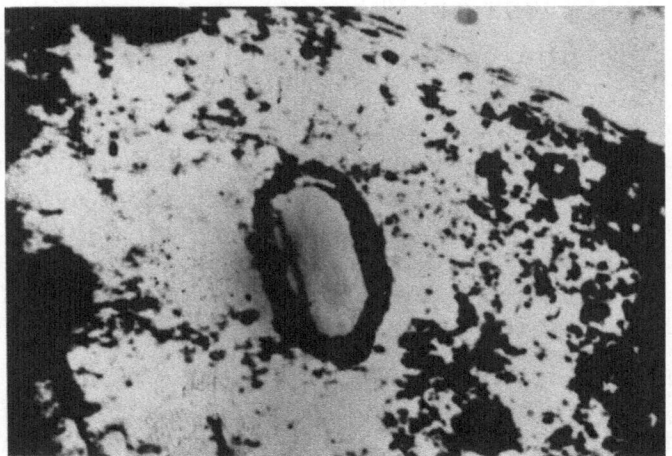

Fig. 74. Formation of dislocation loops around an incoherent
inclusion [136].

4%Cu alloy with θ' inclusions. It is clear from this micrograph that the inclusions intersecting a slip plane bend during deformation and allow dislocations to pass through. However, Nicholson et al. [136] also found cases in which θ' precipitates in Al—Cu alloys and M' precipitates in Al—Zn—Mg alloys act as obstacles which stop dislocations.

Incoherent precipitates are impermeable to dislocations. Nicholson et al. [136] found that slip bands in "overaged" alloys do not intersect incoherent inclusions. Figure 74 is an electron micrograph of the Al—7.5%Zn—2.5%Mg alloy containing an incoherent inclusion surrounded by dislocation loops. These loops were formed, during deformation, from segments of dislocations which failed to travel across incoherent inclusions.

A widely used method of preparing composite materials involves the introduction of inclusions into a matrix by powder metallurgy methods. Oxide particles can be introduced in the way which is employed in the manufacture of SAP (sintered aluminum powder) composites: oxidized powders are compacted and subsequent firing produces approximately isomeric inclusions from the oxide films surrounding the powder particles. Inclusions introduced externally into heterogeneous alloys (composites) almost always have incoherent boundaries.

17. HIGH-TEMPERATURE CREEP OF

DISPERSION-HARDENED ALLOYS

At high temperatures, the diffusion mobility of atoms is high and the kinetics of the creep of dispersion-hardened alloys may, to a considerable degree, be governed by the velocity of migration of inclusions (Sec. 16). The effective velocity of dislocations in a stress field will depend on the number and nature of the inclusions, as well as on the mechanism and velocity of the migration of the inclusions.

At high temperatures, a dislocation may overcome an obstacle either by destroying it or by bypassing it through climb to a different slip plane. The climb mechanism is more likely at high temperatures [137, 138]. In many metals and alloys, this mechanism is the main one and it determines the kinetics of high-temperature creep [139]. However, the climb is unlikely to occur if a dislocation is bound so strongly to an inclusion that the two move together. In this case, the well-known relationship [140] between the rate of deformation $\dot{\varepsilon}$ and the density of moving dislocations N_d,

$$\varepsilon \propto N_d \, bv(\tau, T),$$ (17.1)

can still be used provided $v(\tau, T)$ is understood to be the velocity of a dislocation-inclusion complex averaged out over the various orientations of the Burgers vector.

Thus, by altering the type, composition, and linear dimensions of inclusions and the nature of their boundaries with the matrix, i.e., by altering the velocity of inclusions, we can raise considerably the heat resistance of a composite material even at temperatures at which metals and alloys free of hardening inclusions cannot be used as structural materials.

The presence of foreign inclusions hindering the motion of dislocations can have two very important and, in a sense, opposite consequences.

First, if these inclusions are practically immobile, the high-temperature diffusion creep may become a threshold effect because the pinned dislocations (either single dislocations or those in dislocation walls) cease to be vacancy sources and sinks and, therefore, the diffusion transport of matter is arrested. In this case, such transport can occur only if the external stresses exceed a certain threshold sufficient to enable dislocations to overcome immobile obstacles.

Secondly, the dislocation loops which form around inclusions may act as sources of new dislocations. This process may affect the kinetics of the creep in a heterogeneous system.

At high temperatures, dispersion-hardened alloys soften. This is illustrated in Fig. 75, which gives the temperature dependences of the Brinell hardness of the $Al-Al_2O_3$ composite (SAP), which was cold rolled before being annealed [141, p. 153]. The temperature dependences shown in Fig. 72 indicate that the hardness and the softening temperature rise as the content of the Al_2O_3 disperse particles, which are insoluble in aluminum, in-

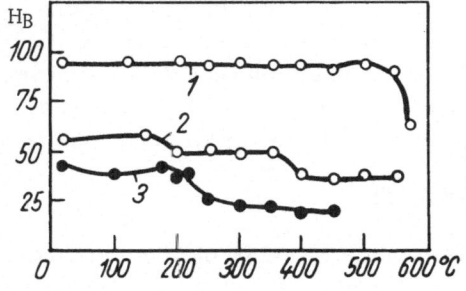

Fig. 75. Temperature dependence of the hardness of the SAP composite [141]: 1) $9\%Al_2O_3$; 2) $7\%Al_2O_3$; 3) $5\%Al_2O_3$.

creases. The material containing 9% Al_2O_3 retains its high value of hardness right up to 550°C, i.e., to a temperature which is only 100°C below the melting point of the matrix. However, above this temperature, the hardness falls rapidly. Similar results have also been obtained for $Cu-Al_2O_3$ composites prepared by powder metallurgy methods [141, p. 119].

The softening of dispersion-hardened composites, particularly those of the SAP type, may be due to the simultaneous effect of several processes. In many real alloys, the high-temperature softening is due to the dissolution of the hardening-phase particles in the matrix. However, we shall be interested primarily in the processes which are due to the migration of the hardening-phase inclusions. Apart from the direct relationship between the rate of deformation and the velocity of inclusions which rises with temperature, a higher mobility of the inclusions which are dragged by migrating grain boundaries or dislocations may reduce the number of hardening-phase particles within grains and blocks because of the clustering of these particles at the boundaries and the coalescence as a result of direct collisions. Microscopic cavities moving together with dislocations may get smaller and disappear so that dislocations become free to move easily.

Migration of Inclusions under the

Action of Forces Exerted by

Dislocations

We have already shown that diffusion fluxes which appear in the interior of the host crystal and on inclusion—host boundaries as a result of inhomogeneous stresses are likely to set in motion not only the inclusions but also the dislocations. In particular, the migration of inclusions may be due to the fields of the elastic stresses around dislocations. Such migration may have a strong influence on the behavior of an inclusion ensemble (for example, it may accelerate the rate of coalescence of inclusions in collisions) and on the behavior of dislocations in heterogeneous systems. Therefore, it is essential to consider the motion of inclusions near dislocations at relatively high temperatures if we wish to understand correctly the mechanisms responsible for the heat resistance of heterogeneous systems and for the high-temperature creep.

A theory of the migration of inclusions in an inhomogeneous field of elastic stresses (which may be generated by dislocations) was presented in Sec. 5. We showed that the velocity of inclusions is proportional to the stress gradient. We established that when the migration is due to diffusion fluxes along an amorphous boundary between an inclusion and the host crystal, the velocity of a spherical inclusion located at a distance from a dislocation line is of the order

$$v \approx \frac{G\omega}{kT} D_{\text{eff}} \frac{b}{r^2}, \qquad (17.2)$$

where

$$D_{\text{eff}} = \frac{a}{R} D_S. \qquad (17.3)$$

If the migration of inclusions is primarily due to diffusion in the interior of the host, we must make the following substitution in Eq. (17.2):

$$D_{\text{eff}} = D. \qquad (17.4)$$

A more detailed calculation yields numerical coefficients in the above formulas and the direction of motion of the inclusions. For example, if the predominant mechanism is boundary diffusion and the boundary between an inclusion and its host is amorphous (so that the tangential stresses vanish), the velocity of an inclusion in the field of an edge dislocation is given by the following formula [171], which applies when r ≫ R:

$$v = -\frac{10}{3\pi} \frac{Gb(1+v)}{1-v} \frac{D_S \omega}{kT} \frac{a}{Rr^2} \left(\vec{e}_r \sin\theta - \vec{e}_\theta \cos\theta\right). \qquad (17.5)$$

Here, $\vec{e}_r$ is a unit vector perpendicular to the dislocation line; $\vec{e}_\theta$ is a unit vector perpendicular to the dislocation line and to the surface of the inclusion; θ is the angle between $\vec{e}_r$ and $\vec{b}$. Thus, depending on the angle θ, an inclusion can move either toward a dislocation or away from it.[†]

We shall now consider some consequences of the migration of inclusions, especially in the case of pinned dislocations. It follows from Eqs. (17.2)-(17.5) that, in a certain range of angles θ, inclusions migrate toward dislocation lines and are gradually precipitated on these lines, forming a type of Cottrell cloud. If initially the inclusions are distributed uniformly over the host crystal, we find from Eqs. (17.2)-(17.5) that at a moment t the number of inclusions per unit length of the dislocation line is of the order of

$$\nu = n_{\mathrm{i}} \left(\frac{G\omega}{kT} D_{\mathrm{eff}}\, b \right)^{2/3} t^{2/3}. \tag{17.6}$$

Here, n_{i} is the number of inclusions per unit volume at $t = 0$.

The precipitation of inclusions on dislocations may have a strong influence on the dislocations. In particular, under external stresses which are insufficient to detach the dislocations from the inclusions, the mobility of the dislocations will be governed by the velocity of migration of the inclusions. This alters also the mechanism and properties of creep, which we shall consider in the next subsection.

Diffusional Creep in Heterogeneous Systems

Diffusional creep, i.e., the deformation of a solid as a result of the diffusion transport of matter, may be a threshold or non-threshold effect, depending on the actual structure of the solid. Phenomenologically, this means that when the stresses are too

---

[†]The motion of inclusions in the elastic fields of dislocations was also considered in [143]. However, the expressions obtained for v differ by a factor $\sim b/r$ from Eq. (17.5). This difference is due to the incorrect identification, in [143], of the whole force exerted by a dislocation on a region around an inclusion with the force responsible for the migration of the inclusion (the diffusion fluxes should be considered in detail, as has been done in [63, 10]).

low to cause plastic deformation, a solid may behave as a Newtonian medium if

$$\dot{\varepsilon} \propto \tau, \tag{17.7}$$

or as a Bingham medium if

$$\dot{\varepsilon} \propto (\tau - \tau^*), \tag{17.8}$$

where $\dot{\varepsilon}$ is the rate of deformation, τ is the applied stress, and
τ^* is the threshold stress governed by the structure of the solid
and the mechanism by which matter is transported.

We shall first consider diffusional creep of the nonthreshold
type. This type of creep can appear even at the very low stresses
which produce directional diffusion fluxes. In accordance with
Eq. (5.6), a directional flux of vacancies responsible for creep
can be expressed in the form

$$\vec{I}_v = D_v \frac{N_0 c}{kT} \vec{\nabla} \mu_v, \tag{17.9}$$

where $\vec{\nabla} \mu_v$ is the gradient of the chemical potential of the vacancies, which depends on the distribution of stresses and the distribution of vacancy sources and sinks.

Diffusional creep may occur in a field of homogeneous but anisotropic stresses. This case is encountered in samples which
are tested for creep by uniaxial tension. In this case, an inhomogeneous distribution of vacancies is subject to the boundary conditions set out in Eq. (5.8). The values of the chemical potential
of the vacancies are different on the free surfaces of a sample and
on the grain or block boundaries oriented in different ways to the
applied stress. Therefore, different parts of such boundaries may
act as vacancy sources and sinks.

According to the Nabarro−Herring−Lifshits model [46-48],
directional vacancy diffusion fluxes appear in each of the structure elements under the influence of the gradient of the chemical
potential of the vacancies $\nabla \mu_v \sim \delta \mu_v / l$, where l is a characteristic
linear dimension of the gradient, i.e., the distance between va-
cancy sources and sinks. The dimension l may be the size of a
mosaic block or the distance between separate dislocations within
such a block. Directional fluxes deform blocks and grains without altering their volume. The deformation of all the structure

elements is coherent, i.e., no discontinuities or accumulations occur at the boundaries. An important feature of diffusional creep is that the vacancy fluxes in structure elements do not produce macroscopic diffusion in the whole sample.

The rate of creep due to the Nabarro–Herring–Lifshits mechanism can easily be estimated from the relationship governing the dimensionless rate of deformation:

$$\dot{\varepsilon} \propto D_v \frac{\partial^2 c_0}{\partial x^2} \approx D_v \frac{\delta c_v}{l^2} , \qquad (17.10)$$

where δc_v is the drop in the vacancy concentration corresponding to a given (constant) drop in the chemical potential:

$$\delta \mu = \delta \tau_{nn} \omega, \qquad (17.11)$$

where τ_{nn} is the normal stress on the surface of a structure element. It follows from Eqs. (17.10), (17.11), and (5.2) that

$$\dot{\varepsilon} \propto \frac{D\omega}{kTl^2} \tau. \qquad (17.12)$$

Equation (17.10) shows that during diffusional creep a crystal behaves as a Newtonian medium whose deformation is a nonthreshold process described by

$$\dot{\varepsilon} = \frac{\tau}{\eta} , \qquad (17.13)$$

where η is the viscosity.

A comparison of Eqs. (17.12) and (17.13) yields a formula for the viscosity of a crystal in which the characteristic dimension of the structure elements is l:

$$\eta \propto \frac{kT}{D\omega} l^2. \qquad (17.14)$$

The basic difference between the Newtonian flow in a crystal and in an amorphous body is that the viscosity of the crystal is not a constant property which depends only on the temperature: it is also a function of the distribution of the vacancy sources and sinks, which are described by the quantity l.

Single dislocations whose Burgers vector has an edge component can act as vacancy sources and sinks in real crystals. The diffusional climb of such dislocations, which results in the lengthening or shortening of half-planes, is also accompanied by the emission or absorption of vacancies. Thus, dislocations which are in nonconservative motion can act as vacancy sources and sinks. Consequently, they can give rise to diffusional creep in crystals.

The motion of a dislocation along a direction forming an angle ψ with the direction of the Burgers vector may result from the diffusion of vacancies or atoms toward the dislocation. If an edge dislocation of unit length absorbs or emits $1/a$ atoms, it can climb a distance of one atomic spacing (a) at right angles to the slip plane.

A sudden displacement of a dislocation segment of length L = na is most unlikely if n $\gg$ 1 because it would require an energy fluctuation of the order of nU_0, where U_0 is the energy needed for the attachment of an atom or a vacancy to the dislocation. The real mechanism of the diffusional climb of dislocations is associated with the absorption of single atoms or vacancies by jogs. These jogs are formed as a result of fluctuations or intersection dislocations with one another. The formation of a jog alters the energy of the system by an amount U_0. Consequently, the number of jogs per unit length of a dislocation line, c_j, is of the order of $\exp(-U_0/kT)$. If the diffusional flux toward a dislocation is distributed uniformly along its length, the average velocity of the diffusional climb of a dislocation is given by

$$v_\perp = v_j c_j \sin \psi, \qquad (17.15)$$

where v_j is the velocity of jogs along the dislocation.

The value of v_j is calculated in [135] for the case when the force F_d, exerted on a unit length of a dislocation line and responsible for the diffusional flux traveling toward it, is weak [$F_d \ll kT/b^2$; this condition is satisfied if $\tau \ll kT/\omega \sim 1$ kJ/cm^3 (10^{10} ergs/cm^3)] and the diffusional climb is due to vacancies. In this case, we obtain the expression

$$v_j \propto \frac{Db}{kT} \frac{F_d}{\sin \psi}. \qquad (17.16)$$

It follows from Eqs. (17.15) and (17.16) that

$$v_\perp \propto Dc_j \frac{b}{kT} F_d \qquad (17.17)$$

or, since $F_d = \tau b$,

$$v_\perp \propto \frac{D\omega c_j}{kTb} \tau. \qquad (17.18)$$

Under certain conditions, dislocation slip does not occur at stresses below the linear creep limit [95] and the deformation is primarily due to the diffusional climb of dislocations. Therefore, we shall consider only those cases of diffusional creep in the absence of inclusions when the slip has practically no effect on the deformation.

If we apply Eqs. (17.1) and (17.18), we obtain

$$\dot{\varepsilon} \propto \frac{D\omega c_j}{kT} N_d \tau. \qquad (17.19)$$

The creep mechanism described by Eq. (17.19) is observed when dislocations can migrate freely in a field of applied stresses without any hindrance from impurity atoms or hardening-phase inclusions.

The rate of the diffusional creep resulting from the motion of dislocations [Eq. (17.1)] may be quite low in real alloys because of the obstacles in the form of impurity clouds or — in the case of dispersion-hardened materials — particles of the hardening phase Formally, the influence of impurities or inclusions represents a reduction in the value of $v(\tau, T)$ given by Eq. (17.1).

The rate of creep as a result of the motion of dislocations together with the surrounding impurity clouds can be estimated by assuming that the velocity of a dislocation is limited by the diffusional mobility of the atoms in the surrounding cloud, irrespec-

Fig. 76. Schematic representation of the motion of a dislocation with the attached impurity atoms.

tive of the mechanism of motion of the dislocation (Fig. 76). If the cloud is not saturated and we can assume that impurity atoms are located at distances L > a and act as discrete pinning centers, we can estimate the rate of creep from the relationship obtained in [135]. It has been established that

$$\dot{\varepsilon} \propto N_{\mathrm{d}} \frac{b}{L} D_1 \sinh \frac{\tau b^2 L}{kT},$$ (17.20)

where $D_1 \propto \nu b^2 \exp(-U_1/kT)$ is the diffusion coefficient of the impurity atoms. In the special case of saturated impurity clouds, when $L \approx b$, we find that, if $\tau b^2 L < kT$, then

$$\dot{\varepsilon} \sim \frac{D_1 \omega}{kT} N_{\mathrm{d}} \tau.$$ (17.21)

This expression is basically similar to Eq. (17.19) but it differs from the latter in the physical meaning of the diffusion coefficient: in Eq. (17.19) D is the self-diffusion coefficient of the host material, whereas in Eq.(17.21) D_1 is the diffusion coefficient of the impurity atoms.

We shall now consider possible reasons why the diffusional creep can be of the threshold type and we shall estimate the threshold stress $\tau*$.

In the foregoing discussion, we have assumed that vacancy sources and sinks which are needed in the diffusional transport of matter are of infinite strength, i.e., that the vacancy emission and absorption processes do not limit the diffusion. Formally, this assumption means that the drop in the chemical potential determined by external stresses [Eq.(17.11)] is used entirely to maintain a vacancy concentration gradient.

However, we may find [144, 145] that dislocation walls or single dislocations are not ideal vacancy sources and sinks. Therefore, a potential difference $\pm\Delta\mu_1$ will be needed between a source and a sink in order to ensure the emission or absorption of vacancies. It follows that a smaller drop in the chemical potential will be available for initiating the vacancy diffusion flux. In this case, the rate of deformation is given by the relationship

$$\dot{\varepsilon} \propto \frac{2D}{kTl^2} (\Delta\mu - \Delta\mu_1).$$ (17.22)

The value of $\Delta\mu_1$ can be determined only if we make some specific assumptions about the structure of the vacancy sources and sinks. We shall estimate $\Delta\mu_1$ for the case when single dislocations or dislocations forming a wall act as sources and sinks. We shall assume that the motion of such dislocations is hindered by immobile obstacles. These obstacles may be very widely dispersed foreign inclusions with coherent boundaries. If these obstacles are distributed along a dislocation at average intervals equal to l, we find that a dislocation may bend by an amount $\sim l/2$ and then become detached from an obstacle if $\Delta\mu$ exceeds the threshold value of $\Delta\mu_1$ given by the expression

$$\Delta\mu_1 \approx \frac{2Gb}{l}\,\omega. \tag{17.23}$$

Since $\Delta\mu \simeq \tau\omega$, Eq. (17.22) can be rewritten in the form

$$\varepsilon \propto \frac{2D\omega}{kTl^2}\left(\tau - \frac{2Gb}{l}\right) \quad \text{for } \tau > \frac{2Gb}{l}. \tag{17.24}$$

Comparing Eqs. (17.8) and (17.24), we obtain the following estimate for the threshold stress

$$\tau^* \approx \frac{2Gb}{l}\,.$$

In the case under consideration, the diffusional creep cannot occur under stresses $\tau \le 2Gb/l$ because vacancy sources and sinks are inactive.

The occurrence of the threshold type of diffusional creep was demonstrated by Sautter and Chen [146]. They investigated creep

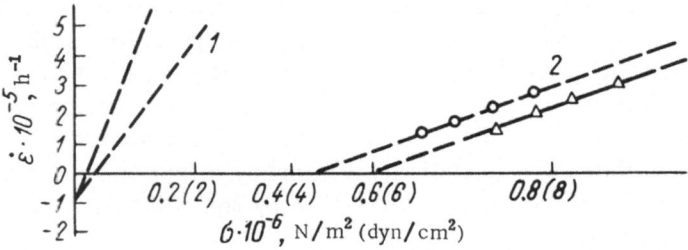

Fig. 77. Dependence of the rate of creep on the applied stress [146]:
1) gold foil; 2) gold foil containing Al_2O_3 inclusions.

in foils of pure gold and gold containing Al_2O_3 particles (Fig. 77). The experiments were carried out at sufficiently low stresses so that the deformation was primarily due to the diffusional transport of matter. It is evident from Fig. 77 that the foils containing Al_2O_3 inclusions exhibited diffusional creep at stresses $\tau > \tau^* = 0.4$-0.5 MN/m^2 or $(4$-$5) \times 10^6$ dyn/cm^2. This value was over an order of magnitude larger than the threshold stress τ' at which the rate of creep in the pure gold was still $\dot\varepsilon = 0$. The value of τ' was governed by the surface tension γ^* : †

$$\tau' \approx \dot\gamma/d.$$

According to the data reported by Sautter and Chen [146], $\tau' \sim 2 \times 10^5$ dyn/cm^2, d $\approx 5 \times 10^{-3}$ and, therefore, the surface tension of gold is $\gamma \sim 100$ N/m (10^3 dyn/cm), which is in agreement with the published values. We should stress that τ' is governed by γ and by the geometry of the sample and it has nothing in common with τ^*.

The value of $\dot\varepsilon$ for dispersion-hardened gold foils was less than the corresponding value for pure gold (Fig. 77), probably because the densities of the mobile dislocations were different.

Dislocation Creep of Heterogeneous

Systems in the Absence of Generation

of New Dislocations

We shall now consider the rate of creep in heterogeneous systems in which dislocations diffuse together with the attached inclusions [171]. We shall initially assume that new dislocations are not generated during this process.

We shall start by calculating the velocity of diffusion of a dislocation and the attached inclusions in the field of external stresse τ. We shall assume that this velocity is controlled by the velocity of inclusions.‡ A segment l of the dislocation, which is the aver-

---

†The values of the surface energy of solids obtained by the "zero creep" method can be overestimated considerably because of the presence of impurities and the possibility that τ^* is taken as equal to τ'.

‡The condition of compatibility of the motion of dislocation—inclusion complexes can be analyzed in the same way as the motion of grain boundary—inclusion systems [Eqs. (19.3)-(19.5)].

age distance between inclusions, is subjected to a force τbl. The same force is exerted on each inclusion by the dislocation. The results given in the "Theory" subsection of Sec. 6 show that the velocity of inclusions (and, consequently, of the dislocation) under the action of a force τbl is of the order of

$$v \propto \frac{\omega}{kT} D_{\text{eff}} \frac{bl}{R^3} \tau. \tag{17.25}$$

For example, if $D_S(a/R) \gg D$, $\omega \sim 10^{-23}$ cm^3, $T \sim 10^3$°K, $D_S \sim 10^{-7}$ cm^2/sec, $a \sim b \sim 3$ Å, $R \sim 3 \times 10^{-6}$ cm, $l \sim 10^{-5}$ cm, $\tau \sim 10$ J/cm^3 (10^8 ergs/cm^3), we find that v ~ 10 Å/sec, i.e., the velocity is quite high.

The rate of creep is given by the formula which follows from Eqs. (17.1) and (17.25):

$$\dot{\varepsilon} \propto N_d b^2 \frac{\omega}{kT} D_{\text{eff}} \frac{l}{R^3} \tau. \tag{17.26}$$

If $N_d \sim 3 \times 10^{10}$ cm^{-2} and the other constants for the same values as given above, it follows from Eq. (17.26) that $\dot{\varepsilon} \sim 10^{-5}$ sec^{-1}.

Equation (17.26) is valid if $\tau < Gb/l$ because at these stresses dislocations are not detached from moving inclusions. The temperature dependence of the rate of creep is then governed by the influence of temperature on the volume or boundary diffusion coefficients.

Dislocations gradually emerge on the surfaces of grains and their density decreases. This means that in this case the creep is not a steady but a gradually weakening process.

If a crystal contains isolated inclusions, new inclusions may be precipitated continuously on moving dislocations. This will reduce the value of l and, consequently, the rate of creep.

New inclusions may be captured by a dislocation as a result of their motion toward dislocations or of their direct collisions with moving dislocations.

If the velocity of dislocations v_d is not too high, a dislocation will attract inclusions from the surrounding cylindrical region whose radius can be estimated from Eq. (17.2):

$$r \propto (At_0)^{1/3}, \tag{17.27}$$

where $A \sim (G\omega/kT)D_{eff}b$. The quantity t_0 in Eq. (17.27) is found from the condition $r \sim vt_0$. We can easily see that in this case

$$\dot{v} \propto \frac{r^2}{t_0} n_i \qquad (17.28)$$

or, after simple transformations,

$$\dot{v} \propto (Av)^{1/2} n_i . \qquad (17.29)$$

This estimate is valid for velocities which satisfy

$$vR \ll (Av)^{1/2} \quad \text{or} \quad v \ll A/R^2. \qquad (17.30)$$

The opposite case of high velocities, $v \gg A/R^2$, applies when $\tau \gg GR/l$. In this case, dislocations become detached from inclusions and the nature of their motion changes.

The accumulation of inclusions at moving dislocations reduces their velocity and, consequently, the rate of creep will decrease. In the later stages of the diffusional creep, all the inclusions may become attached to dislocations.

The inclusions attached to dislocations may grow larger because of the one-dimensional coalescence resulting from the transport of matter between inclusions (due to diffusion along a dislocation pipe). This process may affect the kinetics of creep [172]. A theory of one-dimensional coalescence was developed in [147] on the assumption that the volume of matter enclosed in the inclusions attached to a dislocation of unit length does not vary with time. If this approximation is adopted, it is found that the average size of inclusions $\overline{R}$ and the average distance between them $\overline{l}$ vary with time in the following manner:

$$\overline{R} \propto \left(10^{-1} \theta_0 b^2 D_\perp t\right)^{1/7},$$

$$\overline{l} \propto (b^2 D_\perp t)^{3/7}/\theta_0^{4/7},$$

where $D_\perp$ is the coefficient of diffusion along a dislocation pipe; θ_0 is the volume of matter enclosed in the inclusions attached to a dislocation of unit length.

These approximate formulas for $\overline{R}$ and $\overline{l}$ show that when inclusions move as a result of volume diffusion, the one-dimensional coalescence does not affect the rate of creep because the ratio

l/R^3 of Eq. (17.26) is independent of time. This conclusion follows from the constancy of the volume of matter in the inclusions because $l/R^3 \propto \theta_0^{-1}$. If the inclusions are small and migrate as a result of surface diffusion, we find that l/R^4 in Eq.(17.26) is proportional to $t^{-1/7}$, i.e., the coalescence of inclusions results in some slowing down of the creep because of the growth of the coalesced inclusions, which may become so large that the dominant mechanism changes to volume diffusion and the coalescence-induced time dependence of $\dot{\varepsilon}$ disappears.

A similar analysis can be applied to the process of one-dimensional coalescence in the motion of gas-filled microbubbles attached to a dislocation, when this motion is the result of the transport of matter through the occluded gas.

The climb of dislocations under the action of external forces is only one of many possible mechanisms. Dislocations can move also as a result of the diffusion of excess vacancies. Many real systems contain sources of excess vacancies which become active at temperatures corresponding to the significant diffusion of atoms. These systems include materials which have been irradiated with neutrons or fission fragments at high temperatures, or have been bombarded at low temperatures and then annealed at high temperatures [148]. They include also mixtures of powders of mutually soluble metals, in which diffusion-induced homogenization is due to the unequal partial diffusion coefficients of the components [149].

If a state of vacancy supersaturation $\delta c_v/c_v$ is maintained by a system of vacancy sources and sinks, it is found that dislocations — which can act as vacancy sinks — are subject to a generalized thermodynamic forces. This force is due to the nonequilibrium concentration of vacancies in the crystal lattice, and it is governed by the gradient of the chemical potential of the vacancies near a dislocation whose Burgers vector has an edge component. The force acting on a dislocation segment of length l_d is of the order of

$$ F' \sim \frac{l_d}{a} |\nabla \mu_v| \sim \frac{l_d}{a} kT \frac{|\nabla c_v|}{c_v} . \tag{17.31} $$

Since $\nabla c_v \sim \delta c_v/b$, we find

$$ F' \sim \frac{l_d}{ab} kT \frac{\delta c_v}{c_v} . \tag{17.32} $$

Equations (6.7), (6.8), and (17.32) allow us to determine the velocity of dislocation-dragged inclusions in the field of excess vacancies. If the surface diffusion mechanism is the dominant effect, we find that when $a \approx b$

$$v \propto \frac{\omega l_d}{aR^4} D_S \frac{\delta c_v}{c_v} .$$ (17.33)

In the volume diffusion case, we obtain

$$v \propto \frac{\omega l_d}{a^2 R^3} D \frac{\delta c_v}{c_v} .$$ (17.34)

If $\omega \sim 10^{-23}$ cm^3, $l_d \sim 10^{-4}$ cm, $D_S \sim 10^{-5}$ cm^2/sec, $\delta c_v / c_v \sim 10^{-2}$, we find that Eq. (17.33) yields $v \sim 10^{-26}/R^4$ cm/sec. Therefore, small inclusions of $R \sim 10^{-5}$ cm, dragged by dislocations climbing in a medium of excess vacancies, can move at velocities as high as $\sim 10^2$ Å/sec.

It is interesting to consider the diffusional climb of a dislocation together with the attached cavities. Such motion should alter the volume of the cavities. The change is due to the fact that such a climb is accompanied by the absorption or emission of vacancies, depending on the direction of motion. In this case, the cavities attached to a moving dislocation can act as natural vacancy sources and sinks (Fig. 78).

The cavities which act as vacancy sources decrease in size. If we assume that each spherical cavity of radius R supplies vacancies to a dislocation segment located between two neighboring cavities and that the contribution of other vacancy sources can be ignored, the law which describes the change in the volume Ω of a cavity which is absorbed only by the associated dislocation can be written in the form

$$\frac{d\Omega}{dt} = -v_\perp b l_d .$$ (17.35)

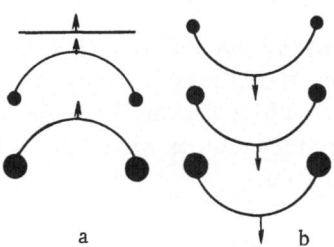

Fig. 78. Changes in the dimensions of cavities moving together with the dislocation: a) increase; b) decrease.

a   b

Equation (17.35) expresses the equality of the changes in the volumes of the cavity and of the matter in the associated half-plane.

It follows from Eqs. (17.35) and (17.33) that, if $\delta c_v/c_v = \text{const}$,

$$R^7 = R_0^7 - \psi t, \qquad \psi = \frac{7}{4\pi} \omega l_d^2 D_s \cdot \frac{\delta c_v}{c_v} . \qquad (17.36)$$

A cavity of initial radius R_0 disappears completely in the time interval during which the associated dislocation travels

$$x^* \propto v_\perp t^* \propto \frac{R_0^3}{b l_d} . \qquad (17.37)$$

The disappearance of cavities frees the dislocation from their retarding influence. Consequently, softening may occur. This softening is similar to that which occurs when precipitates in a supersaturated solid solution are dissolved when the temperature is raised. However, there is an important difference between these two processes: in contrast to the dissolution of inclusions, the disappearance of a pore is the result of motion and not of a change in the equilibrium conditions at the inclusion—host interface.

This mechanism of the disappearance of cavities may affect the softening mechanism if the path traveled by a cavity before its disappearance is not very long (less than the average linear size of a block l).

It follows from Eq. (17.37) that

$$\frac{R}{b} \propto \left(\frac{x^* l_d}{b^2} \right)^{1/3} .$$

If $x^* \sim l_c \sim 10^{-3}$ cm, it is found that $R/b \sim 10^3$. This estimate shows that the cavities with $R < 10^{-5}$ cm disappear before a diffusing dislocation reaches the boundary of a mosaic block.

Apart from the cavities whose volume decreases during their motion, there may be cavities which grow (Fig. 77b). This process should additionally slow down the creep because a cavity which increases in size slows down the motion of the associated dislocation.

Thus, the motion of dislocation—cavity complexes may be accompanied by the softening or hardening of an alloy. The relative importance of these mechanisms depends on the extent to which cavities retard the diffusional climb of dislocations.

Dislocation Creep of Heterogeneous

Systems in the Case of Generation

of New Dislocations

We shall now consider the diffusional creep of dispersion-hardened systems in which new dislocations are generated. The nature of the source of the dislocations is unimportant and will not be specified [171].

In this case, the kinetics of creep should be quite different in the initial and final stages. In the initial stage, the randomly distributed sources of dislocations may be assumed to be independent. Then, the dislocation density will increase with time and the creep will therefore be accelerated. In the steady-state stage, the interaction between the expanded dislocation loops establishes a dynamic equilibrium in respect of the dislocation density, which corresponds to some constant rate of creep. In both stages, the rate of creep depends on the interaction between dislocation loops and second-phase inclusions.

In the initial stage, the dislocation loops, which appear around sources, capture inclusions during their motion. If we calculate the diffusion velocity of the head dislocation decorated by inclusions, we must bear in mind that this dislocation is acted upon by a stress $n\tau$, where $n = 2L\tau/Gb$ is the number of loops in a given tangle. This velocity can be estimated using Eq. (17.25):

$$v \propto \frac{\omega}{kT} D_{\text{eff}} \frac{bl}{R^3} n\tau \propto \frac{\omega D_{\text{eff}}}{kTv} \frac{L}{R^3 G} \tau^2. \qquad (17.38)$$

It is assumed in the above formula that $l = v^{-1}$.

Equations (17.29) and (17.38) define the functions $\dot{\nu}(t)$ and $v(t) = \frac{d}{dt} L(t)$. After a sufficiently long time, these equations reduce to

$$
\left.
\begin{aligned}
v &\propto \frac{\omega}{kT} D_{\text{eff}} \frac{1}{P^2 b G^3} \tau^4, \\
\dot{v} &\propto \frac{\omega}{kT} D_{\text{eff}} \frac{n_i}{PG} \tau^2.
\end{aligned}
\right\}
\tag{17.39}
$$

In Eq. (17.39), $P = (4/3)\pi R^3 n_i$ is the volume concentration of inclusions. Substituting Eq. (17.38) into Eq. (17.1) and bearing in mind that $N_d \propto MN l \propto (M\tau / Gb)(vt)^2$ (M is the density of dislocation sources), we find that

$$
\dot{\varepsilon} \propto M \left(\frac{\omega}{kT} D_{\text{eff}} \right)^3 \frac{1}{P^6 b^3} \frac{\tau^{13}}{G^{10}} t^2. \tag{17.40}
$$

It is worth noting that, according to Eq. (17.40), the rate of creep is proportional to the thirteenth (!) power of the stress and that the activation energy is three times as large as the activation energy associated with D_{eff}.

In later stages, the interaction between different dislocation tangles becomes important. The stresses acting on the head dislocation may become sufficiently high for the detachment of this dislocation from inclusions or for bypassing the inclusions by the Orowan mechanism. Moreover, a considerable redistribution of inclusions may result from the drag exerted by dislocations in inhomogeneous stress fields.

When the stage of steady-state creep is reached, a dynamic equilibrium is established between the numbers of created and annihilated dislocations and annihilated dislocations, and the processes of inclusion redistribution should be accompanied by homogenization. We shall consider one of the simplest models of steady-state creep: we shall assume that dislocations of opposite sign are distributed uniformly in a crystal and that they are separated by distances l_d from one another, forming Taylor's network. In this case, the dislocation sources are separate segments of the network. The dislocation density $N_d \sim 1/l_d^2$ is found from the requirement that the average stresses Gb/l_d generated by dislocations at a distance $\sim l_d$ should compensate the external stresses τ and should prevent the creation of new dislocations, i.e., $N_d \sim (\tau/Gb)^2$. Dislocations of opposite sign may meet and interfere destructively as a result of their motion through a distance $\sim l_d$ in the random stress field. The resultant reduction in the number of dislocations is compensated by the creation of new dislocations, which re-establishes the original density N_d.

The concentration of inclusions on newly formed dislocations is low but it gradually increases as a result of the capture of inclusions during the motion of dislocations over distances $\sim l_d$. If we assume that inhomogeneous stress fields ensure continuous homogenization of the distribution of inclusions, we can calculate the rate of capture of inclusions from Eq.(17.29). Since, according to Eq. (17.38), $v \propto 1/\nu$, we find that in the case of motion in a field of average stresses $\sim \tau$,

$$ v \propto G^{1/3} \left(\frac{\omega D_{\text{eff}}\ b}{kT} \right)^{2/3} n_i^{2/3} \frac{\tau^{1/3}}{R} t^{2/3} . \tag{17.41} $$

If we substitute this expression ($\nu \propto 1/l$) in Eq. (17.25) for the velocity of diffusing dislocations, we can find the time $\bar{t}$ needed to travel the average distance $\sim l_d$ between dislocations of opposite sign (the lifetime of dislocations) and the average velocity of such motion $\bar{v}$:

$$ \bar{t} \sim \frac{l_d}{v} , \qquad \bar{v} \sim \frac{1}{l_d^2} \frac{D_{\text{eff}}}{kT} \frac{\omega}{P^2 G} \frac{b}{} \tau^2 . \tag{17.42} $$

Substituting Eq. (17.42) for $\bar{v}$ in Eq. (17.1) and bearing in mind that $N_d \propto 1/l_d^2 \propto (\tau/Gb)^2$, we can estimate the rate of steady-state creep:

$$ \varepsilon \propto \frac{D_{\text{eff}}\ \omega}{kT} \frac{1}{b^2 p^2} \frac{\tau^6}{G^5} . \tag{17.43} $$

Thus, the temperature dependence of the rate of creep is the same as that of the effective (volume or boundary) diffusion coefficient which governs the mobility of inclusions. The rate of creep is proportional to τ^6, where τ represents the external stresses. The dependence on the volume concentration of inclusions P is given by the factor $1/P^2$. The radius of inclusions affects only the boundary diffusion mechanism for which $D_{\text{eff}} \propto 1/R$. These quantitative results may be modified somewhat by a more detailed statistical theory.

18. SWELLING

The phenomenon known as swelling is the increase in volume of reactor materials (in particular, fissionable materials) under the influence of a neutron flux. This increase in volume occurs during high-temperature irradiation or during heating following low-temperature irradiation. Swelling prevents the efficient utilization of nuclear fuels and the achivement of "high burnup" because of the loss of mechanical strength by the swollen fuel elements.

Neutron-induced nuclear reactions in fuel elements and reactor materials generate various products, including inert gases (helium, neon, krypton, and xenon). In particular, krypton and xenon are gaseous fragments of fissioning uranium nuclei. Since the solubility of inert gases in fuel materials (for example, uranium dioxide) is low, the excess gas forms bubbles. The nucleation and growth of these bubbles is responsible for swelling [150, 152]. The amount of gas which accumulates in an irradiated material may be very considerable. In fuel elements, the volume of the gas (at P = 1 atm) in the bubbles is of the same order of magnitude as the volume of the element.

The mechanism of swelling is not simple. Macroscopic swelling may be due to many simultaneous processes whose relative importance will depend on the properties of the material, the temperature, the irradiation conditions, etc.

The mechanism of growth of gas-filled bubbles is of basic importance in swelling. Initially, the bubbles are nucleated as a result of the diffusion of the gas atoms toward various lattice defects such as microcracks, dislocation jogs, dislocation loops, etc. The atoms of a practically insoluble gas diffuse mainly along interstices. If swelling occurs during high-temperature irradiation, the gas diffuses continuously toward bubbles.

The accumulation of a gas in bubbles generates pressure, which is responsible for the deformation of the material in question. Depending on the gas pressure and the temperature, the creep which deforms a given material may be due to either threshold or nonthreshold mechanisms.

The threshold creep mechanisms are active if the stresses generated by the gas pressure exceed a certain threshold value

$\tau*$ necessary for dislocations to overcome the obstacles which hinder their motion. The threshold stress $\tau*$ is governed by the actual interaction between moving dislocations and obstacles and by the way in which these obstacles are overcome. The deformation of a medium under the influence of a stress $\tau > \tau*$ can be described phenomenologically as viscous flow in a Bingham medium, so that the rate of deformation is $\dot{\varepsilon} \propto (\tau - \tau*)$. If the stresses are $\tau < \tau*$, the plastic deformation of the swelling material, which accompanies the growth of gas-filled bubbles, may be due to a nonthreshold diffusion mechanism (Sec. 17).

A theory of high-temperature swelling in the presence of a source of gas atoms is developed in [152-154]. An allowance is made in this theory for the diffusional coalescence of bubbles interacting via a balanced diffusion field. However, it is assumed in [152-154] that at all stages of the process the gas bubbles remain static.

Under real conditions, the processes described above may be accompanied by the coalescence of bubbles colliding directly during their motion in a force field or as a result of the random Brownian motion. In this case, swelling is due to the fact that the volume of a bubble formed by the coalescence of two colliding bubbles is larger than the sum of the original volumes. This can be demonstrated by considering equilibrium bubbles, in which the gas pressure is compensated by the Laplace pressure (it is assumed that the gas in the bubbles is ideal – Sec. 10).

When two gas-filled equilibrium bubbles coalesce, we sum not their volumes (which would be the case for solid particles or liquid droplets) but their surface areas.† In view of this, the volume of the new (larger) bubble is greater than the sum of the volumes of the original bubbles. If the radii of the coalescing bubbles are $R_1 = R$ and $R_2 = bR$, it follows that the radius of the new bubble is

$$R = R_1 (1 + b^2)^{1/2} \tag{18.1}$$

---

†It is readily shown that the energy of a gas in an equilibrium bubble (w) is equal to its surface energy. Since $w = n_i \frac{3}{2} kT$, where $n_i = (P_g / kT)(\frac{4}{3} \pi R^3)$ is the number of gas molecules in a bubble, it follows that, for $P_g = 2\gamma / R$, $w = 4\pi R^2 \gamma \propto R^2$ [Eq. (10.5)]. Therefore, the summation of the energies of atoms in two coalescing bubbles leads to the summation of the surface areas of these bubbles.

and, consequently, the relative increase in the volume of the bubbles resulting from their coalescence is

$$\frac{\Delta\Omega}{\Omega} = \frac{\Omega - (\Omega_1 + \Omega_2)}{\Omega_1 + \Omega_2} = \frac{(1 + b^2)^{3/2}}{1 + b^3} - 1. \tag{18.2}$$

We can readily show that $\Delta\Omega/\Omega$ has its maximum value for $b = 1$, i.e., when two bubbles of the same radius merge to form a larger bubble. In this case, $\Delta\Omega/\Omega \approx 0.4$. However, this change may be somewhat greater if the radii of the coalescing bubbles are so small that the gas enclosed in them is not ideal but is described by the van der Waals equation. This occurs in equilibrium bubbles whose radii are of the order of 10^{-7}-10^{-5} cm. Coalescence reduces the departure from the ideal properties of the gas because the pressure in the new bubble is reduced.

The coalescence of two equilibrium gas-filled bubbles is favored by the energy considerations because the entropy of the gas increases as a result of the increase in volume (the surface and the internal energies of the enclosed gas remain constant).

Obviously, the number of gas atoms n_i in a bubble is much smaller than the number of atoms n_a in the solid which originally occupied the same volume as the bubble. It follows from Eq. (10.5) that in an equilibrium bubble filled with an ideal gas the relevant numbers of atoms are

$$n_i = \frac{8\pi R^2 \gamma}{3kT}, \quad n_a = \frac{4\pi}{3} \frac{R^3}{\omega}. \tag{18.3}$$

It is clear from Table 3 [155] that the degree of occupancy of an equilibrium bubble by gas atoms, $n_i/n_a = 2\gamma\omega/kTR$, decreases with increasing radius. This means that the increase in the volume resulting from the coalescence of bubbles must result in the swelling of a sample. The dependences of the number of atoms in a cavity on its radius R, calculated for helium and xenon in copper (Fig. 79), show that the gas in bubbles of radii $R < 10^{-5}$ cm is not ideal. The departure from the ideal properties increases with decreasing radius R but, as pointed out earlier, the collision-induced coalescence of cavities reduces this departure.

The coalescence in an ensemble of moving inclusions was considered earlier for the case when inclusions migrate under the influence of a temperature gradient or the Brownian motion (Sec. 10). In the former case, the velocity of small bubbles is propor-

tional to $1/R$ in accordance with Eq. (10.1). The coalescence kinetics is then described by Eqs. (10.13)-(10.16).

However, the drag exerted on bubbles by moving block or
grain boundaries, or by dislocations, may play an important role
in swelling. In this case, the forces $\vec{F}$ acting on inclusions can
be assumed to be approximately independent of the radius. Then,
the dependence of the velocity of an inclusion on its radius will
differ considerably from the dependence given by Eq. (10.1) and
it will be described by $v \propto F/R^4$ of Eq. (6.8) or $v \propto F/R^3$ of Eq.
(6.7), which apply to migration under the influence of surface and
volume diffusion fluxes.

The kinetics of coalescence applicable to these $v(R)$ dependences can be investigated on the basis of the similarity principle
(Sec. 10). If we solve the transport equation (10.7) in the same
way as in the derivation of Eqs.(10.13)-(10.16) and if we bear in
mind that the order of homogeneity h is -2 or -1 for Eqs. (6.8)
and (6.7), we can readily show that the average radius of the bubbles $\overline{R}(t)$ and their total relative volume $V(t)$ can be expressed −
in terms of the average initial radius R_0 and the initial number of
bubbles − by the formulas [155]

$$\overline{R} \propto R_0 (t/t^*)^{1/4}, \quad V(t) \propto \frac{kT}{\gamma} N_i R_0 (t/t^*)^{1/4}, \quad t^* \propto \frac{\gamma R^4}{D_S a\omega F N_i}, \quad (18.4)$$

if the motion of the bubbles is governed by the surface diffusion
fluxes, and by the formulas

$$\left. \begin{array}{c} \overline{R} \propto R_0 (t/t^*)^{1/3}, \quad V(t) \propto \frac{kT}{\gamma} N_i R_0 (t/t^*)^{1/3}, \\[2mm] t^* \propto \frac{\gamma R_0^3}{D\omega F N_i}, \end{array} \right\} \quad (18.5)$$

TABLE 3. Swelling as a Result of Annealing after
Removal of Pressure

R, cm	n_i	n_a	n_a/n_i	$\Delta\Omega/\Omega$, % for 10^{-3} % at.
10^{-7}	10^2	$4 \cdot 10^2$	4	0.4
10^{-6}	$5 \cdot 10^4$	$4 \cdot 10^5$	8	0.8
10^{-5}	$8 \cdot 10^6$	$4 \cdot 10^8$	50	5.0
10^{-4}	$8 \cdot 10^8$	$4 \cdot 10^{11}$	500	50.0

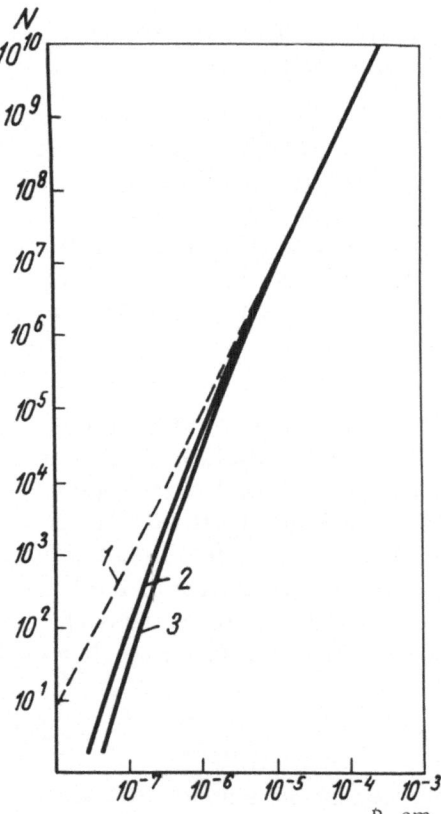

Fig. 79. Dependences of n_a, n_i, and n_i' on the radius R of helium- or xenon-filled bubbles in copper [155]: 1) ideal gas; 2) van der Waals He gas; 3) van der Waals Xe gas.

if the motion of the bubbles is governed by volume diffusion fluxes. Here, N_i is the total number of atoms of an ideal gas in 1 cm³; t* is the characteristic time.

The value of t* can also be estimated on the basis of elementary considerations of the type used, for example, in gas-kinetic theory. The estimate can be obtained by assuming that $t^* \sim l^0/v$, where $l^0 \sim 1/R^2 N_B$ is the mean free path of an inclusion before its collision with another inclusion (N_B is the number of bubbles per unit volume).

These estimates of the kinetics of swelling as a result of the coalescence of migrating gas bubbles apply to the case when total amount of gas in a sample does not vary with time. However, such a situation is realized only if the gas is introduced at moderate temperatures and the sample is then annealed at temperatures at which the diffusional mobility is sufficient for inclusions to be significantly displaced in a measurable time. Under real conditions, the coalescence occurring during irradiation is accompanied by the appearance of new gas atoms. This may be allowed for by assuming, for simplicity, that the rate of generation of gas atoms $\dot{N}_i$ is constant. Then, Eq. (10.5) transforms to

$$N_B(0)\,\overline{R}^2 = \frac{3kT}{8\pi\gamma}\,\dot{N}_i\,t. \tag{18.6}$$

In this case, the formulas describing the time dependences of $\overline{R}$ and $V(t)$ are of the form

$$\left.\begin{array}{l} \overline{R} \propto \left(\dot{N}_i\,\dfrac{D_S\,a\omega}{\gamma}\,Ft^2\ln t\right)^{1/4}, \\[3mm] V(t) \propto \left(\dfrac{\dot{N}_i}{\gamma}\right)^{5/1} kTa\left(D_S\,F\ln t\right)^{1/1}\,t^{3/2}, \end{array}\right\} \tag{18.7}$$

if the mass transport is the result of surface diffusion, and

$$\left.\begin{array}{l} \overline{R} \propto \left(\dfrac{\dot{N}_i}{\gamma}\,D\omega Ft^2\right)^{1/3}, \\[3mm] V(t) \propto \left(\dfrac{\dot{N}_i}{\gamma}\right)^{1/3} kTa\,(DF)^{1/3}t^{5/3}, \end{array}\right\} \tag{18.8}$$

if the mass transport is due to volume diffusion in the host substance.

The force responsible for the migration of bubbles does not remain constant for a long time if it is due to the motion of boundaries separating structure elements, or is due to the relaxation of stresses accompanied by the motion of dislocations. These two processes reduce the size of the moving structure elements (grain boundaries or dislocations), the average force causing their motion and, consequently, the velocity of the gas-filled bubbles. This comment is valid at least in the case when the annealing, during which the bubbles migrate, is not accompanied by irradiation, which would generate new dislocations.

These considerations and estimates relating to the kinetics of swelling can be compared with experimental results. The most reliable results are obtained if the comparison is made with the experimental data relating to the swelling that is observed during

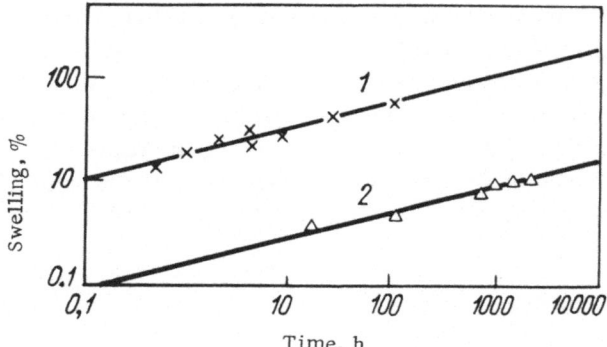

Fig. 80. Dependences of the swelling of beryllium (1) and uranium (2) during isothermal annealing after neutron irradiation [150, 156].

the high-temperature annealing of samples previously irradiated at moderate temperatures.

We may assume that in the initial stage of the swelling process the forces are constant, the gas bubbles are small (r $\sim$ 10^{-5}–10^{-6} cm), and the surfaces of the bubbles are sufficiently clean so that the transport of matter takes place primarily by surface diffusion. Therefore, the experimental results [155] corresponding to this stage of the process can be described by the formulas in Eq. (18.4). According to these formulas, the time dependences $\overline{R}$

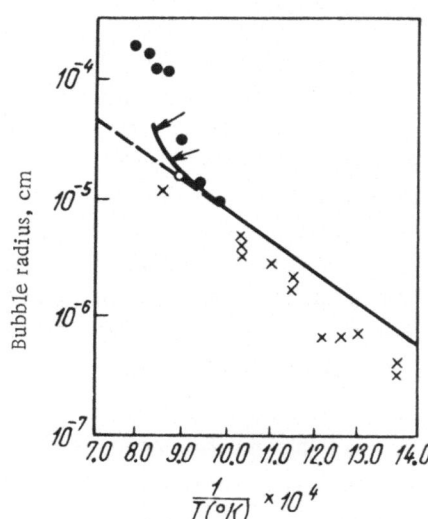

Fig. 81. Temperature dependence of the average radius of helium-filled bubbles in copper [151].

and $V(t)$ are governed by the factor $t^{1/4}$, and the temperature dependences by the factor $\exp(-U_S/kT)$, where U_S is the activation energy of surface self-diffusion. Figure 80 shows the dependences of the quantity $V(t)$ on the duration of the isothermal annealing of beryllium and uranium† previously irradiated with neutrons. Neutron bombardment produced helium filled bubbles in beryllium, whereas the bubbles in uranium were filled with krypton and xenon. The slope of the dependences plotted in Fig. 80 are approximately 1/4, in agreement with Eq. (18.4).

Figure 81 shows the temperature dependence of the average radius of bubbles in copper which are filled with helium atoms [151]. The slope of the dependence plotted in the coordinates ln R and $1/T$ shows that $U_S \approx 2.7$ eV. This value exceeds $U_S \approx 2.1$ eV, obtained from a study of self-diffusion on the surface of copper by the mass transport method [20].

These experimental results support the model used in calculations relating to swelling, and they show that the force F, responsible for the migration of bubbles, remains practically constant under the conditions in the experiments discussed here.

The importance of the contribution of the coalescence of gasfilled bubbles to the swelling of a sample was confirmed experimentally in studies of silver tubes [185] in which gas bubbles were generated as a result of chemical reduction of oxides with hydrogen, which produced water vapor.

Impurities in the host substance also exert a considerable influence on the swelling kinetics. We have assumed that gas-filled bubbles can migrate freely, so that their velocity $v \propto F$. However, it is known that insoluble impurities forming inclusions may inhibit the free motion of those bubbles which are attached to such inclusions. The force binding bubbles to inclusions is due to the fact that the surface energy of a combined bubble—inclusion system can be smaller than for the bubble and the inclusion separately The binding may be strong enough to prevent the bubble from being dislodged by the force F.

An estimate of the force of interaction F* between a bubble of radius R and an inclusion whose radius is at least R yields the

---

†The results given in [150, 156] and cited in [155].

expression

$$F^* \propto R(\gamma - \gamma'), \tag{18.9}$$

where γ' is the surface energy of the matter in the bubble at the bubble–host interface.

The force F^* which binds a gas-filled bubble to an inclusion is proportional to R. This means that when the force F is independent of the radius of the bubble, we should find that the bubbles bound to inclusions are mainly those whose radius is $R > F/(\gamma - \gamma')$.

Thus, in the presence of impurities (inclusions), the process of coalescence may stop at the stage when practically all the remaining bubbles are bound to impurity (inclusion) particles and are effectively immobile. In this situation, the mobility will be governed by the mobility of the bubble–impurity complexes, which may be much lower than the mobility of bubbles. If N_B^* is the number of bubbles stabilized in this way, their radius R^* and the maximum swelling $V^*(t)$ is now given by the following relationships, which are obtained from Eq. (10.5):

$$R^* \approx \left(\frac{3 N_i kT}{8 \pi \gamma N_B^*} \right)^{1/2}, \qquad V(t) \propto \frac{1}{10} \left(\frac{N_i kT}{\gamma N_B^*} \right)^{3/2} N_B^*. \tag{18.10}$$

This enhancement of the resistance to swelling, resulting from the presence of an insoluble phase, was observed in uranium containing admixtures of aluminum and iron or disperse particles of the high-temperature β phase, which formed during quenching.

In the foregoing discussion, we assumed that the coalescence of two equilibrium gas-filled bubbles occurs when these bubbles come in direct contact during their migration. This assumption is based on the observation that the coalescence of bubbles is favored by energy considerations and is not a threshold process. Nevertheless in real crystals, particularly in nuclear fuel elements, the coalescence of gas-filled bubbles may be a threshold effect because of the positive adsorption of surface-active impurities on the surfaces of the bubbles (these impurities are often present in the host substance). A specific energy may be required to break through a narrow layer between two adjacent bubbles when this layer contains such impurities. Consequently, an energy threshold may result from adsorption and this may inhibit the coalescence of adjacent cavities (Sec. 10).

In view of the possible influence of the adsorption of impurities on the coalescence of gas-filled bubbles, we may assume that small amounts of surface-active substances would prevent the coalescence of bubbles and, consequently, reduce the swelling of a sample.

A considerable influence on the experimentally observed swelling is also exerted by the external pressure P^0. When an allowance is made for this pressure, the condition for the equilibrium of the gas in a bubble follows from Eq. (2.34):

$$\left(\frac{2\gamma}{R} + P^0 \right) \frac{4}{3} \pi R^3 = n_i kT. \tag{18.11}$$

It is clear from Eq. (18.11) that the pressure of the gas in a bubble contained within a sample subjected to an external pressure P^0 is $[1 + (P^0 R/2\gamma)]$ times higher than the gas pressure in the same bubble in the absence of P^0.

The swelling of a sample under a pressure P^0 can be determined from the condition $V(t) = N_B (t)(4\pi/3)R^3$, i.e., when Eq. (18.11) is used, we find that

$$V_{\max}(t) = \frac{N_i kT}{2\gamma/R + P^0}. \tag{18.12}$$

If $P^0 \gg 2\gamma/R$, the maximum swelling V_m is $V_m \approx N_i kT/P^0$.

The influence of pressure on the dimensions of bubbles may be considerable if $P^0 \geq 2\gamma/R$, i.e., when $P^0 \approx 100$ bar (10^2 atm), provided $R \approx 10^{-5}$ cm. However, this pressure cannot appreciably affect the diffusion constants of a metal and, therefore, the velocity of bubbles and the kinetics of their coalescence will not change greatly. This means that, until the mobility of bubbles becomes negligibly small because of the continuous increase in their size, the bubbles in a sample subjected to an external pressure will be the same size as those in a sample which is annealed with the application of this pressure. This will occur in spite of the fact that the gas pressure in the former bubbles will be higher. Consequently, annealing after the removal of an external pressure will give rise to an additional swelling since each of the bubbles will increase in size as a result of the relaxation of the associated stresses and the arrival of vacancies.

TABLE 4. Atoms in Equilibrium Gas Bubbles

Run	Pressure, kN/m^2 (kgf/cm^2)	Time, h	$V(t)$, %
1	4(400)	3	2.0
	4(400)	24	2.0
	0(0)	24	10.6
2	4(400)	3	4.1
	0(0)	24	11.3

The influence of an external pressure on the swelling of a sample annealed at a high temperature after low-temperature irradiation was investigated experimentally by Churchman et al. [150]. They studied uranium irradiated with neutrons. The highest hydrostatic pressure P^0 in their experiments was 4 kN/m^2 (400 kgf/cm^2).

Typical results obtained by Churchman et al. are given in Table 4, which lists the changes in the values of $V(t)$ after various anneals at 810°C, carried out consecutively. The results presented in Table 4 show that annealing after the removal of pressure strongly enhances the swelling.

19. RECRYSTALLIZATION AND SINTERING

The migration of second-phase inclusions and of pores may affect the process of secondary recrystallization and the kinetics of the final stage of sintering in porous bodies.

We shall consider some of the consequences of the interaction between grain boundaries and inclusions in real polycrystalline bodies. In Sec. 11 we considered the results of experiments which were carried out specially in order to determine the relationships governing the interaction between boundaries and inclusions. These experiments demonstrated that a moving boundary can drag inclusions. If the force responsible for the motion of a boundary is small, inclusions are capable of stopping a boundary. That is why the inhibiting effect of second-phase inclusions on secondary recrystallization is manifested clearly in the final stage of the process when the linear dimensions of grains are large and the force responsible for the motion of boundaries is insufficient to overcome the inhibiting effect of the inclusions. The velocity of boundary—inclusion systems during the later stages of recrys-

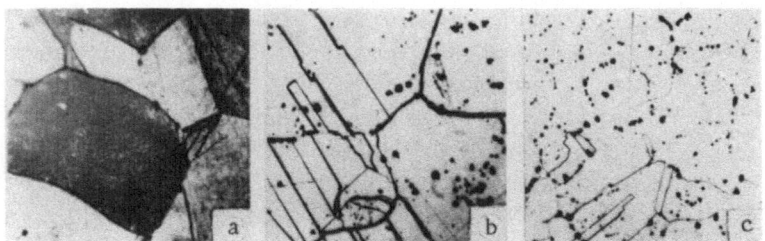

Fig. 82. Influence of pressure on the recrystallization of porous copper at T = 1050°C, applied for t = 2 h. ×500. a) P = 10 MN/m²
(100 atm); b) P = 0.3 MN/m² (3 atm); c) P = 0.1 MN/m² (1 atm)
[161].

tallization are governed by the velocity of the inclusions, which may be zero.

The inhibiting effect of inclusions on the kinetics of secondary recrystallization has been clearly observed in experiments on many materials containing inclusions in the form of oxide particles [126], second-phase inclusions [127], or pores [157-159].

The results of these experiments show that the dimensions of grains which are stabilized during the later stages of the secondary recrystallization and the rates of growth of these grains decrease with increasing volume concentration of the inclusions in the recrystallized matrix. This is illustrated clearly by studies of the kinetics of recrystallization of samples in which an inclusion may dissolve when some changes are made in the recrystallization annealing conditions.

Beck et al. [160] investigated the recrystallization of the binary aluminum-manganese alloy with 1.1% Mn and they found that a

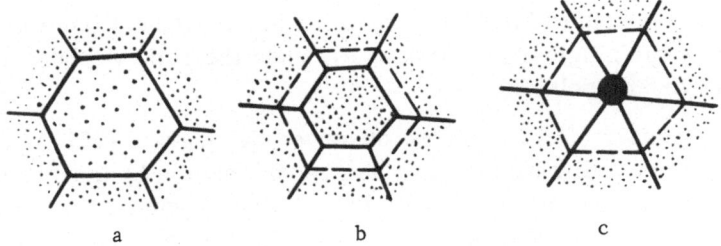

Fig. 83. Formation of impurity clusters as a result of secondary recrystallization [92]: a, b, c, successive stages.

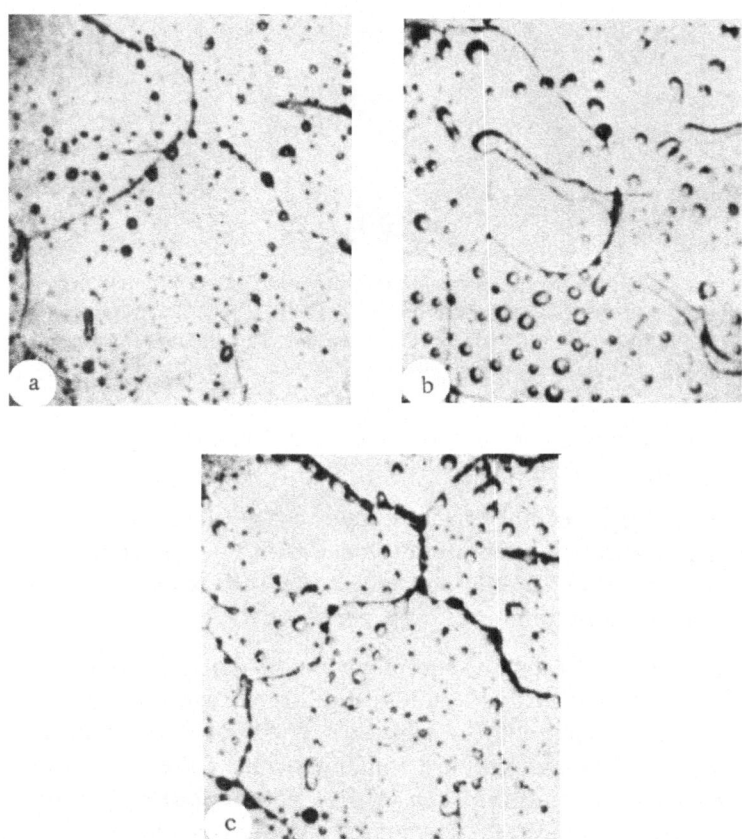

Fig. 84. Recrystallization-type of migration of grain boundaries in polycrystalline camphor under the influence of gas-filled bubbles moving in the field of a temperature gradient $| \nabla T | = 6.1$ deg/cm. ×33.

sudden increase in the rate of growth and the maximum size of the grains occurred at 650°C when the alloy changed from the two-phase to the single-phase type, i.e., when the second-phase inclusions were dissolved.

A similar effect was also observed by Geguzin [161] in a study of the recrystallization of porous copper under a hydrostatic pressure, which was known to favor the healing of pores. In the case of a polycrystalline sample, the application of pressure reduced the number of inclusions hindering the motion of the grain boundaries and, consequently, favored secondary recrystallization in a porous body. This is illustrated clearly in Fig. 82.

The drag of inclusions by grain boundaries migrating during secondary recrystallization may give rise to the accumulation of second-phase particles. Clusters of such particles form at the points where the boundaries meet and they move toward the centers of small grains which shrink during the recrystallization. This process has been observed experimentally and is shown schematically in Fig. 83.

The motion of grain boundaries during recrystallization may also be due to their drag by inclusions migrating in an external force field. This effect was observed clearly in polycrystalline camphor [175], in which gas-filled bubbles migrated in the field of a temperature gradient (Fig. 84).

In discussing the interaction of inclusions with boundaries and its influence on the kinetics of the recrystallization coarsening of grains, we must consider two aspects: the velocity of the combined motion of a boundary and the dislocations pinned to it, and the conditions under which such combined motion can occur.

The first aspect is concerned with the rate of recrystallization coarsening of grains in a medium containing insoluble inclusions.

Let us consider some consequences of the drag of pores by grain boundaries during the recrystallization of porous bodies, particularly compacted powders whose porosity during the later stages of sintering is entirely due to the presence of closed isolated pores.

A moving boundary which drags a pore pinned to it distorts the pore in such a way that a chemical potential gradient appears along the surface of the pore. This gradient gives rise to the transport of matter from the front of the pore to the rear wall of the pore, i.e., it gives rise to the migration of a pore along a grain boundary (Fig. 85) [162].

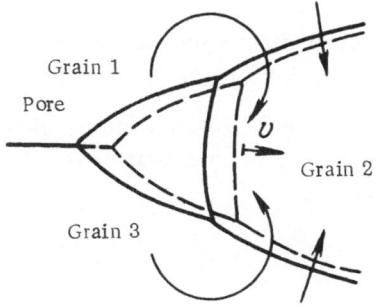

Grain 1
Pore
Grain 2
Grain 3

Fig. 85. Motion of a pore dragged by moving grain boundaries. The arrows show the directions of flow of the vacancies [162].

In the case of pore-free polycrystalline bodies, the kinetics of recrystallization grain growth can be described satisfactorily by a relationship of the type

$$\left(L_t^2 - L_0^2 \right) \propto t, \tag{19.1}$$

where L_0 and L_t are, respectively, the average linear size of the grains at the beginning of the secondary recrystallization process and at a moment t. It is assumed that the motion of a grain boundary is due to such recrystallization.

Equation (19.1) is derived on the assumption that the velocity of a grain boundary toward its center of curvature is proportional to the curvature, i.e., that relationships such as Eq. (11.8) apply. We note that the law (19.1) describes best the results of experiments on high-purity metals. In metals of technical grade, Eq. (19.1) applies only in the range of temperatures at which the dissolution of the second-phase inclusions is possible.

In the presence of inclusions and particularly in the presence of pores, Eq. (19.1) should not be obeyed because the motion of the boundaries during recrystallization is inhibited by the presence of inclusions. The kinetics of the recrystallization growth of grains separated by boundaries which carry inclusions or pores depends on the mechanism of the transport of matter from the front to the rear of a migrating inclusion (pore).

We shall provide an elementary description of the kinetics of secondary recrystallization in a crystal in which pores are located at grain boundaries. The velocity of a pore-carrying grain boundary is given by Eq. (11.10)-(11.12) and it can be written in the form (which applies to different diffusion mechanisms):

$$v_g \sim \frac{1}{R^n} e^{-\frac{U_1}{kT}}. \tag{19.2}$$

We shall now consider the conditions of the simultaneous motion of grain boundaries and pores or inclusions [163-165].

If a boundary together with second-phase inclusions move at a velocity v, i.e., if the velocity of the boundary equals that of the inclusions, we find that

$$B_g F_g = B_i F_i, \tag{19.3}$$

where B_g and B_i are the mobilities of, respectively, a grain boundary and an inclusion; F_i is the force exerted on the inclusion by the boundary; F_g' is the force exerted on unit area of the boundary, calculated with allowance for the inhibiting effect of the inclusions.

The force F_g' is determined by the force [see Eq. (11.8)] acting on a boundary during recrystallization (F_g), the force F_i, and the average number of inclusions attached to the boundary $N = 1/\rho^2$:

$$F_g' = F_g - N F_i .$$

(19.4)

It follows from Eqs. (19.3) and (19.4) that the velocity of a complex comprising a grain boundary and the inclusions pinned to it is given by

$$v = F_i \tilde{B} = F_i \frac{B_g B_i}{B_g + N B_i} = F_i \frac{B_i}{1 + N \frac{B_i}{B_g}}$$

(19.5)

The quantity $\tilde{B}$ is the reduced mobility of the boundary–inclusion complex. We can easily see that for different limiting relationships between B_i and B_g the mobility of the complex is determined either by the mobility of the inclusions or by the mobility of the grain boundary.

If $N B_i / B_g \ll 1$, the velocity of a boundary depends only on its own intrinsic mobility. This inequality applies when the number of inclusions pinned to a boundary is very small ($N \to 0$) or when inclusions are so small that $B_g \gg N B_i$. This case corresponds to free recrystallization, i.e., to recrystallization which is not inhibited by pores or inclusions.

If $N B_i / B_g \gg 1$, the reduced mobility is $\tilde{B} \approx B_g / N$, i.e., it is governed by the mobility and the number of inclusions or pores. In this case, the rate of recrystallization grain growth dL/dt depends strongly on the dimensions of the pores or inclusions and on the mechanism responsible for the transport of pores or inclusions. Obviously, we have

$$\frac{dL}{dt} \approx v = F_i \tilde{B} = F_i \frac{B_g}{N} .$$

(19.6)

We shall discuss Eq. (19.6) by applying it to a porous polycrystalline material.

Studies of the role played by pores in inhibiting the motion of a grain boundary are very important for the following reasons. It is known that the high-temperature creep of metals and alloys is accompanied by the cavitation process, i.e., by the formation of microcavities located mainly on dislocations and grain boundaries. The cavitation process is particularly strong in reactor materials which suffer radiation damage and, therefore, recrystallization grain growth in such materials is influenced strongly by the inhibiting effect of pores. The role played by pores is also very important in the recrystallization of porous powder compacts, which can be regarded as a mixture of two phases, one being the metal in question and the other the pores.

If we wish to use Eq. (19.6) to determine the law L(t), we must know the relationship between the average linear dimension of a grain and the pore radius.

In the process of secondary recrystallization, the dimensions of moving pores do not remain constant. The pores increase as a result of coalescence and of direct collisions. The coalescence of an ensemble of fine pores located on a grain boundary may be much faster than the coalescence in the interior of a grain because of the higher diffusion permeability of the boundaries. If this process is determined primarily by the diffusion along grain boundaries, its kinetics resembles that of the two-dimensional diffusion coalescence of foreign particles located on the surface of a solid [87]. It follows from [87] that in the course of two-dimensional coalescence, the average radius of a pore should vary with time in accordance with the law $\overline{R} \propto (D_{gb} t)^{1/4}$, where D_{gb} is the grain boundary diffusion coefficient. Direct collisions between pores can occur when two boundaries meet. Such collisions may result in the growth of a pore, as shown schematically in Fig. 86.

Fig. 86. Enlargement of inclusions as a result of disappearance of a grain during recrystallization. The arrows indicate the direction of motion of grain boundaries.

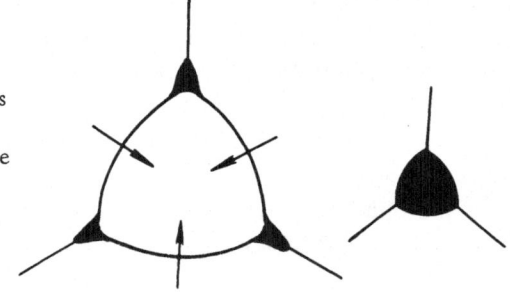

We may expect the dimensions of growing pores and growing grains to be correlated. It follows from [163] that this correlation can be represented very approximately in the form $R \propto L$.

The mobility of pores B_g, which occurs in Eq. (19.6), can be represented — using the formulas in Sec. 6 and the assumption that $R \propto L$ — in the following form:

$$B_g = \frac{\text{const}}{L^n} e^{-U_i/kT} \qquad (19.7)$$

In Eq. (19.7), the exponent is $n = 3$ for the volume diffusion mechanism and for the diffusional transport of matter across a gaseous medium under a constant pressure ($P = \text{const}$); $n = 4$ for the surface diffusion mechanism; $n = 2$ for the diffusional transport of matter across a gaseous phase when the gas pressure in the pore is $P = 2\gamma/R$. The quantity U_i in Eq. (19.7) is the activation energy of the relevant diffusion process.

Equations (19.6) and (19.7) yield the following time dependence of the average size of grains in a porous medium during recrystallization:

$$L_t^\beta - L_0^\beta \propto \frac{e^{-\bar{U}_i/kT}}{N} t. \qquad (19.8)$$

where $\beta = n + 1$.

Equation (19.8) applies to that stage of secondary recrystallization in a porous body when the boundaries migrate together with the associated pores. It follows from general considerations that, for a certain ratio between the pore and the grain-boundary mobilities and between the forces applied to the pores and the boundary, a migrating boundary can detach itself from the pores and leave them behind in the interior of the grain.

If we use Eq. (19.4), we can easily find the relationship between v_g and v_i in the form

$$\varkappa = \frac{v_i}{v_g} = \frac{K}{\theta - 1}, \qquad (19.9)$$

where $K = B_i/NB_g$ and $\theta = F_g/NF_i$. It follows from Eq. (19.9) that when $\varkappa \ll 1$, i.e., when $\theta - 1 \gg K$, a grain boundary should become detached from the associated pores, whereas when $\varkappa \gg 1$, i.e., when $\theta - 1 \ll K$, the boundary should migrate together with the pores. Such motion may occur under various conditions: the limiting cases of this motion are represented by $K \gg 1$ when the velocity of the boundary–pore system is governed by the mobility of the boundary, and $K \ll 1$ when this velocity is governed by the mobility of the pores.

Under isothermal conditions and for a specified grain size, the values of θ and K depend solely on the pore radius.

The criteria given above and the values of the quantities occuring in them can be used to determine the dimensions of the grains and pores at which the motion of a grain boundary with its associated pores is controlled either by the mobility of the boundary or by the mobility of the pores. We can also find the conditions under which a moving boundary becomes detached from its associated pores.

We note that the coalescence of pores located on a boundary may occur as a result of diffusion, i.e., due to the diffusional transport of matter between pores through a balanced diffusion field (two-dimensional coalescence), or as a result of direct collisions in the Brownian motion. These two processes alter the average distances between the pores ($\bar{l}$), their average radius $\bar{R}$), and, therefore, they can alter the velocity of a system consisting of pores and a grain boundary.

When pores migrate as a result of volume diffusion, the velocity of such a system is determined by the law $v \propto \bar{l}^2/\bar{R}^3 \approx 1/\theta_0$ [see Eq. (11.11)], where θ_0 is the total volume of pores per unit area of the grain boundary. If the two-dimensional coalescence occurs under conditions such that θ_0 remains constant, the coalescence cannot alter the velocity of the system. However, if pores migrate under the influence of surface diffusion, so that $v \propto \bar{l}^2/\bar{R}^4 \approx 1/\theta_0\bar{R}$ [see Eq. (11.12)], the two-dimensional diffusional coalescence should slow down the boundary–pore system so that its velocity becomes $v \approx 1/\bar{R} \approx t^{1/4}$. Estimates show that such deceleration of a boundary during secondary recrystallization may occur at high temperatures (below the melting point) when the diffusional mobility is sufficient for the grain boundaries to migrate a considerable distance $\bar{R}$ during the recrystallization.

The influence of the two-dimensional Brownian coalescence (Sec. 6) on the motion of a boundary–pore system may be sig-

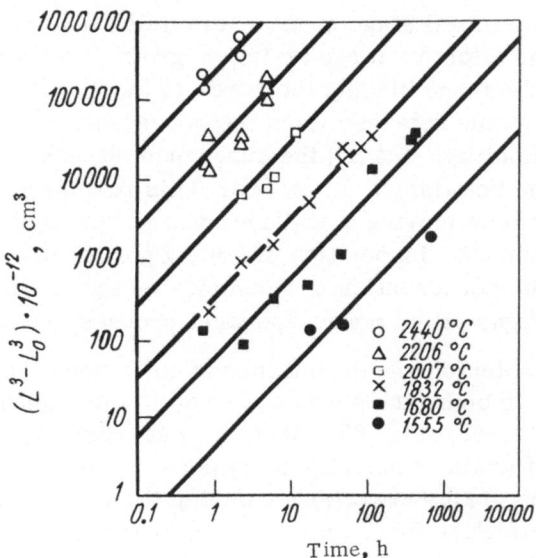

Fig. 87. Changes in the dimensions of grains during the secondary recrystallization of UO_2 compacts [167].

nificant only if the linear dimensions of the pores are very small: $\leq 5 \times 10^{-7}$ cm.

The recrystallization of porous bodies has been studied experimentally by many workers [158-167]. We shall compare the experimental results given in [166-167] with our theory of the kinetics of grain growth.

These investigations were concerned with the sintering of porous compacts of UO_2 and Al_2O_3, and it was found that Eq. (19.8) was obeyed with n = 2. Figure 87 shows the dependence of $\log(L_t^3 - L_0^3)$ on t, plotted using the data on the UO_2 compacts [166]. The temperature dependence of the rate of recrystallization growth of UO_2 grains yielded the activation energy of this process, U_i = 500 kJ/mole (124 kcal/mole), which is close to the heat of evaporation of UO_2, 600 kJ/mole (138 kcal/mole) [168]. Evidently, in the case of UO_2 and Al_2O_3 compacts, the migration of pores pinned to moving boundaries resulted from the transport of matter across the gaseous phase subject to the condition that the inert-gas pressure in the pores was equal to the Laplace pressure resulting from the curvature of the pore surfaces ($P \propto 1/R$). The inert gas filling the pores was insoluble in the host material.

During the initial stage of the recrystallization of a porous body, when the effective force acting on grain boundaries is large and exceeds the force binding the pores to the boundary, the boundaries become detached from the pores, and, consequently, have a high mobility. During the subsequent stages, the number of pores at the boundaries increases. This results either from collisions between moving boundaries and pores located within grains, or from the dissociation of hollow pores into vacancies followed by the condensation of vacancies on the grain boundaries. In the case of gas-filled pores, the first process predominates.

A nonmonotonic time dependence of the number of pores located at grain boundaries was observed in the experiments on porous copper [158] (Fig. 88). It is evident from Fig. 88 that the detachment of grain boundaries from pores occurred only during one cycle. The grains were larger during the next cycle and this reduced the effective force acting on the grain boundaries. The force was then insufficient to detach the boundaries from the pores.

The migration of pores dragged by grain boundaries gives rise to certain special features in the later stages of the sintering of porous bodies. During the stages of the formation of closed equilibrium gas-filled pores, the gas pressure is compensated by the Laplace pressure and the densification should stop. This happens because near an equilibrium pore the excess concentration of vacancies is zero and, consequently, the vacancies neither migrate toward the pore nor away from it. Diffusional coalescence does not occur in an ensemble of such pores. However, coalescence can still occur because of the collisions between moving pores. As in the case of swelling, the pore volume increases as a result of the coalescence and the density of a compact should decrease.

The swelling of compacts in the late stages of sintering was observed experimentally by Mansour and White [169]: they in-

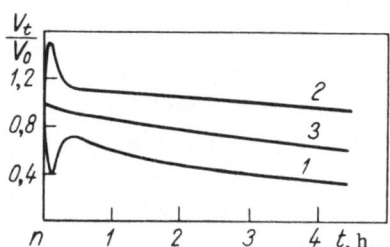

Fig. 88. Time dependence of the relative volume of pores in cyclically treated copper (V_t is the volume of pores at a time t, V_0 is the initial volume of pores, T = 700°C) [158]: 1) pores on grain boundaries; 2) pores within grains; 3) pores on boundaries and within grains.

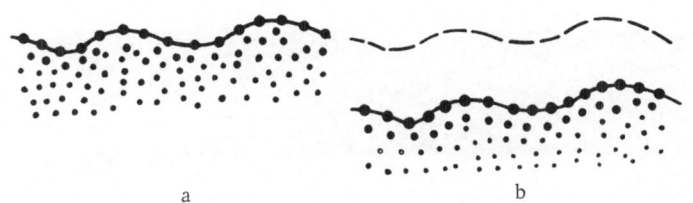

Fig. 89. Enlargement of inclusions on a moving boundary: a, b) successive stages.

vestigated the densification of UO_2 powder compacts in which the pores were filled with an inert gas. Isothermal annealing increased the porosity from ε = 0.02 to 0.05. This increase in porosity was accompanied by an increase in the average pore size. A similar phenomenon was observed during the sintering of uranium oxide compacts[†] and in the later stages of the sintering of UO_2–PuO_2 mixtures.[‡]

The migration of grain boundaries during recrystallization increases the average size of pores, gas-filled bubbles, and second-phase inclusions located at these boundaries. This may be a consequence of two different processes. In one of these processes, a boundary dragging its associated pores meets pores located in the interiors of the grains, which are then captured by the moving boundary. During such motion, the dimensions of the pores may increase as a result of direct collisions between the internal pores captured by the boundary and the diffusional coalescence of pores located on a boundary. This mechanism is facilitated by the fact that the density of the pores at the boundary increases during its motion and by the fact that vacancies can flow between the pores as a result of grain-boundary diffusion. The increase in the size of the pores associated with a moving boundary is shown schematically in Fig. 89. The postulated process is supported by the photograph in Fig. 90, which shows a moving grain boundary in a copper film. Helium-filled bubbles can be seen ahead of or on the boundary and the bubbles near the boundary are larger than those far from the boundary.

[†]The work of Lafontaine and Vandem Benden cited in [163].
[‡]The work of Harrison, Foster, and Russell cited in [163].

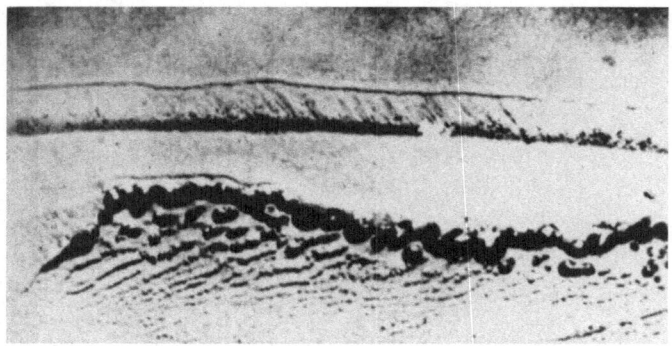

Fig. 90. Accumulation of helium bubbles in front of a moving
grain boundary in copper. ×800.

The size of pores may increase during recrystallization for
different reasons. Collisions between pores may occur in different boundaries of a grain, which shrinks during recrystallization.
This process is shown schematically in Fig. 83.

We must also consider two further effects which are associated
with the migration of pores and which should be manifested in sintering.

In Sec. 17, we drew attention to the fact that in a crystal supersaturated with vacancies to the extent $\delta c_v / c_v$, dislocations ex-
perience a force F' given by Eq. (17.31). If a chain of pores is
located along a dislocation, the pores may be healed and become
sources of vacancies. This would naturally maintain the state
of supersaturation with vacancies in the region adjoining the dislocation.

The force resulting from vacancy supersaturation can set a
dislocation in motion and this dislocation may drag pores behind
it. Thus, the healing of pores located on a dislocation may result from the motion of dislocation—pore systems. In this process, the rate at which pores heal and the velocity of motion of
the system must be matched.

In the case of a nonequilibrium distribution of pores along a
dislocation line, the motion of the dislocation—pore system should
bend the dislocation. If the dislocation is pinned by immobile obstacles the pore-carrying segment between these obstacles will be
bent. When a certain radius of curvature, depending on the pore

radius and on the distance between the obstacles, is reached, the dislocation ceases to act as the source of the vacancies emitted by the pore. In this case, the volume of the pore may decrease as a result of the loss of vacancies to other sinks.

This process is schematically similar to that shown in Fig. 75 but there is a difference: the force which sets a dislocation—pore system in motion is not external but results from the healing of pores. The energy necessary for the combined motion of pores together with a dislocation is acquired from the free surface energy released by the pores which are healed.

The other process can be described as follows. Surface diffusion fluxes generally do not result in densification of a porous body. The porosity decreases as a result of the volume diffusion fluxes directed toward pores. This applies to closed pores located far from any external surface of a powder compact. On the other hand, the pores located near a free surace may migrate toward it as a result of surface diffusion (Sec. 8) and, having reached such a surface, they can collapse and produce some densification. Thus, densification may take place without volume diffusion.

The mechanism of densification as a result of the migration of pores to free surfaces may be important in the case of thin porous plates. The "floating up" of gas-filled pores is reported in [17]. However, under the conditions of the experiments described in [17], the motion of pores was not due to the force of attraction of a free surface but to a temperature gradient ∇T.

Adjacent pores may converge during sintering. In the case of large pores, the diffusion—elastic interaction between them is mainly due to diffusion fluxes in the interior of the host. If the radius of one of the pores is considerably larger than the radius of the other pore, the velocity of the smaller pore in a perfect crystal can be found from Eq. (8.17), which applies to the interaction of a pore with a planar boundary. Such migration is always accompanied by a change in the pore volume. The ratio of the velocity of transition of a pore to the rate of change of its radius can be deduced from Eqs. (8.17) and (8.15). This ratio is of the order of

$$\left(\frac{v}{\dot{R}}\right) \sim \left(\frac{R}{L}\right)^2 \quad (L > R). \tag{19.10}$$

The ratio $v/\dot{R}$ is of the same order of magnitude in the case of pores of similar radii (provided $R \ll L$).

If the pores are located in a crystal containing a large number of dislocations and the distance between these dislocations is much less than the pore radius, the migration may result from viscous flow. In this case, the rate of change of the distance between two pores is again given by the approximate formulas (19.10) [73].

We note that, in the absence of an interaction between pores located in a homogeneous viscous medium, the pores may converge as a result of the densification in the sintering process. The densification of a uniformly distributed system of pores is uniform throughout the porous medium and, therefore, any two pores should approach each other at a relative velocity $v_1 - v_2$ proportional to the distance between their centers,

$$\vec{v}_1 - \vec{v}_2 = \frac{1}{3}\,\dot{u}_{ll}\,\vec{r}_{12}, \qquad (19.11)$$

where $\dot{u}_{ll}$ is the rate of deformation of a porous crystal as a result of densification. This densification-induced migration of pores is additional to the motion resulting from the diffusion-elastic interaction or from the Brownian motion.

The kinetics of the approach of macroscopic pores during their healing was investigated in synthetic NaCl single crystals [173], which contained pores generated accidentally during the growth of the crystals. Some of these crystals had closely spaced

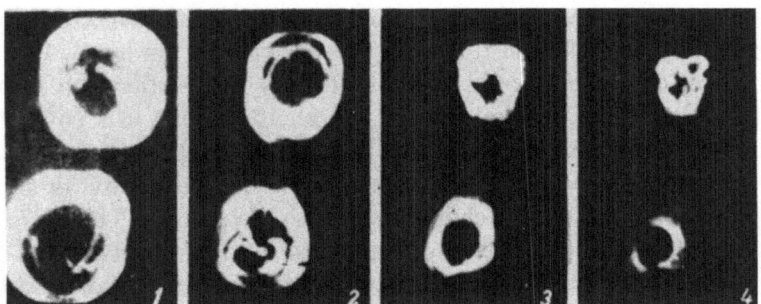

Fig. 91. Successive stages (1-4) in the healing of two neighboring pores in NaCl single crystal [73]. $L_0 = 1.66 \times 10^{-2}$ cm, $R_0 = 5.38 \cdot 10^{-2}$ cm, T = 750°C, P = 2 MN/m^2 (20 kgf/cm^2).

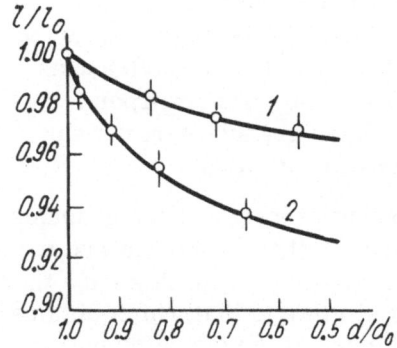

Fig. 92. Dependence $l/l_0 = \varphi\,(d/d_0)$ (the continuous curves are calculated and the points are the experimental results): 1) $L_0 = 1.33 \times 10^{-1}$ cm; $R_0 = 1.83 \times 10^{-1}$ cm; 2) $L_0 = 1.66 \times 10^{-2}$ cm; $R_0 = 5.38 \times 10^{-2}$ cm.

pairs of pores and ensembles of pores of linear dimensions R $\sim$ 10^{-1}-10^{-2} cm. The experiments were carried out on samples containing approximately isomeric pores. Since the crystals were transparent, the changes in the linear dimensions of the pores could be evaluated easily. The healing occurred at temperatures of 700-750°C in an autoclave under an argon pressure of 20-30 kgf/cm². Under these conditions, the host crystal became deformed as a result of viscous flow.

Figure 91 shows the successive stages of the healing of pores in one of the samples. The experimental results obtained for two closely spaced pores could be represented conveniently in the dimensionless coordinates L/L_0 and d/d_0 (d = 2R). The results of the integration of Eq. (19.10) could be represented conveniently in terms of these coordinates and a suitable comparison could be made with the experimental data on the rate of pore healing.

It is evident from Fig. 92 that Eq. (19.10) describes quite satisfactorily the actual kinetics of the approach of the centers of the pores. We can see that the dependence plotted in terms of the coordinates L/L_0 and d/d_0 is independent of the rate of the process and is governed solely by the initial size of the pores and their relative positions.

20. EVAPORATION AND CONDENSATION.

SMOOTHING OF THE SURFACE RELIEF

The processes of evaporation, condensation, and mass transfer along a macroscopically distorted surface of a crystal should be accompanied by directional migration of elements of the surface microrelief, such as steps and kinks. This migration is the consequence of the role played by steps and kinks which act as sources and sinks of atoms participating in mass transport. Foreign-phase particles (for example, oxide granules) on the surface may hinder the directional migration of steps.

The migrating steps may be bound to foreign-phase granules if $\gamma_{12} < \gamma_1 + \gamma_2$ (the subscript 1 refers to the crystal matrix and the subscript 2 to the foreign granule material). In this case the situation on the surface is basically similar to that which obtains in the bulk of a dispersion-hardened crystal during deformation. Here, the surface steps act as analogs of dislocations and the foreign-phase granules as analogs of disperse inclusions which hinder the motion of dislocations in the bulk of the crystal [186].

We shall consider first a macroscopically smooth surface in contact with supersaturated or unsaturated vapor ($|\Delta P/P_0| \neq 0$). If a step migrates under the influence of the difference between the chemical potentials $|\Delta\mu| = kT|\Delta P/P_0|$ near or far from the steps, a segment of the step held back by two granules is bent into the shape of an arc of a circle of radius ρ. Consequently, the granules experience a force

$$F_g = \frac{\gamma_1 h l_g}{\rho} = \frac{h l_g}{\omega} \Delta\mu, \qquad (20.1)$$

where h is the height of the step and l_g is the distance between the two granules. If the radii of the granules are such that $R \ll l_g$, we find that $\rho_{min} \approx l_g/2$. Obviously, if $|\Delta\mu| > \gamma_1\omega/\rho_{min}$, the step bends so that $\rho > \rho_{min}$ and it overcomes the pinning effect of the granules. If $|\Delta\mu| < \gamma_1\omega/\rho_{min}$, the step migrates together with the granules provided F_g is less than the force necessary to detach the step from the granules.

The velocity of a system which consists of a step and attached granules is given by a formula similar to Eq. (19.5):

$$v = F_{s'}\widetilde{B} = F_{s'}\frac{B_g B_{s'}}{B_g + n_g B_{s'}}, \qquad (20.2)$$

where B_g and $B_{s'}$ are, respectively, the mobilities of the granules and the step; $n_g = 1/l_g$ is the linear density of the granules attached to the step; $F_{s'} = (h/\omega)\Delta\mu$ is the force acting on a unit length of the step.

The mobility of the granules in the case of the bulk or volume (subscript b) and surface (subscript s) diffusion mechanisms is given by the formulas which follow from Eqs. (6.7) and (6.8):

$$B_{gb} \simeq \frac{D}{kT}\left(\frac{a}{R}\right)^3, \quad B_{gs} \simeq \frac{D_s}{kT}\left(\frac{a}{R}\right)^4 \qquad (20.3)$$

By definition, the mobility of a step is $B_{s'} = v_{s'}/F_{s'}$, where $v_{s'} = b\Delta\mu$, and, consequently, $B_{s'} = b(\omega/h)$. The constant b depends on the mechanism responsible for the migration of atoms away from the step [187]. When the predominant mechanism is the surface diffusion, we have $b_s = (2D_s'/kT)(a/h\lambda_s)$, where λ_s is is the mean free path of atoms in the surface diffusion mechanism and D' is the coefficient of diffusion along the surface of the crystal in question. If the migration of a step is limited by the rate of diffusion of atoms which evaporate and form the vapor above the crystal, we find that $b_v = (D_v/k^2T^2)(\omega d/h\delta)P_0$, where D_v is the diffusion coefficient in the vapor; P_0 is the equilibrium vapor pressure, δ is a characteristic linear size which depends on the experimental geometry and which governs the rate of evaporation, and d is the distance between steps. Using these values of b_s and b_v, we can rewrite the formulas for the mobility of a step as follows:

$$B_{s's} = \frac{2D'_s}{kT}\frac{a^4}{\lambda_s h^2}, \quad B_{s's} = \frac{D_v P_0}{h^2}\left(\frac{\omega}{kT}\right)^2\frac{d}{\delta} \qquad (20.4)$$

The motion of a system which consists of a step and the associated foreign-phase granules can be described qualitatively by means of Eqs. (20.1) and (20.4).

The rate of evaporation or condensation from or on a macroscopically smooth surface of a crystal along which steps migrate together with the associated granules can therefore be described by the formula

$$i_g = \frac{vh}{\omega d} = \left(\frac{h}{\omega}\right)^2\frac{kT}{d}\left|\frac{\Delta P}{P_0}\right|\tilde{B}, \qquad (20.5)$$

where $|\Delta P/P_0|$ is the relative supersaturation or undersaturation. A comparison of Eq. (20.5) with a similar formula describing "free" evaporation or condensation, i.e., the case when the migration of steps is not hindered in any way, shows that the presence of obstacles in the form of granules may slow down the evaporation or condensation by a factor $K = \tilde{B}/B_{s'}$.

By way of example, we shall estimate K in the special case when $B_g/n_g B_s$, $\ll$ 1 and the mobility of the step–granule system is governed by the mobility of the granules: $\tilde{B} = B_g/n_g$. If we assume that the migration of the granules and the step is governed by the surface diffusion, we find that

$$K = \frac{D_s}{2D_s'} \frac{\lambda_s h^2 l_g}{R^4}.$$

If $D_s/2D_s' \approx 10^{-2}$, $R \approx 10^{-5}$ cm, $l_g \approx 10^{-4}$ cm, $\lambda_s \approx 10^{-3}$ cm, and $h \approx 10^{-6}$ cm, we find that K = 0.1.

The kinetics of smoothing of a macroscopic surface irregularity can be considered similarly bearing in mind that the force acting on a step depends on the local curvature of the surface. A special feature of the processes on an irregular surface is the possibility of formation of steps at distances $d < \lambda_s$. In this case we cannot ignore their diffusional interaction and, consequently, an allowance must be made for the possibility of the formation of arrays of steps [187]. If $d > \lambda_s$, the hindering influence of foreign-phase granules on the kinetics of smoothing is described, as in the case of evaporation and condensation, by the factor K = $\tilde{B}/B_s$.

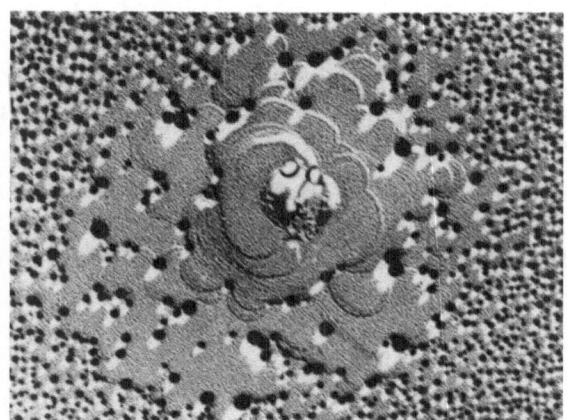

Fig. 93. Evaporation nucleus on the surface of an LiF single crystal carrying gold granules. The migration of steps is seen to be accompanied by the drag of the granules and their coalescence as a result of collisions. ×12,000.

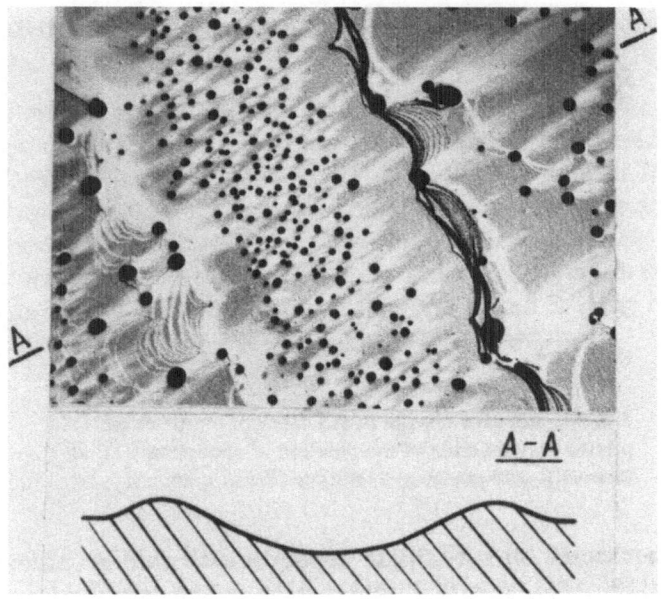

Fig. 94. Migration of steps on a surface with macro-scopic irregularities is accompanied by the accumulation of granules in regions of maximum curvature. ×12,000.

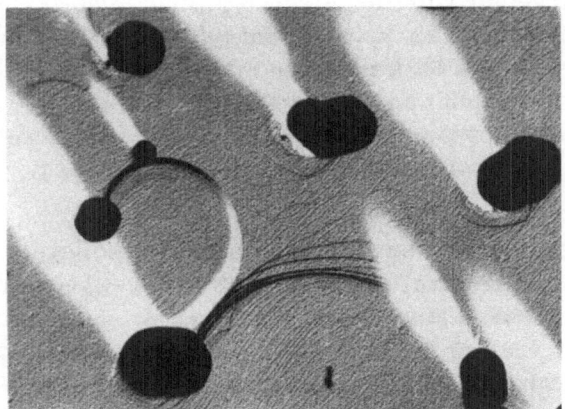

Fig. 95. Formation of "pedestals" under granules bypassed by moving steps. A "pedestal" is made visible by the gradation of the shadow near the granule (the shadow was produced by evaporation of a suitable material). ×12,000.

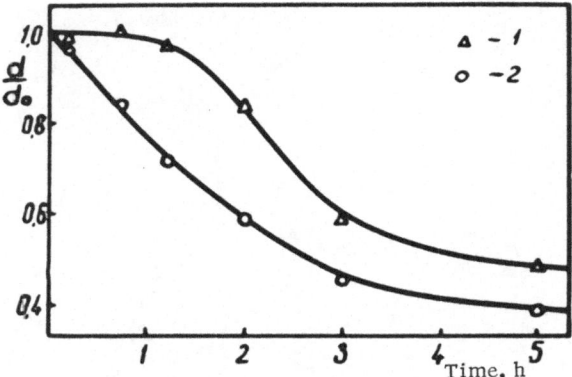

Fig. 96. Relative change in the depth of scratches on NaCl plotted as a function of the duration of annealing: 1) surface with gold granules; 2) surface free of granules.

Experiments on LiF single crystals with gold granules on the surface [186, 188] provided information on the interaction between steps and foreign-phase particles.

Figure 93 shows an evaporation nucleus with bent steps and gold particles. These particles coalesced by collisions while moving together with the steps. It is evident from this figure that the steps together with the attached granules moved at a velocity of $\sim 10^{-7}$ cm/sec. If we assume that the velocity of the step—granule system was limited by the migration of the granules, we can apply Eqs. (20.2) and (20.4) to show that, because $\Delta P/P_0 \approx$ 0.1, the motion of the granules was governed by the surface diffusion mechanism characterized by the coefficient $D_s \approx 10^{-9}$ cm^2/sec.

The force exerted on a step depends on the local curvature of the surface. Therefore, near "humps" where the positive curvature is greatest, the granules and the steps move rapidly toward the peak of a "hump" where coalescence results in increase of the size of the granules. This can be seen clearly in Fig. 94.

If the value of $\Delta \mu$ is large, the migrating steps may bypass a granule leaving it behind on a "pedestal" of height equal to the sum of the heights of all the steps which have bypassed this granule (Fig. 95). This phenomenon resembles the formation of dislocation loops around an inclusion surrounded by many dislocations (Fig. 74).

The experimental results [188] show that the kinetics of the migration of step—granule systems is complicated by the one-dimensional (along the steps) [147] and two-dimensional [87] diffusional coalescence of granules and the coalescence as a result of collisions. Eventually these processes may reduce the retarding influence of the granules on the kinetics of mass transport. This was observed clearly in studies of the healing of scratches on the surface of NaCl (Fig. 96) [181].

Literature Cited

1. A. A. Chernov, Zh. Eksp. Teor. Fiz., 31:709 (1956).
2. W. A. Tiller, J. Appl. Phys., 34:2757, 2763 (1963).
3. M. V. Speight, J. Nucl. Mater., 13:207 (1964).
4. J. Biersack, Nukleonik, 8:439 (1966).
5. W. Diez and J. Biersack, Phys. Status Solidi, 32:247 (1969).
6. G. W. Greenwood and M. V. Speight, J. Nucl. Mater., 10:140 (1963).
7. P. G. Shewmon, Trans. AIME, 230:1134 (1964).
8. E. Ya. Mikhlin, Fiz. Tverd. Tela, 6:2819 (1964).
9. I. M. Lifshits, A. M. Kosevich (Kossevich), and Ya. E. Geguzin, J. Phys. Chem. Solids, 28:783 (1967).
10. M. A. Krivoglaz, A. M. Masyukevich, and K. P. Ryaboshapka, Fiz. Metal. Metalloved., 23:200 (1967).
11. J. Biersack and W. Diez, Phys. Status Solidi, 27:139 (1968).
12. M. A. Krivoglaz and M. E. Osinovskii, Fiz. Metal. Metalloved., 23:988 (1967).
13. M. A. Krivoglaz, Fiz. Metal. Metalloved., 28:577 (1969).
14. M. A. Krivoglaz and M. E. Osinovskii, in: Structure and Properties of Metals (Metal Physics Series) [in Russian], No. 31, Naukova Dumka, Kiev (1970), p. 45.
15. V. B. Fiks, Ionic Conduction in Metals and Semiconductors [in Russian], Nauka, Moscow (1969).
16. R. S. Barnes and D. J. Mazey, Proc. Roy. Soc. London, A275:47 (1963).
17. J. K. Williamson and R. M. Cornell, J. Nucl. Mater., 13:278 (1964).
18. J. Taylor, J. Nucl. Mater., 13:301 (1967).
19. J. Y. Choi and P. G. Shewmon, Trans. AIME, 224:589 (1962).
20. J. Y. Choi and P. G. Shewmon, Trans. AIME, 230:123 (1964).
21. Ya. E. Geguzin, in: Surface Diffusion and Spreading [in Russian], Nauka, Moscow (1969), p. 5.
22. Y. Adda, G. Brebec, N. V. Doan, and M. Gerl, in: Proc. Intern. Symp. on Thermodynamics, Vienna, 1965, Vol. 2, publ. by International Atomic Energy Agency, Vienna (1966), p. 255.
23. G. G. Lemmlein, Dokl. Akad. Nauk SSSR, 85:325 (1952).

24.   W. D. Kingery and W. H. Goodnow, in: Ice and Snow — Properties, Processes, and Applications (Proc. Conf. Massachusetts Institute of Technology, Cambridge, Mass., 1962), publ. by MIT Press, Cambridge, Mass. (1963), p. 237.

25.   P. Hoekstra, T. E. Osterkamp, and W. F. Weeks, J. Geophys. Res., 70:5035 (1965).

26.   J. D. Harrison, J. Appl. Phys., 36:3811 (1965).

27.   J. D. Harrison and W. A. Tiller, J. Appl. Phys., 34:3349 (1963).

28.   J. D. Harrison, J. Appl. Phys., 36:326 (1965).

29.   G. G. Lemmlein, Dokl. Akad. Nauk SSSR, 78:685 (1951).

30.   A. I. Bublik and B. Ya. Pines, Dokl. Akad. Nauk SSSR, 87:215 (1952).

31.   P. Hoekstra and R. D. Miller, "The movement of water film between glass and ice," US Army Cold Regions Res. and Eng. Laboratory Research Rep. (1965).

32.   R. F. Chaiken, D. J. Sibbett, J. E. Sutherland, D. K. van de Mark, and A. Wheeler, J. Chem. Phys., 37:2311 (1962).

33.   W. Seith and H. Wever, Z. Elektrochem., 57:893 (1953).

34.   V. B. Fiks, Fiz. Tverd. Tela, 1:16 (1959).

35.   P. P. Kuz'menko, Ukr. Fiz. Zh., 7:117 (1962).

36.   J. Bardeen and C. B. Herring, in: Imperfections in Nearly Perfect Crystals (Proc. Symp., Pocono Manor, Pa., 1950), publ. by Wiley, New York (1952), p. 261.

37.   K. Compaan and Y. Haven, Trans. Faraday Soc., 52:786 (1956).

38.   P. S. Ho and H. B. Huntington, J. Phys. Chem. Solids, 27:1319 (1966).

39.   Ya. I. Frenkel', Zh. Eksp. Teor. Fiz., 16:29 (1946).

40.   B. Ya. Pines, Zh. Tekh. Fiz., 16:737 (1946).

41.   B. Ya. Pines, Usp. Fiz. Nauk, 52:501 (1954).

42.   Ya. E. Geguzin and I. M. Lifshits, Fiz. Tverd. Tela, 4:1326 (1962).

43.   Ya. E. Geguzin, Macroscopic Defects in Metals [in Russian], Metallurgizdat, Moscow (1962).

44.   I. M. Lifshits and V. V. Slezov, Zh. Eksp. Teor. Fiz., 35:479 (1958).

45.   I. M. Lifshits and V. V. Slezov, Fiz. Tverd. Tela, 1:1401 (1959).

46.   F. R. N. Nabarro, Report Conf. on Strength of Solids, University of Bristol, 1947, publ. by the Physical Society, London (1948), p. 75.

47.   C. J. Herring, J. Appl. Phys., 21:437 (1950).

48.   I. M. Lifshits, Zh. Eksp. Teor. Fiz., 44:1349 (1963).

49.   I. M. Lifshits and V. B. Shikin, Fiz. Tverd. Tela, 6:2780 (1964).

50.   S. T. Konobeevskii, Zh. Eksp. Teor. Fiz., 13:200 (1943).

51.   B. Ya. Lyubov and N. S. Chastov, Dokl. Akad. Nauk SSSR, 84:939 (1952).

52.   M. A. Krivoglaz, Fiz. Metal. Metalloved., 17:161 (1964).

53.   L. I. Lur'e, Spatial Problems in Theory of Elasticity [in Russian], GITTL, Moscow (1955).

54.   J. D. Eshelby, "Continuum theory of lattice defects," Solid State Phys., 3:79 (1956).

55.   M. A. Krivoglaz, A. M. Masyukevich, and K. P. Ryaboshapka, Fiz. Metal. Metalloved., 24:1129 (1967).

56.   E. M. Baroody, J. Appl. Phys., 38:4893 (1967).

57. R. Kelly and E. Ruedl, Phys. Status Solidi, 13:55 (1966).
58. M. E. Gulden, J. Nucl. Mater., 23:30 (1967).
59. E. Ruedl and R. Kelly, J. Nucl. Mater., 16:89 (1965).
60. C. W. Tucker, Jr., and F. J. Norton, J. Nucl. Mater., 2:329 (1960).
61. R. Kelly and F. Brown, Acta Met., 13:169 (1965).
62. A. A. Bauer, D. L. Morrison, F. R. Winslow, J. Bugl, and T. S. Elleman, J. Nucl. Mater., 13:75 (1964).
63. M. A. Krivoglaz and M. E. Osinovskii, Fiz. Metal. Metalloved., 24:805 (1967).
64. M. A. Krivoglaz, in: Structure and Properties of Metals (Metal Physics Series) [in Russian], No. 31, Naukova Dumka, Kiev (1970), p. 37.
65. M. A. Krivoglaz and M. E. Osinovskii, Fiz. Metal. Metalloved., 28:3 (1969).
66. C. Herring, in: The Physics of Powder Metallurgy (ed. by W. E. Kingston), McGraw-Hill, New York (1950), Ch. 8.
67. W. W. Mullins, J. Appl. Phys., 28:333 (1957).
68. A. A. Chernov, Kristallografiya, 1:583 (1956).
69. F. A. Nichols, J. Nucl. Mater., 27:137 (1968).
70. N. S. Chastov, in: Problems in Metallography and Physics of Metals [in Russian], No. 5, Metallurgizdat, Moscow (1958), p. 595 [Tr. Tsent. Nauch.-Issled. Inst. Chem. Metall., No. 5, p. 595 (1958)].
71. J. D. Eshelby, Acta Met., 3:487 (1955).
72. M. E. Osinovskii, in: Diffusion Processes in Metals (Metal Physics Series) [in Russian], No. 25, Naukova Dumka, Kiev (1968), p. 55.
73. Ya. E. Geguzin and A. S. Dzyuba, Fiz. Tverd. Tela, 5:891 (1963).
74. M. Smoluchowski, in: Brownian Motion [Russian translation], ONTI, Moscow (1939), p. 332.
75. S. Chandrasekhar, Stochastic Problems in Physics and Astronomy [Russian translation], IL, Moscow (1947).
76. S. V. Pshenai-Severin, Dokl. Akad. Nauk SSSR, 94:865 (1954).
77. D. L. Swift and S. K. Friedlanger, J. Colloid. Sci., 19:621 (1964).
78. G. M. Hidy, J. Colloid. Sci., 20:123 (1965).
79. R. S. Barnes, J. Nucl. Mater., 11:135 (1964).
80. M. V. Speight, J. Nucl. Mater., 12:216 (1964).
81. E. E. Gruber, J. Appl. Phys., 38:243 (1967).
82. Ya. E. Geguzin, Yu. I. Boiko, and L. N. Paritskaya, Fiz. Metal. Metalloved., 24:418 (1967).
83. Ya. E. Geguzin, S. S. Simeonov, and É. A. Éivazov, Fiz. Tverd. Tela, 9:1440 (1967).
84. F. A. Nichols, J. Appl. Phys., 37:2805 (1966).
85. F. A. Nichols, Acta Met., 15:365 (1967).
86. R. S. Barnes, Phil. Mag., 5:635 (1960).
87. Ya. E. Geguzin, Yu. S. Kaganovskii, and V. V. Slezov, J. Phys. Chem. Solids, 30:1173 (1969).
88. D. McLean, Grain Boundaries in Metals, Oxford University Press (1957).
89. T. Gladman, Proc. Roy. Soc., London, A294:298 (1966).
90. M. Hillert, Acta Met., 13:227 (1965).
91. M. V. Speight and G. W. Greenwood, Phil. Mag., 9:683 (1964).

92.   M. F. Ashby and R. M. Centamore, Acta. Met., 16:1081 (1968).

93.   M. F. Ashby and I. G. Palmer, Acta Met., 15:420 (1967).

94.   L. N. Larikov, in: Problems in Physics of Metals and Metallography [in Russian], No. 11, Izd. AN UkrSSR, Kiev (1960), p. 161.

95.   B. Ya. Pines, E. E. Badiyan, and V. P. Khizhkovyi, Fiz. Tverd. Tela, 5:2859 (1963).

96.   K. K. Ziling and A. I. Grankin, Izv. Vyssh. Ucheb. Zaved., Fizika, No. 11, p. 157 (1968).

97.   K. K. Ziling, Fiz. Metal. Metalloved., 22:931 (1966).

98.   E. F. Koch and K. T. Aust, Acta Met., 15:405 (1967).

99.   A. Napolitano, P. B. Macedo, and E. G. Hawkins, J. Amer. Ceram. Soc., 48:613 (1965).

100.   E. H. Fontana and W. A. Plummer, Phys. Chem. Glasses, 7:139 (1966).

101.   M. A. Krivoglaz and A. A. Smirnov, Theory of Orderable Alloys [in Russian], Fizmatgiz, Moscow (1958).

102.   J. R. Manning, Phys. Rev., 125:103 (1962).

103.   L. S. Darken, Trans. AIME, 175:184 (1948).

104.   J. Bardeen and C. Herring, in: Atom Movements, American Society for Metals, Cleveland (1951), p. 87.

105.   A. D. le Claire, in: Progress in Metals Physics [Russian translation], No. 1, Metallurgizdat, Moscow (1956), p. 224.

106.   F. Bailly, Y. Marfaing, G. Cohen-Solal, and J. Melngailis, J. Phys., 28:573 (1967).

107.   J. Frenkel, Kinetic Theory of Liquids, Oxford University Press (1946).

108.   K. Lehovec, J. Chem. Phys., 21:1123 (1953).

109.   J. D. Eshelby, C. W. A. Newey, P. L. Pratt, and A. B. Lidiard, Phil. Mag., 3:75 (1958).

110.   I. M. Lifshits and Ya. E. Geguzin, Fiz. Tverd. Tela, 7:62 (1965).

111.   T. B. Grimley, Proc. Roy. Soc., London, A201:40 (1950).

112.   M. A. Krivoglaz, Úkr. Fiz. Zh., 14:3 (1969).

113.   M. A. Krivoglaz, Ukr. Fiz. Zh., 14:43 (1969).

114.   A. M. Kosevich, Fiz. Tverd. Tela, 7:451 (1965).

115.   M. A. Krivoglaz, Fiz. Tverd. Tela, 10:3348 (1968).

116.   Ya. E. Geguzin and N. N. Ovcharenko, Dokl. Akad. Nauk SSSR, 163:621 (1965).

117.   Ya. E. Geguzin, N. N. Ovcharenko, S. S. Simeonov, and A. I. Gvozdikov, Dokl. Akad. Nauk SSSR, 193:1286 (1970).

118.   Ya. E. Geguzin and Yu. S. Kaganovskii, Kolloid. Zh., 30:681 (1968).

119.   Ya. E. Geguzin and L. N. Paritskaya, Dokl. Akad. Nauk SSSR, 141:833 (1961).

120.   Ya. E. Geguzin, V. I. Solunskii, and L. M. Reznik, Fiz. Tverd. Tela, 7:802 (1965).

121.   A. M. Kosevich and I. G. Margvelashvili, Ukr. Fiz. Zh., 14:289 (1969).

122.   Ya. E. Geguzin, N. N. Ovcharenko, and É. A. Éivazov, Ukr. Fiz. Zh., 13:580 (1968).

123.   Ya. E. Geguzin and V. I. Solunskii, 10:3670 (1968).

124.   Ya. E. Geguzin and V. I. Solunskii, Dokl. Akad. Nauk SSSR, 156:644 (1964).

125.   Ya. E. Geguzin and S. S. Simeonov, Fiz. Tverd. Tela, 12:911 (1970).

126. Z. Jeffries and R. S. Archer, in: The Science of Metals, McGraw-Hill, New York (1924), p. 95.
127. J. E. Burke, Metals Technol., Vol. 15, No. 7 (Tech. publ. 2472), 19 pp. (1948).
128. Ya. E. Geguzin, Physics of Sintering [in Russian], Nauka, Moscow (1967).
129. J. M. Hedges and J. W. Mitchell, Phil. Mag., 44:223 (1953).
130. W. C. Dash, J. Appl. Phys., 27:1193 (1956).
131. S. Amelinckx, Phil. Mag., 1:269 (1956).
132. S. Amelinckx, W. van der Vorst, R. Gevers, and W. Dekeyser, Phil. Mag., 46:450 (1955).
133. S. Amelinckx, in: Dislocations and Mechanical Properties of Crystals (ed. by J. C. Fisher, W. G. Johnston, R. Thomson, and T. Vreeland, Jr.), Wiley, New York (1957), p. 3.
134. F. Coulomb and J. Friedel, in: Dislocations and Mechanical Properties of Crystals (ed. by J. C. Fisher, W. G. Johnston, R. Thomson, and T. Vreeland, Jr.), Wiley, New York (1957), p. 555.
135. J. Friedel, Dislocations, Pergamon Press, Oxford (1964).
136. R. B. Nicholson, G. Thomas, and J. Nutting, Acta Met., 8:172 (1960).
137. J. Weertman, J. Appl. Phys., 26:1213 (1955).
138. J. Weertman, J. Appl. Phys., 28:362 (1957).
139. O. D. Sherby, Acta Met., 10:135 (1962).
140. A. H. Cottrell, Dislocations and Plastic Flow in Crystals, Oxford University Press (1953).
141. Collection: Investigations of Refractory Alloys [in Russian], Vol. 9, Izd. AN SSSR, Moscow (1962).
142. N. H. Nachtrieb, Advan. Phys., 16:309 (1967).
143. R. W. Weeks, S. R. Pati, M. F. Ashby, and P. Barrand, Acta Met., 17:1403 (1969).
144. M. F. Ashby, Scripta Met., 3:843 (1969).
145. M. F. Ashby, Scripta Met., 3:837 (1969).
146. F. K. Sautter and E. S. Chen, in: Oxide Dispersion Strengthening, Proc. Second Bolton Landing Conf., Bolton Landing, N. Y., 1966, publ. by Gordon and Breach, New York (1968), p. 495.
147. V. V. Slezov and D. M. Levin, Fiz. Tverd. Tela, 12:1748 (1970).
148. Ya. E. Geguzin and I. V. Vorob'eva, Ukr. Fiz. Zh., 14:1499 (1969).
149. Ya. E. Geguzin, Usp. Fiz. Nauk, 61:217 (1957).
150. A. T. Churchman, R. S. Barnes, and A. H. Cottrell, J. Nucl. Energy, 7:88 (1958).
151. J. G. Ball, J. Inst. Metals, 84:239 (1956).
152. V. V. Slezov and V. B. Shikin, Fiz. Tverd. Tela, 6:7 (1964).
153. V. M. Agranovich and É. Ya. Mikhlin, At. Energ., 12:385 (1962).
154. L. P. Semenov, At. Energ., 15:404 (1963).
155. R. S. Barnes, J. Nucl. Mater., 11:135 (1964).
156. J. B. Rich, G. P. Walters, and R. S. Barnes, J. Nucl. Mater., 4:287 (1961).
157. Ya. E. Geguzin and L. N. Paritskaya, Fiz. Metal. Metalloved., 12:900 (1961).
158. Ya. E. Geguzin and L. N. Paritskaya, Porosh. Met., No. 5, p. 20 (1962).
159. É. V. Dekhtyareva, in: Experimental Techniques in Technical Mineralogy and Petrography [in Russian], Nauka, Moscow (1966), p. 212.

160.   P. A. Beck, M. L. Holzworth, and P. R. Sperry, Trans. AIME, 180:163 (1949).
161.   Ya. E. Geguzin, Dokl. Akad. Nauk SSSR, 135:829 (1960).
162.   W. D. Kingery and B. Francois, J. Amer. Ceram. Soc., 48:546 (1965).
163.   F. A. Nichols, J. Appl. Phys., 37:4599 (1966).
164.   R. J. Brook, J. Amer. Ceram. Soc., 52:56 (1969).
165.   F. A. Nichols, J. Amer. Ceram. Soc., 51:468 (1968).
166.   J. R. MacEwan, J. Amer. Ceram. Soc., 45:37 (1962).
167.   R. L. Coble, J. Appl. Phys., 32:793 (1961).
168.   R. J. Ackermann, P. W. Gilles, and R. J. Thorn, J. Chem. Phys., 25:1089 (1956).
169.   N. A. L. Mansour and J. White, Powder Met., No. 12, p. 108 (1963).
170.   Ya. E. Geguzin and S. S. Simeonov, Ukr. Fiz. Zh., 16:495 (1971).
171.   Ya. E. Geguzin, M. A. Krivoglaz, and K. P. Ryaboshapka, Fiz. Metal. Metal-
       loved., 31:23 (1971).
172.   Ya. E. Geguzin and D. M. Levin, Fiz. Metal. Metalloved., 32:670 (1971).
173.   Ya. E. Geguzin, S. S. Simeonov, and V. M. Mostovoi, Fiz. Tverd. Tela, 13:100
       (1971).
174.   M. E. Osinovskii, Fiz. Tverd. Tela, 13:3648 (1971).
175.   T. R. Anthony and R. A. Sigsbee, Acta Met., 19:1029 (1971).
176.   P. S. Ho, J. Appl. Phys., 41:64 (1970).
177.   L. E. Willertz and P. G. Shewmon, Met. Trans., 1:2217 (1970).
178.   T. R. Anthony and H. E. Cline, Phil. Mag., 22:893 (1970).
179.   T. R. Anthony and H. E. Cline, J. Appl. Phys., 42:3380 (1971).
180.   M. N. Botvinko and M. A. Krivoglaz, Ukr. Fiz. Zh., 16:621 (1971).
181.   G. B. Gibbs, Mater. Sci. Eng., 2:269 (1967-8).
182.   R. Raj and M. F. Ashby, Met. Trans., 2:1113 (1971).
183.   F. A. Nichols, Acta Met., 20:207 (1972).
184.   S. R. Pati and P. S. Maiya, Acta Met., 19:807 (1971).
185.   H. R. Patil and H. B. Huntington, J. Appl. Phys., 42:3916 (1971).
186.   Ya. E. Geguzin, V. V. Kalinin, and Yu. S. Kaganovskii, Dokl. Akad. Nauk
       SSSR, 206:580 (1972).
187.   A. A. Chernov, Usp. Fiz. Nauk, 73:277 (1961).
188.   Ya. E. Geguzin, V. V. Kalinin, and Yu. S. Kaganovskii, Proc. Conf. on Crys-
       tal Growth [in Russian], Tsekhkadzor (1972).